31. Colloquium der Gesellschaft für Biologische Chemie
14.—19. April 1980 in Mosbach/Baden

Biological Chemistry of Organelle Formation

Edited by
Th. Bücher W. Sebald H. Weiss

With 114 Figures

Springer-Verlag
Berlin Heidelberg New York 1980

Professor Dr. Dr. TH. BÜCHER
Institut für Physiologische Chemie
Physikalische Biochemie und Zellbiologie
der Universität München
Goethestraße 33
8000 München 2/FRG

Priv. Doz. Dr. W. SEBALD
Gesellschaft für Biotechnologische Forschung mbH
Mascheroder Weg 1
3300 Braunschweig-Stöckheim/FRG

Priv. Doz. Dr. H. WEISS
Europäisches Laboratorium für Molekularbiologie
Meyerhofstraße 1
6900 Heidelberg/FRG

ISBN 3-540-10458-5 Springer-Verlag Berlin Heidelberg New York
ISBN 0-387-10458-5 Springer-Verlag New York Heidelberg Berlin

Library of Congress Cataloging in Publication Data. Gesellschaft für Biologische Chemie. Biological chemistry of organelle formation/31. Bibliography: p. Includes index. 1. Cell organelles-Congresses. 2. Cytochemistry-Congresses. 3. Developmental cytology-Congresses. I. Bücher, Theodor. II. Sebald, W., 1941-. III. Weiss, Hanns, 1940-. IV. Title. [DNLM: 1. Chloroplasts-Congresses. 2. Mitochondria-Congresses. 3. Cytogenetics-Congresses. 4. Genetics, Biochemical-Congresses. W 3 MO 87 31st 1980 b/QH 605 M 894 1980 b] QH 611. G 47 1980 574.87 80-28246.

Printing and binding: Brühlsche Universitätsdruckerei, Giessen.
2131/3130-543210

Preface

Eukaryotic cells contain a plurality of organelles distinguished by
their specific membranes and contents. Their biogenesis occurs by
growth and division of preexisting structures rather than de novo.
Mitochondria and chloroplasts, which appear to be descended from
prokaryotic ancestors, have retained some DNA and the biosynthetic
capability for its expression. They synthesize, however, only a few
of their proteins themselves. Most of their proteins are synthesized
on free ribosomes in the cytoplasm and are only assembled in the
correct membrane after synthesis is complete. The biogenesis of
peroxisomes and glyoxysomes also appears to occur by an incorporation
of proteins synthesized first in the cytoplasm. Other organelles,
the Golgi complex, lysosomes, secretory vesicles, and the plasma
membrane, are formed in a different manner. Their proteins are
assembled in the membrane of the endoplasmic reticulum during translation by bound ribosomes and they must then be transported to the
correct membrane.

The 1980 Mosbach Colloquium was one of the first attempts to discuss
the biogenesis of the various organelles in biochemical terms. This
was appropriate since the crucial problems now center on the search
for signals and receptors that dictate the site of assembly, the
route taken, and the final location of a particular organelle protein.
The assembly of prokaryotic membranes and the membrane of an animal
virus were also discussed, since these simpler systems might shed light
on the biogenesis of organelles in eukaryotes.

Acknowledgments. The organizers are grateful to the Gesellschaft
für Biologische Chemie and its chairman, Prof.Dr. H. G. Wittmann, for
actively supporting the colloquium. Special thanks are due to Prof.
Dr. E. Auhagen and Prof. Dr. H. Gibian for their help in the organization and to the Deutsche Forschungsgemeinschaft and all persons and
institutions who provided the necessary funds to make this meeting
a successful one.

October, 1980   TH. BÜCHER
                                                           W. SEBALD
                                                           H. WEISS

Contents

Contributors

You will find the addresses at the beginning of the respective contribution

ALT, J.   97
ANDERSON, S.   11
BANKIER, A.T.   11
BAR-NUN, S.   147
BARRELL, B.G.   11
BARTLETT, S.G.   113
BETZ, H.   175
BISANZ, C.   97
BOGORAD, L.   87
BORST, P.   27
BUSE, G.   59
CHEN, E.   11
CHUA, N.-H.   113
COULSON, A.R.   11
CZAKO-GRAHAM, M.   147
DAYHOFF, M.O.   71
DE BRUIJN, M.H.L.   11
DRIESEL, A.J.   97
DROUIN, J.   11
EPERON, I.C.   11
GAROFF, H.   221
GORDON, K.   97
GRATZL, M.   165
GROSSMAN, A.R.   113
HASILIK, A.   207
HEARNE, P.   245
HERRMANN, R.G.   97
HERZOG, V.   119
HILDEBRANDT, J.W.   97
HIROTA, Y.   245
ITO, K.   245
JOLLY, S.O.   87
KLEIN, U.   207
KREIBICH, G.   147
LAZAROW, P.B.   187
LINK, G.   87

MC INTOSH, L.   87
MOK, W.   147
NACK, E.   147
NIERLICH, D.P.   11
NOBREGA, F.G.   1
NOKELAINEN, M.   245
OKADA, Y.   147
PONTICELLI, A.   245
POULSEN, D.   87
REHM, H.   175
ROBBI, M.   187
ROE, B.   11
ROSENFELD, M.G.   147
SABATINI, D.D.   147
SANGER, F.   11
SCHEDEL, R.   97
SCHMIDT, G.W.   113
SCHREIER, P.H.   11
SCHWARTZ, R.M.   71
SCHWARTZ, M.   235
SCHWARZ, Z.   87
SEARS, B.B.   97
SEYER, P.   97
SHIO, H.   187
SIMONS, K.   221
SMITH, A.J.H.   11
STADEN, R.   11
STEINMETZ, A.   87
TZAGOLOFF, A.   1
VON FIGURA, K.   207
WERNER, S.   43
WICKNER, W.   245
WINTER, P.   97
WLASCHEK, M.   97
YOUNG, I.G.   11

Structure and Nucleotide Sequence of the Cytochrome B Gene in Yeast Mitochondrial DNA

A. Tzagoloff and F. G. Nobrega[1]

Introduction

The mitochondrial respiratory chain contains two distinct cytochromes of the b type (b_k and b_t). Cytochromes b_k and b_t differ in their redox potentials and absorption properties in the visible spectrum (Table 1). Both cytochromes are part of the coenzyme QH_2-cytochrome c reductase and have been implicated to function on the main pathway of the electron transport chain (Rieske, 1976). Despite the fact that cytochrome b has been isolated from fungal (Weiss and Ziganke, 1974) and mammalian (VonJagow et al., 1978) sources, the relationship of the purified protein to cytochromes b_k and b_t has not been satisfactorily resolved and until recently it was not clear whether they represent two different or a single heme protein.

Table 1. Properties of mitochondrial b type cytochromes

| Name | Absorption maxima of reduced band | | | | Midpoint potential (mV) |
| | 23°C | | | 77 K | |
	α	β	γ	α	
b_k or b_{562}	562	532	430	560	+30 to +65
b_t or b_{566}	566	538	432	562.5	-30 to -50

Chance et al. (1970); Rieske (1976)

During the past several years genetic and molecular analyses of the apocytochrome b gene have helped to enlarge our understanding of the structure of this important member of the respiratory chain. In this communication we would like to review some of the information that has emerged from studies of cytochrome b-deficient mutants of yeast and to present some more recent data on the organization and DNA sequence of the gene. The combined genetic and biochemical evidence indicates that cytochromes b_k and b_t are products of a single mitochondrial gene. The gene is composed of three exons which code for a protein with a molecular weight of 44,000. The specific heme content of purified yeast cytochrome b further suggests that there are two hemes associated with the protein. Based on this evidence we propose cytochromes b_k and b_t to be a single protein with two separate heme-binding sites.

[1]Department of Biological Sciences, Columbia University, New York, N.Y. 10027, USA

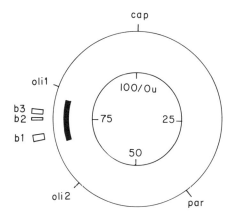

Fig. 1. Physical map of yeast mtDNA. The positions of the antibiotic resistance loci *oli1*, *oli2* (oligomycin), *cap* (chloramphenicol), and *par* (paromycin) are indicated on the *outer circle*. The *open boxes* show the location of the cytochrome b exons and the *darkly drawn arc* of the segment of DNA retained in DS400/A12. The map units are marked on the *inner circle*

Genetic of Cytochrome b

Weiss et al. (1973) first demonstrated cytochrome b of *Neurospora crassa* to be synthesized on mitochondrial ribosomes. The addition of chloramphenicol, a known inhibitor of mitochondrial ribosomes, to a suspension of cells, completely blocked the incorporation of labeled amino acids into a mitochondrial translation product whose migration on SDS gels was coincident with cytochrome b. Similar results have been obtained with yeast (Katan et al. 1976b) and it is now thought that the mitochondrial origin of cytochrome b may be general for all eukaryotic organisms.

In addition to being translated in mitochondria, cytochrome b is also a gene product of mitochondrial DNA (mtDNA). This became evident when cytoplasmic mutants of *Saccharomyces cerevisiae* were isolated with specific deficiences in coenzyme QH_2-cytochrome c reductase (Tzagoloff et al., 1975). The respiratory-deficient phenotype was often correlated with either the complete absence of the apoprotein or the presence of new low-molecular weight polypeptides structurally related to cytochrome b (Tzagoloff et al. 1976; Claisse et al. 1978; Alexander et al. 1979).

Several laboratories have carried out extensive genetic studies of the yeast cytoplasmic mutants deficient in cytochrome b. Recombinational analysis and petite deletion mapping conclusively established all the mutations to be confined to a region of mtDNA between the *oli1* and *oli2* resistance loci (Slonimski and Tzagoloff 1976; Tzagoloff et al. 1976; Slonimski et al. 1978; Haid et al. 1979; Alexander et al. 1979). The location of the cytochrome b region on the wild-type mtDNA of yeast is shown in Figure 1. Furthermore, depending on the strain of yeast studied, the mutations were found to fall into a number of different genetic loci variously designated as *cob* or *box*. Each locus represented a clustering of genetically linked mutational sites. The fact that crosses of mutants from different loci produced wild-type recombinants with frequencies approaching values of nonlinked markers, suggested that genetic determinants of cytochrome b are scattered over a long region of the mitochondrial genome (Tzagoloff et al. 1976; Slonimski et al., 1978). This was demonstrated more directly by physical mapping of the cytochrome b mutations (Slonimski et al. 1978; Alexander et al. 1979). In the D273-10B strain reported here, the minimal distance between the most distal cytochrome b mutations has been estimated to be 3.5 kilobases (kb) (Nobrega and Tzagoloff 1980a). In other strains

of yeast such as KL14-4A (Grivell et al. 1979) the cytochrome b gene
is even longer, spanning some 7 kb.

Detergent solubilized apocytochrome b has an apparent molecular weight
of 30,000 measured either on SDS gels (Katan et al. 1976a,b; Lin et
al. 1978) or on molecular sieving columns (Weiss and Ziganke 1975).
A colinear gene coding for a protein of this size should, therefore,
be 800 - 900 nucleotides long. The earlier evidence for the clustering
of cytochrome b mutations in genetically unlinked loci together with
the discrepancy in the size of the protein and the physical length of
the gene was interpreted to indicate that it might have a mosaic structure similar to other eukaryotic split genes.

The presence of intervening sequences in the cytochrome b gene has
been confirmed by electron microscopic visualization of RNA:DNA hybrids (Grivell et al. 1979) and by the DNA sequence of the gene (Nobrega and Tzagoloff 1980b). Grivell et al. (1979) reported discontinuous hybridization of mtDNA to a mitochondrial RNA species presumed to
be the cytochrome b messenger. The single-stranded DNA loops observed
in such hybrids agreed well with the predicted number and sizes of introns based on the genetic and physical mapping data. The KL14-4A
strain used in the studies of Grivell et al. (1979) was found to have
four introns totaling some 6 kb of DNA. As shown below, the cytochrome
b gene of the D273-10B strain has two introns with lengths of 1.4 and
0.7 kb.

Enrichment of the Cytochrome b Gene in ρ⁻ Clones

To sequence the cytochrome b gene, we isolated ρ⁻ clones whose mitochondrial genomes were deleted for all known markers except those of
cytochrome b. The clones were all derived from the respiratory competent haploid D273-10B previously shown by Morimoto and Rabinowitz
(1979) to lack two long inserts in the cytochrome b region of wild-
type mtDNA. In this strain the overall length of the gene is approximately 3.5 kb shorter than in the so-called "long" strains.

The most genetically complex clone used in our studies (DS400/A12) was
established to have a tandemly repeated segment of mtDNA with a repeat
length of 7.6 kb. The genotype of DS400/A12 was consistent with the
presence of the entire gene. Furthermore, restriction analysis of the
DNA indicated that the DS400/A12 segment spanned the region of wild-
type mtDNA from 69 to 80 units where the gene had been previously
mapped.

Since the DS400/A12 genome was too long to be sequenced directly, a
series of less complex clones was isolated by sequential mutagenesis
of DS400/A12 with ethidium bromide. The genetic properties of some of
these derivative clones are listed in Table 2. The clones were selected
primarily for their differential retention of various cytochrome b
markers in the *cob1* and *cob2* loci. Based on the genotypes and unit
lengths of their genomes, the clones represented a broad range of com-
plexities. Some of the simpler clones (N series) exhibited single marker retention with unit lengths of only several hundred base pairs.

Physical Map of the Cytochrome b Gene in *S. cerevisiae* D273-10B

A complete physical map of the cytochrome b region has been constructed
by digestion of the mtDNA's of DS400/A12 and of the less complex clones
with combinations of restriction endonucleases (Nobrega and Tzagoloff

Table 2. Genetic Properties of ρ⁻ clones

| Clone | *oli1* | *cob1* | | | | | *cob2* | | | | | Unit length of mtDNA |
		M7-40	M8-53	M13-101	M6-200	M8-181	M9-228	M33-119	M21-71	M17-162	M10-152	(kb)
DS400/A12	−	+	+	+	+	+	+	+	+	+	+	7.6
DS401	+	−	−	−	−	−	−	−	−	−	−	5.4
DS400/N1	−	−	−	−	−	−	−	−	−	−	+	0.44
DS400/M8	−	−	−	−	−	−	+	+	+	+	+	1.6
DS400/M4	−	−	−	−	+	+	+	+	+	+	+	3.5
DS400/N9	−	−	−	−	−	−	+	+	+	+	−	0.94
DS400/N7	−	−	−	−	−	−	+	−	−	−	−	0.68
DS400/N23	−	−	−	−	−	+	−	−	−	−	−	0.3
DS400/N2	−	+	−	−	−	−	−	−	−	−	−	0.58
DS400/M11	−	+	+	+	−	−	−	−	−	−	−	3.5

Plus and minus signs indicate the presence or absence of the genetic marker in the ρ⁻clone. (Nobrega and Tzagoloff 1980a)

1980a). The results of these analyses were useful in obtaining not only the restriction map of DS400/A12 itself, but also in aligning the
segments of the low-complexity clones on the overall map and in assign-
ing fairly precise physical limits to the cytochrome b mutations present in the various *cob* mutants. The orientation of DS400/A12 with
respect to the wild-type map was deduced from the occurrence of HincII,
BamHI, BglII, HhaI, and EcoRI sites that had been mapped on the mtDNA
of D273-10B (Morimoto and Rabinowitz 1979). The deletion endpoints of
the DS400/A12 segment were established by comparing the restriction
maps and nucleotide sequences of DS400/A12 and DS401. Since the latter
clone had a genome that partially overlapped with DS400/A12 at the
oli1 proximal end, the sequence divergence between the two mtDNA's provided the exact nucleotides where the deletions had been initiated.

The physical maps of the different ρ⁻ genomes and the locations of
the cytochrome b markers are depicted in Figure 2. The DS400/A12 segment includes the sequence between 69.3 and 80.2 units. The DS400/M11
and M4 genomes represent approximately equal halves of the DS400/A12
segment. The low complexity N clones span different subregions of
DS400/A12. All the markers previously assigned to the *cob1* locus are
located between 74.7 and 76.3 units while the *cob2* markers fall within
the region of 71.5 and 72.9 units. The maximal separation of the most
distal *cob1* (M7-40) and *cob2* (M10-150) mutations is 3.4 kb, confirming
earlier findings on the physical size of the gene in this strain.

Nucleotide Sequence of the Apocytochrome b Gene

The nucleotide sequence of the segment retained in DS400/A12 was obtained by sequencing the genomes of the low-complexity clones and also
of preparative restriction fragments of the DS400/A12 itself. All the
sequences were determined by the chemical modification method of Maxam

5

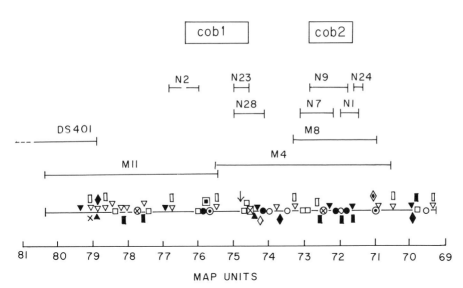

Fig. 2. Physical maps of DS400/A12 and of low-complexity mtDNA's. The restriction sites are indicated by the following symbols: HpaII (Δ), HaeII ( ), SacII (X), TaqI (◆), MboII (□), MboI (O), EcoRI (⊗), BamHI (■), HphI ( ), AluI (●), HinfI (▲), EcoRII (↓), BglII (x), HincII (⊙), HindIII (◊), HhaI (◈). The physical limits of the *cob1* and *cob2* mutations are indicated by the *upper boxes*. (Nobrega and Tzagoloff 1980a)

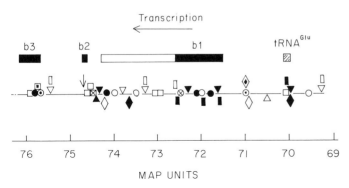

Fig. 3. Physical location of the cytochrome b exons and of the glutamic acid tRNA gene. The symbols used for the restriction sites are the same as in Fig. 2. The three cytochrome b exons *(b1, b2, b3)* are depicted by the *solid bars*. The extension of the reading frame in the first intron is indicated by the *open bar*. The direction of transcription of genes is shown by the *arrow*. (Nobrega and Tzagoloff 1980b)

and Gilbert (1977). An analysis of the DNA sequence has revealed the presence of two genes in the DS400/A12 mtDNA. One of the genes, located at 70 map units consists of a 72 nucleotide long sequence that can be folded into a tRNA structure with a 5'-UUC-3' anticodon. This gene codes for the glutamic acid tRNA previously mapped in the *cob* region of wild-type mtDNA by Wesolowski and Fukuhara (1979). The second gene occurs in the region from 71.4 to 76.3 units. This gene is 3308 nucleotides long and contains a number of reading frames, three of which

```
    -100
5'-TAATTAATAATATATATTTATATATTTTTTATTAATTAATATATATAAAA TATTAGTAATAAATAATATTATTAATATTTTATAAATAAATAATAATAAT


      fMet Ala Phe Arg Lys Ser Asn Val Tyr Leu Ser Leu Val Asn Ser Tyr Ile Ile Asp Ser Pro Gln Pro Ser Ser
   +1 ATG GCA TTT AGA AAA TCA AAT GTG TAT TTA AGT TTA GTG AAT AGT TAT ATT ATT GAT TCA CCA CAA CCA TCA TCA


      Ile Asn Tyr Trp Trp Asn Met Gly Ser Leu Leu Gly Leu Cys Leu Val Ile Gln Ile Val Thr Gly Ile Phe Met
  +76 ATT AAT TAT TGA TGA AAT ATG GGT TCA TTA TTA GGT TTA TGT TTA GTT ATT CAA ATT GTA ACA GGT ATT TTT ATG


      Ala Met His Tyr Ser Ser Asn Ile Glu Leu Ala Phe Ser Ser Val Glu His Ile Ile Arg Asp Val His Asn Gly
 +151 GCT ATG CAT TAT TCA TCT AAT ATT GAA TTA GCT TTT TCA TCT GTT GAA CAT ATT ATA AGA GAT GTG CAT AAT GGT


      Tyr Ile Leu Arg Tyr Leu His Ala Asn Gly Ala Ser Phe Phe Phe Met Val Met Phe Met His Met Ala Lys Gly
 +226 TAT ATT TTA AGA TAT TTA CAT GCA AAT GGT GCA TCA TTC TTT TTT ATG GTA ATG TTT ATG CAT ATG GCT AAA GGT


      Leu Tyr Tyr Gly Ser Tyr Arg Ser Pro Arg Val Thr Leu Trp Asn Val Gly Val Ile Ile Phe Ile Leu Thr Ile
 +301 TTA TAT TAT GGT TCA TAT AGA TCA CCA AGA GTA CTA TTA TGA AAT GTA GGT GTT ATT ATT TTC ATT TTA ACT ATT


      Ala Thr Ala Phe Leu Gly Tyr Cys Cys Val Tyr Gly Gln Met Ser His Trp Gly Ala Thr Val Ile Thr Asn Leu
 +376 GCT ACA GCT TTT TTA GGT TAT TGT TGT GTT TAT GGA CAG ATG TCA CAT TGA GGT GCA CTA GTT ATT ACT AAT TTA


      Phe Ser Ala Ile Pro Phe Val Gly Asn Asp Ile Val Ser Trp Leu Trp Gly Gly Phe Ser Val Ser Asn Pro Thr
 +451 TTC TCA GCA ATT CCA TTT GTA GGT AAC GAT ATT GTA TCT TGA TTA TGA GGT GGG TTC TCA GTA TCT AAC CCT CTA


      Ile Gln Arg Phe Phe Ala Leu His Tyr Leu Val Pro Phe Ile Ile Ala Ala Met Val Ile Met His Leu Met Ala
 +526 ATC CAG AGA TTC TTT GCG TTA CAT TAT TTA GTA CCT TTT ATC ATT GCT GCA ATG GTT ATT ATG CAT TTA ATG GCA


      Leu His Ile His Gly Ser Ser Asn Pro Leu Gly Ile Thr Gly Asn Leu Asp Arg Ile Pro Met His Ser Tyr Phe
 +601 TTA CAT ATT CAT GGT TCA TCT AAT CCA TTA GGT ATT ACA GGT AAT TTA GAT AGA ATT CCA ATG CAT TCA TAC TTT


      Ile Phe Lys Asp Leu Val Thr Val Phe Leu Phe Met Leu Ile Leu Ala Leu Phe Val Phe Tyr Ser Pro Asn Thr
 +676 ATT TTT AAA GAT TTA GTA ACT GTT TTC TTA TTT ATG TTA ATT TTA GCA TTA TTT GTA TTC TAT TCA CCT AAT ACT

                                                     +2174
      Leu Gly Gln                                    His Pro Asp Asn Tyr Ile Pro Gly Asn Pro Leu Val Thr Pro Ala
 +751 TTA GGT CAA (---- 1414 nucleotides----) CAT CCT GAT AAC TAT ATT CCT GGT AAT CCT TTA GTA ACA CCA GCA

                                                     +2958
      Ser Ile Asp                                    Val Pro Glu Trp Tyr Leu Leu Pro Phe Tyr Ala Ile Leu Arg Ser
+2219 TCT ATT GAT (---- 730 nucleotides ----) GTA CCT GAA TGA TAC TTA TTA CCA TTC TAT GCT ATT TTA AGA TCT


      Ile Pro Asp Lys Leu Leu Gly Val Ile Thr Met Phe Ala Ala Ile Leu Val Leu Leu Val Leu Pro Phe Thr Asp
+3003 ATT CCT GAT AAA TTA TTA GGA GTT ATT CTA ATG TTT GCA GCT ATT TTA GTA TTA TTA GTT TTA CCA TTT ACT GAT


      Arg Ser Val Val Arg Gly Asn Thr Phe Lys Val Leu Ser Lys Phe Phe Phe Phe Ile Phe Val Phe Asn Phe Val
+3078 AGA AGT GTA GTA AGA GGT AAT ACT TTT AAA GTA TTA TCT AAA TTC TTC TTC TTT ATC TTT GTA TTC AAT TTC GTA


      Leu Leu Gly Gln Ile Gly Ala Cys His Val Glu Val Pro Tyr Val Leu Met Gly Gln Ile Ala Thr Phe Ile Tyr
+3153 TTA TTA GGA CAA ATT GGA GCA TGC CAT GTA GAA GTA CCT TAT GTC TTA ATG GGA CAA ATC GCT ACA TTT ATC TAC


      Phe Ala Tyr Phe Leu Ile Ile Val Pro Val Ile Ser Thr Ile Glu Asn Val Leu Phe Tyr Ile Gly Arg Val Asn
+3228 TTC GCT TAT TTC TTA ATT ATT GTA CCT GTT ATC TCT ACT ATT GAA AAT GTT TTA TTC TAT ATC GGT AGA GTT AAT


      Lys TER
+3303 AAA TAA TATATAATTAAATTAATACATAGATATAATA -3'
```

Fig. 4. Nucleotide sequences of the cytochrome b exons. The sequence shown is that of the nontranscribed strand. The amino acid sequences encoded in the exons are shown above the DNA sequence. There is an ambiguity of one amino acid *(underlined)* at each of the two exon junctures (Nobrega and Tzagoloff 1980b)

have been identified to be exons of the cytochrome b gene. Each exon
can be translated into a protein sequence averaging 50% homology with
the cytochrome b encoded in bovine mtDNA (B. Barrell and F. Sanger,
pers. comm.).

The structure of the yeast gene is shown in Figure 3. The sequences
constituting the exons together with the amino acid sequences are presented in Figure 4. The first exon (b1) starts with an AUG initiator
at approximately 71.4 units. The exon is 756 - 759 nucleotides long,
although the reading frame itself continues for an additional 1180
nucleotides before terminating with an ochre codon. The second exon
(b2) consists of only 48 - 54 nucleotides and lies 1136 nucleotides
downstream from the end of exon b1. The third exon (b3) is 349 - 361
nucleotides long and is separated from exon b2 by 730 nucleotides.
The cytochrome b gene is flanked by A+T-rich sequences that are characteristic of spacer regions in yeast mtDNA. The amino acid sequence
encoded in the gene indicates a hydrophobic protein with stretches of
25 - 35 amino acids without a single charged residue. Yeast cytochrome
b is composed of 384 amino acids and from its composition has a molecular weight of 44,000. This value is 30% higher than estimated by
SDS gel electrophoresis (Katan et al. 1976; Lin et al. 1978).

As mentioned above, the first intron in the cytochrome b gene contains
a long reading frame capable of coding for a protein with a molecular
weight of 45,000. This sequence has a somewhat lower G+C content (22%)
than most of the known mitochondrial genes (27% - 33% G+C). Its amino
acid composition suggests a rather basic protein with an unusually
high content of lysine. It is unclear at present whether this intron
sequence is transcribed and translated into a bona fide protein. It
is of interest that a long reading frame has also been found in the
first intron of the cytochrome b gene of the KL14-4A strain (Jacq et
al. 1980) and the intron sequence of the 21S ribosomal RNA gene (Dujon
1980). These observations have raised the intriguing possibility that
the intron sequences code for enzymes required for the maturation of
the cytochrome b messenger (Jacq et al. 1980; Church and Gilbert 1980).

Organization of the Cytochrome b Gene in Other Strains of Yeast

Restriction maps of the cytochrome b region indicate the presence in
some strains of two inserts with a total length of approximately 3.5
kb. Each insert has been shown to correspond to an intervening sequence
within the structural gene of aprocytochrome b (Grivell et al. 1979).
A comparison of the restriction maps of D273-10B and Kl14-4A suggests
the crucial difference to be in the part of the gene that codes for
the amino terminal half of the protein (Fig. 5). Whereas the first
252 - 253 amino acid residues in D273-10B are encoded in a single exon,
in KL14-4A this sequence is fragmented into three smaller exons. The
extra exons and introns account for the larger number of genetic loci
(box) reported in the "long" strains (Slonimski et al. 1978; Haid et
al. 1979; Alexander et al. 1979).

The organizational variations in the gene in closely related strains
of yeast raises some interesting questions concerning the role of the
intron sequences in the expression of cytochrome b. There is no reason
to believe that the primary structure of yeast cytochrome b differs
among the strains examined. Nor is there evidence for strain-specific
differences in the regulation of cytochrome b biosynthesis. Although
it cannot be excluded that the introns may have some subtle regulatory
role, their presence may simply reflect an active mitochondrial re-

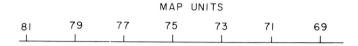

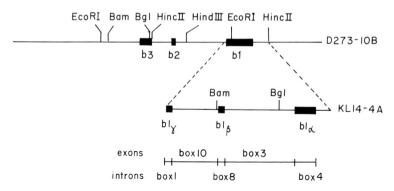

Fig. 5. Structure of the cytochrome b gene in *S. cerevisiae* D273-10B and KL14-4A.
Only those restriction sites which have been mapped on the wild-type mtDNA are
shown. The cytochrome b exons are drawn as *solid bars*. The correspondence between
the first three exons in KL14-4A and the genetic loci *(box)* is shown in the lower
part of the figure

combinational activity which results in the transposition of certain
sequences. The insertion of a sequence into a gene is probably tolerated, provided it can be efficiently removed at the RNA level.

Is There a Single Cytochrome b with Two Heme Prosthetic Groups?

All the known mutations leading to a deficiency of cytochrome b (b_k
and b_t) have been mapped in the *oli1-oli2* span of the genome. The complete sequence of this region of DNA (Nobrega and Tzagoloff 1980a,b),
indicates the presence of a single gene which finds its counterpart in
mammalian mtDNA (B. Barrell et al., this vol.). The reading frame found
in the first intron is unlikely to code for a second cytochrome b on
the following grounds. (1) The amino acid sequence encoded by the intron indicates a basic protein which does not agree with the known properties of mitochondrial cytochrome b. (2) No major mitochondrial product with a high content of lysine is detected in in vivo labeling experiments (G. Coruzzi, unpubl. data).

The existence of only one cytochrome b gene in mtDNA implies that cytochromes b_k and b_t are either a single polypeptide with two heme groups
or that the same protein with a single heme by virtue of different en-
vironments gives rise to the characteristic properties of the two cytochromes. According to the former interpretation, the spectral and redox potential differences of the two cytochromes are the result of two
distinct heme binding sites on the same polypeptide.

The presence of two heme attachment sites on the same protein is supported by the properties of purified yeast cytochrome b. The coenzyme
QH_2-cytochrome c reductase complex of yeast mitochondria has been fractionated on DEAE-cellulose and hydroxylapatite in the presence of detergents to yield a spectrally pure cytochrome b with a heme content

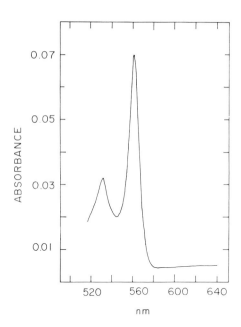

Fig. 6. Absorption spectrum of purified yeast mitochondrial cytochrome b. The concentration of protein was 0.042 mg per ml. (F. Foury, unpubl. data)

ranging from 51 - 54 nmol per mg protein (Fig. 6). The protein migrates as a single band on SDS gels with only trace amounts of contamination by a low molecular weight polypeptide. The heme content of the purified cytochrome indicates a maximum molecular weight of 19,000 - 20,000, assuming one heme per polypeptide chain. Since the molecular weight of cytochrome b calculated from the amino acid sequence is 44,000, the specific heme content is most consistent with the presence of two hemes per polypeptide chain.

Acknowledgments. This research was supported by Grant HL22174 from the National Institutes of Health, United States Public Health Service.

References

Alexander NJ, Vincent RD, Perlman PS, Miller DH, Hanson DK, Mahler HR (1979) Genetic and biochemical characterization of some mutant types affecting apocytochrome b and cytochrome oxidase. J Biol Chem 254:2471-2479

Chance B, Wilson DF, Dutton PL, Erecinska M (1970) Energy-coupling mechanisms in mitochondria: Kinetic spectroscopic and thermodynamic properties of an energy-transducing form of cytochrome b. Proc Natl Acad Sci USA 66:1175-1182

Church GM, Gilbert W (1980) Yeast mitochondrial intron products required in trans for RNA splicing. In: Joseph DR, Schultz J, Scott WA, Werner R (eds) Mobilization and reassembly of genetic information. Academic Press, London New York, in press

Claisse ML, Spyridakis A, Wambier-Kłuppel ML, Pajot P, Slonimski PP (1978) Analysis of proteins translated from the *box* region. In: Bacila M, Horecker BL, Stoppani AOM (eds) Biochemistry and genetics of yeast: Pure and applied aspects. Academic Press, London New York, pp. 369-390

Dujon B (1980) Nucleotide sequences of the intron and flanking exons of the mitochondrial 21S rRNA gene of yeast strains with different alleles at the *omega* and *rib1* genetic loci. Cell submitted for publication

Grivell LA, Arnberg AC, Boer PH, Borst P, Bos JL, van Bruggen EFJ, Groot GSP, Hecht
    NB, Hensgens LAM, van Ommen GJB, Tabak HF (1979) Transcripts of yeast mitochondrial
    DNA and their processing. In: Cummings DJ, Borst P, Dawid IB, Weissman SM, Fox
    CF (eds) Extrachromosomal DNA: ICN-UCLA symposia on molecular and cellular biology,
    vol XV. Academic Press, London New York, pp 305-324
Haid A, Schweyen RJ, Bechmann H, Kaudewitz F, Solioz M, Schatz G (1979) The mito-
    chondrial COB region in yeast codes for apocytochrome b and is mosaic. Eur J
    Biochem 94:451-464
Jacq C, Lazowska J, Slonimski PP (1980) Sur un nouveau mecanism de la regulation
    de l'expression genetique. CR Acad Sci Paris Ser D t290:1-4
Katan MB, van Harten-Loosbroek N, Groot GSP (1976a) The cytochrome bc_1 complex of
    yeast mitochondria. Eur J Biochem 65:95-105
Katan MB, van Harten-Loosbroek N, Groot GSP (1976b) The cytochrome bc_1 complex of
    yeast mitochondria: Site of translation of the polypeptides in vivo. Eur J Biochem
    70:409-417
Lin LF, Clejan L, Beattie DS (1978) The synthesis of cytochrome b on mitochondrial
    ribosomes in Baker's yeast. Eur J Biochem 87:171-179
Maxam A, Gilbert W (1977) A new method for sequencing DNA. Proc Natl Acad Sci USA
    74:560-564
Morimoto R, Rabinowitz M (1979) Derivation of the fine structure map of strain
    D273-10B and comparison with a strain (MH41-7B) differing in genome size. Mol
    Gen Genet 170:25-48
Nobrega FG, Tzagoloff A (1980a) Complete restriction map of the cytochrome b region
    on mitochondrial DNA in *Saccharomyces cerevisiae* D273-10B. J Biol Chem submitted
    for publication
Nobrega FG, Tzagoloff A (1980b) DNA sequence and organization of the cytochrome b
    gene in *Saccharomyces cerevisiae* D273-10B. J Biol Chem submitted for publication
Rieske JS (1976) Composition and function of complex III of the respiratory chain.
    Biochim Biophys Acta 456:195-247
Slonimski PP, Tzagoloff A (1976) Localization in yeast mitochondrial DNA of mutations
    expressed in a deficiency of cytochrome oxidase and/or coenzyme QH_2-cytochrome c
    reductase. Eur J Biochem 61:27-41
Slonimski PP, Pajot P, Jacq C, Foucher M, Perrodin G, Kochko A, Lamouroux A (1978)
    Genetic, physical and complementation maps of the *box* region. In: Bacila M,
    Horecker BL, Stoppani AOM (eds) Biochemistry and genetics of yeast: Pure and
    applied aspects. Academic Press, London New York, pp 339-368
Tzagoloff A, Akai A, Needleman RB, Zulch G (1975) Cytoplasmic mutants of *Saccharomyces
    cerevisiae* with lesions in enzymes of the respiratory chain and in the mitochondrial
    ATPase. J Biol Chem 250:8236-8242
Tzagoloff A, Foury F, Akai A (1976) Genetic loci on mitochondrial DNA involved in
    cytochrome b biosynthesis. Mol Gen Genet 149:33-42
VonJagow G, Schagger H, Engel WD, Machleidt W, Machleidt I, Kolf HJ (1978) Beef
    heart complex III: Isolation and characterization of cytochrome b. FEBS Lett 91:
    121-125
Weiss H, Ziganke B (1974) Cytochrome b in *Neurospora crassa* mitochondria: Site of
    translation of the heme protein. Eur J Biochem 41:63-71
Weiss H, Sebald W, Schwab AJ, Kleinow W, Lorenz B (1973) Contribution of mitochondrial
    and cytoplasmic protein synthesis to the formation of cytochrome b and cytochrome
    aa_3. Biochimie 55:815-821
Wesolowski M, Fukuhara H (1979) The genetic map of transfer RNA genes in yeast mito-
    chondria: Correction and extension. Mol Gen Genet 170:261-275

Sequence of Mammalian Mitochondrial DNA

B. G. Barrell, S. Anderson, A. T. Bankier, M. H. L. De Bruijn, E. Chen, A. R. Coulson,
J. Drouin, I. C. Eperon, D. P. Nierlich, B. Roe, F. Sanger, P. H. Schreier, A. J. H. Smith,
R. Staden, and I. G. Young[1]

Introduction

The human mitochondrial (mt) genome consists of a closed circular duplex DNA approximately 10×10^6 daltons and has been the most intensely
studied animal mt genetic system. The positions of the origin of replication of H strand synthesis (Crews et al. 1979), the 12S and 16S
ribosomal RNA genes (Robberson et al. 1972) and 19 tRNA genes (Angerer
et al. 1976) have been located on the genetic map shown in Figure 1.
A number of discrete products of mitochondrial protein synthesis have
been demonstrated and three of them identified as subunits 1, 2 and 3
of the cytochrome oxidase complex (Hare et al. 1980). In comparison
with other mito-systems, genes for up to four subunits of the ATPase
complex, one of the cytochrome bc_1 complex and possibly for a ribosomal protein would be expected to be present (see review by Borst
1977). Both strands are thought to be completely transcribed symmetrically from a point near the origin of the H strand synthesis (Aloni
and Attardi 1971; Murphy et al. 1975). These transcripts are then processed to give the rRNAs, the tRNAs and a number of polyadenylated but
not capped mRNAs (Attardi et al. 1979). Both the L and H strands have
been shown to be coding with the L strand containing the sense sequence
of the rRNA genes, most of the tRNA genes and most of the stable polyadenylated mRNAs.

We have obtained a complete sequence of human mt DNA and a near complete sequence of bovine mt DNA using the dideoxynucleotide chain termination method of Sanger et al. (1977) and Sanger and Coulson (1978).
Initially restriction fragments (MboI, EcoRI/HindIII) of human mt DNA
obtained from a single placenta were cloned in the plasmid pBR322 by
Drouin (1980) and single stranded template DNA was prepared according
to the exonuclease III procedure of Smith (1979). For the bovine mt
DNA clones were similarly prepared from an EcoRI/BamHI digest. These
fragments were then isolated and further digested with enzymes that
recognize tetrameric sequences, and these digests were "shotgunned"
into the single stranded M13 vector M13mp2 (Gronenborn and Messing
1978; Schreier and Cortese 1979; Sanger et al. 1980). These clones
were sequenced using a "universal primer" complementary to the region
adjacent to the point of insertion. The sequences obtained by this
random method were collated and analysed using the computer programs
of Staden (1977, 1978, 1979, 1980a,b).

[1]MRC Laboratory of Molecular Biology, Hills Road, Cambridge, CB2 2QH, England

12

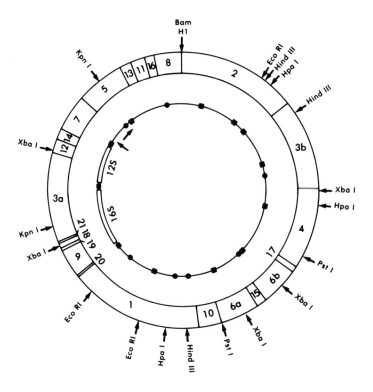

Fig. 1. The genetic map of the human mtDNA genome (Angerer et al. 1976; Ojala and
Attardi 1977) aligned with the MboI restriction enzyme map (Drouin 1980). The origin
and direction of H strand synthesis is shown. The L strand sequence is anticlockwise
and c des for the 12S and 16S rRNAs and tRNAs marked ■ (complementary to H strand).
The ⁻ strand codes for the tRNAs marked ● (complementary to L strand)

Different Genetic Codes

One of the first results of the sequence analysis of human mt DNA was
the identification of the cytochrome oxidase subunit II gene (COII
gene). By comparison of the DNA sequence with the complete amino acid
sequence of the bovine COII determined by Steffens and Buse (1979) we
were able to show that the termination codon UGA coded for tryptophan
as well as UGG in human mitochondria and also to predict that AUA was
methionine as well as AUG and not an isoleucine codon (Barrell et al.
1979). The use of AUA as a methionine codon has recently been confirmed
by comparison of the bovine COII gene with the bovine protein sequence
(I.G. Young and S. Anderson, unpublished). More recently we have evidence for further coding changes in mammalian mt in that AGA and AGG
are probably termination codons and not arginine and that AUA and possibly AUU can serve as initiation codons as well as AUG. The evidence
for these changes is discussed in the sections entitled "Protein Coding
Genes" and "Initiation and Termination of Translation". In a similar
comparison with the yeast COII gene, Macino et al. (1979) and Fox (1979)
were also able to show that UGA is a tryptophan codon in yeast mt DNA.
However, the human and yeast mitochondrial genetic codes are not identical, AUA being isoleucine as in the "universal" genetic code. In addition another change has been found in the yeast mitochondrial genetic

Table 1. Different mitochondrial genetic codes

Codon	Genetic code			
	Mammalian mt	Yeast mt	*Neurospora* mt	"Universal"
UGA	Trp	Trp	Trp	Term.
AUA	Met	Ile	n.d.[b]	Ile
CUA[a]	Leu	Thr	Leu	Leu
AGA/G	Term.	Arg.	n.d.[b]	Arg

[a]In yeast mitochondria it is probable that all of the codon family CUN are threonine

[b]Codon response not determined

code which is not found in that of human mitochondria, in that CUA is a threonine codon and not leucine. This was established by comparison of the yeast mt ATPase subunit 9 protein sequence (Sebald and Wachter 1978) with the sequence of its gene (Hensgens et al. 1979; Macino and Tzagoloff 1979) and the sequence analysis of the yeast mitochondrial tRNA gene (Li and Tzagoloff 1979). Sequence analysis of *Neurospora* mt tRNAs indicates that this genetic system also has a different genetic code similar to but not identical with either mammalian or yeast (Heckman et al. 1980). Thus in the three mt genetic systems so far studied — mammalian, yeast and *Neurospora* — all three have genetic codes which differ not only from the "universal" genetic code but also from each other. These codes are summarised in Table 1.

Organisation of the Genome

Figure 2 shows an attempt to understand the organisation of the human mitochondrial genome. Apart from small differences in the number of bases in the intergenic regions it is likely that the bovine map is the same from the sequence so far determined. The lack of genetics, the difficulties in identifying and sequencing the mainly hydrophobic, N-terminally blocked mt proteins, the paucity of RNA sequence data and the novel features found in the organisation and expression of mammalian mt DNA necessitates that this map is conjectural and based mainly on the DNA sequence data. Thus any mistake in the DNA sequence, e.g. the addition or deletion of a single base in a reading frame, could alter the map. However, the pattern emerging from the sequence and the comparison with the bovine mt DNA sequence suggests that the map is essentially correct.

Origins of Replication

The start points for replication have been located in the HeLa cell mt DNA sequence for the H strand (Crews et al. 1979) and for the L strand in mouse L cells (Martens and Clayton 1979). In the case of the H strand origin, the sequence at the 5' end of the stable 7S DNA of the D loop has been determined, but there is no sequence available at the 3' end of this DNA. The size of this 7S DNA has been estimated to be approximately 680 bases. The distance between the 5' end of the 7S DNA and the proline tRNA gene is 736 bases so that most, if not all, of this region corresponds to the 7S DNA.

14

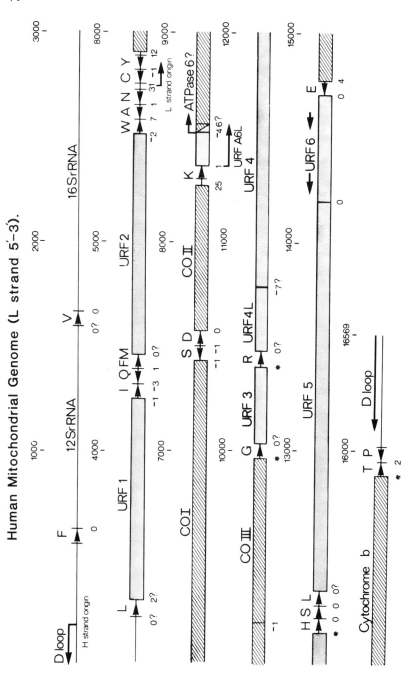

Fig. 2. Gene map of human mtDNA deduced as described in the text. URFs are unidentified reading frames. All protein coding genes are L strand coded except URF 6. tRNA genes are denoted by the one-letter amino acid code and are either L strand coded (>) or H strand coded (<). Numbers above the genes show the scale in base pairs and below the predicted number of bases between genes. * denotes no termination codon in the DNA sequence

The origin of L strand synthesis is located in or near a small 31 base intergenic region between the H strand coded Cys and Asn tRNA genes by comparison of the sequence determined for mouse L cells with the human DNA sequence (Martens and Clayton 1979; Sanger et al. 1980). In both sequences this region contains a stable hairpin loop structure which probably serves as a start point for L strand synthesis.

Ribosomal RNA Genes

The positions of the ribosomal RNA genes and the tRNA genes flanking these sequences have been established on the map shown in Figure 1 (Robberson et al. 1972; Angerer et al. 1976). Sequence analysis of the 5' ends of the 12S and 16S ribosomal RNAs by Attardi et al. (1979) has allowed these ends to be precisely located in the DNA sequence (Eperon et al. 1980). The three tRNA genes in this region have been identified in the DNA sequence using the computer program TRNA (Staden 1980a) and identified by their anticodon sequences. This establishes the order of genes tRNA^{Phe}-12S rRNA-tRNA^{Val}-16S rRNA-tRNA^{Leu}. There are no non-coding nucleotides between the genes for tRNA^{Phe} and the 12S rRNA and similarly between tRNA^{Val} and the 16S rRNA. The 3' sequences of the ribosomal RNAs are not known so it is not possible to say whether there is a similar situation existing between the 3' ends of the rRNA genes and the tRNA^{Val} and tRNA^{Leu} genes. However, we consider it likely that these genes are also contiguous or at least very close on the basis of the sizes of the rRNAs. Also, homology at the sequence preceding the tRNA^{Val} gene with sequences present at the 3' end of other small subunit rRNAs suggests that these genes are very close. Thus it is quite possible that an enzyme or enzymes similar to RNase P could process the primary transcript with single cuts between these genes and release the three tRNAs and two ribosomal rRNAs without the need for further trimming.

No evidence exists for a 5S type rRNA molecule in mitochondrial ribosomes nor has any similar sequence been found in the DNA sequence. The small "3.5S" RNA found in the mt tRNA fraction has been suggested to be a mt 5S rRNA equivalent (Dubin et al. 1974). We believe this RNA molecule to be a tRNA and this will be discussed later in the section on "Different tRNA structures".

Protein Coding Genes

The economy of gene organisation found in the rRNA gene region is also seen in the rest of the molecule which is coding for proteins and tRNAs. Only a very small proportion of the DNA has no predicted coding function, the intergenic distances ranging from only a few bases to butt joints, i.e. no non-coding bases and, in a few cases, to small overlaps between genes.

Where there are significant distances between tRNA genes one or two open reading frames, i.e. potential protein coding genes, are found that span all or nearly all the inter-tRNA regions. These are shown in Figure 2; the labelled regions are those reading frames which have been identified as a particular gene and the stippled regions are unidentified reading frames (URFs). The gene for cytochrome oxidase subunit I (COI) has been identified from an N-terminal amino acid sequence of the bovine protein determined by J. Walker (personal communication), the COII gene from the complete amino acid sequence of Steffens and Buse (1979) and COIII from an N-terminal amino acid sequence of the bovine protein (Buse, personal communication) and also by homology with the yeast mt COIII gene (Tzagoloff, personal communication). The genes for cytochrome b and tentatively ATPase 6 were identified by homology with the corresponding yeast genes (Tzagoloff, personal communication).

In two reading frames ATPase 6? and URF4 the reading frame is not continuous between the flanking genes. In both cases a short reading frame is found starting at the tRNA^{Lys} and tRNA^{Arg} genes respectively, and then a larger reading frame starts up and continues to the next gene. It is not known whether these reading frames both specify a short and long polypeptide product or whether this break in the reading frame is due to an intervening sequence. There is always the possibility that it may be due to mistakes in the DNA sequence. In all reading frames the bovine mt DNA sequence is approximately 70% homologous, whereas the predicted amino acid sequence is 75% homologous. The differences in the DNA sequence are mainly due to third position changes in the codons of the predicted reading frame, making it extremely likely that we have correctly identified unknown protein coding genes. The homology with the yeast ATPase 6? gene is high at the 3' end of the gene with very little homology at the 5' half of the gene, but it is not sufficient to say where the mammalian gene starts. This low homology may indicate that this region codes for a similar protein and not ATPase 6, but in any case the identification of the yeast ATPase 6 gene is tentative. All possible protein genes are coded on the L strand, with one exception — URF6, as are the majority of the tRNA genes and the two rRNA genes.

In some cases the predicted ATG initiation codon is preceded by ATA methionine codons and in URF3 the only possible initiation codon is an ATA which is preserved in the bovine sequence followed by high amino acid homology. Thus it is likely that AUA may be an intiation codon as well as AUG in this genetic system. URF1 in bovine mt is predicted to begin with ATA, but in human mt there is an ATT codon (isoleucine) in the equivalent position. Thus it is possible that AUU might serve as an initiation codon as well as AUG and AUA in mammalian mitochondria.

The identification of URF3 as well as the other protein genes is further supported by the map of the polyadenylated limit transcription products determined by Attardi and co-workers (personal communication). This map corresponds with the sizes and positions of URF1, URF2, COI, COII, URFA6L and ATPase 6?, COIII, URF3, URF4L and URF4, URF5 and the H strand URF6 region and cytochrome b. Although there are large L strand transcripts covering the region of the H strand coded URF6, it is not yet known whether these are further processed to yield a mRNA nearer the size of URF6 to translate this predicted protein.

This map predicts that a minimum of 11 - 13 proteins are coded by mammalian mitochondria with the predicted molecular weights shown in Table 2. Some of the reading frames are larger than any known mitochondrially coded proteins. The reason for this is not clear and could be due to intervening sequences or protein processing or, more likely, inaccuracy in the estimation of molecular weight of hydrophobic proteins on SDS-acrylamide gels. The number and sizes of these predicted proteins are in fairly good agreement with those estimated by gel electrophoresis (J.E. Walker, personal communication). The function of the 6 - 8 unidentified reading frames is not known. The small ATPase 9 gene coded by yeast mt (Hensgens et al. 1979; Macino and Tzagoloff 1979) does not seem to be coded by mammalian mt DNA, there being no homologous sequences with the bovine ATPase 9 protein sequence (Wachter, personal communication). It is likely that at least one of the unidentified

Table 2. Sizes of mammalian mt proteins predicted from the DNA sequence

Name	Length	Molecular weigth
Cytochrome c oxidase subunit I	513	57,000
Cytochrome c oxidase subunit II	227	25,500
Cytochrome c oxidase subunit III	261	30,000
Cytochrome b	380	42,700
ATPase subunit 6?	226	24,800
Unidentified reading frame 1	318	35,600
Unidentified reading frame 2	347	38,900
Unidentified reading frame A6L	68	7,900
Unidentified reading frame 3	115	13,200
Unidentified reading frame 4L	98	10,700
Unidentified reading frame 4	459	51,400
Unidentified reading frame 5	603	66,600
Unidentified reading frame 6	174	18,600

reading frames specifies a ribosomal protein, as is the case in yeast mitochondria. Preliminary analysis of HeLa cell mt ribosomal proteins indicates that possibly three ribosome associated proteins are synthesized by the mitochondrion (J.E. Walker, personal communication).

Initiation and Termination of Translation

The processing model discussed for the ribosomal RNAs and the mRNAs predicts that the mRNAs either start with an initiation codon or else have only a few non-coding bases prior to the initiation codon at their 5' end. This would rule out a prokaryotic mechanism of translation initiation and this is borne out by the observation that there is no complementarity between the predicted end of the 12S rRNA and the sequences before the predicted initiation codons (Eperon et al. 1980). Thus it is likely that the first AUG/A (and possibly AUU) codon in the mRNA is recognised as the initiation codon for translation, as is thought to be the case for AUG in eukaryotic mRNA (Kozak 1978). There is a precedent in a prokaryote for a mRNA beginning with an initiation codon which is that of the bacteriophage λ prm transcript (Walz et al. 1976).

Whereas some reading frames terminate at or close to the junction with other genes with TAA or TAG codons, others do not, and the reading frame continues into the next gene and sometimes beyond. For several reasons we think that this does not happen and two hypotheses are discussed below to allow termination of these reading frames at the junction with the next gene.

AGA and AGG Are Termination Codons

Arginine is normally coded by six codons, CGN and AGA/G. All codons are used in the reading frames shown in Figure 2 except the arginine codons AGA and AGG. These codons are only found in human mt DNA in the

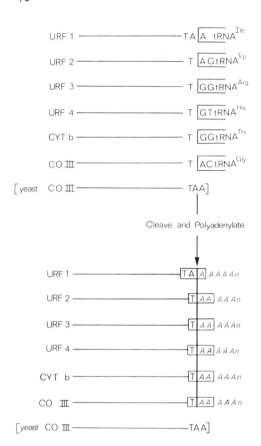

Fig. 3. The transcriptional processing and polyadenylation model for termination of translation

frame, either as the last codon before the next gene or when the reading frame continues into the next gene. In the case of COI and URF6 the last codon before the tRNA^{Ser} and URF5 genes respectively are AGA and AGG. There is very high amino acid homology at the 3' end of URF6 between human and bovine mitochondria, and significantly the bovine reading frame ends with the termination codon TAA in exactly the position where the human reading frame is postulated to end in AGG. We do not have the bovine DNA sequence for the end of COI, but the bovine cytochrome b has an AGA as the penultimate codon to the tRNA^{Thr} gene at its 3' end. The composition of a C-terminal peptide of bovine cytochrome b (Machleidt, personal communication) supports the use of this AGA as a termination codon. Human cytochrome b does not have a termination codon at this junction and a different model for termination of translation for this human gene and other genes is discussed later. The use of AGA/G as termination codons is further supported by the fact that we have been unable to find a mt tRNA gene for these codons. The number of tRNA genes and how they read the mt genetic code is discussed in a later section.

A Transcriptional Processing and Polyadenylation Model for Termination of Translation

The map of the stable polyadenylated transcription products (Attardi, personal communication) agrees very well with the reading frame map shown in Figure 2. Although the ends of the mRNAs have not been precisely located in the DNA sequence the transcription map agrees with the prediction from the DNA sequence in that the tRNA genes or sequences very close to them are the signals for processing the primary transcript. As discussed for the processing of the rRNA and associated tRNA genes, an RNase P type enzyme could cleave the primary transcript with single clips to yield the mature rRNA and tRNA sequences. If, as seems likely, a similar situation exists for the processing of the protein genes and associated tRNA genes, then a single clip at the exact 5' terminus of the tRNA genes would, in the case of URF1 and URF2, destroy the termination codon for the preceding protein gene or leave it with no termination codon at all in the case of COIII, URF3, URF4 and cytochrome b. It is possible that there are different pathways for processing the primary transcript so as to avoid the complication of the overlapping termination codons with the tRNA genes (URF1/ tRNA^{Ile} and URF2/tRNA^{Trp}). However, we think that this is unlikely from the observation that a single clip exactly before the first base of the tRNA gene followed by polyadenylation of the RNA preceding the tRNA gene will recreate the termination codon in the case of URF1, create a different termination codon in the case of URF2 and create termination codons in phase for those reading frames which have no termination codons in the DNA sequence at the junctions with the next gene. This model is illustrated in Figure 3. A similar mechanism could operate for the production of the two separate mRNAs for ATPase6? and COIII, which have overlapping termination codons and initiation codons respectively. Here a single clip before the ATG of COIII destroying the ATPase6? termination codon, followed by polyadenylation of the ATPase6? mRNA would recreate the TAA termination codon for ATPase6?. However, as we have found no tRNA gene in this region, it is not clear what the signal for the processing would be.

In the case of COIII the model predicts that a termination codon not present in the DNA sequence would be created in the position shown in Figure 3, which is in exactly the same place as that present in the DNA sequence for the yeast COIII gene (Tzagoloff, personal communication). The high amino acid homology of this predicted COIII C-terminus with that of yeast COIII as shown below lends further support for the model.

Human	L D V	V	W L F L Y V	S I	Y W W G S
Yeast	L D V	I	W L F L Y V	L F	Y W W G V

However, the correctness of the model and, if correct, how generally it is used in mammalian mitochondria, will depend on sequence analysis of the 3' ends of the mRNAs at the junction with the poly(A).

tRNA Genes

Twenty-two tRNA genes have been located in the human sequence and their position and identity confirmed in the bovine sequence, and also by direct sequence analysis of the bovine tRNAs. This may represent the full mt complement of tRNAs except that there is evidence for a methionine tRNA in addition to the formylatable species (Aujame and Freeman 1979). This tRNA specifying internal methionine would be predicted to have the anticodon UAU, which would respond to the mt methionine codons AUG and AUA. This anticodon would also be predicted for the formyl methionine initiator tRNA from the results of the comparison of the bovine and human sequences which suggest that AUA is an initiation codon as well as AUG (and possibly AUU). The only tRNA^{Met} gene that we have identified with certainty is probably an initiator tRNA by comparison with other initiator tRNAs, having the characteristic three G:C base

SECOND LETTER

FIRST LETTER	U	C	A	G	THIRD LETTER
U	UUU Phe 11	UCU 32	UAU Tyr 46	UGU Cys 5	U
	UUC Phe 140	UCC Ser 99	UAC Tyr 89	UGC Cys 17	C
	UUA Leu 73	UCA Ser 83	UAA Ter –	UGA Trp 93	A
	UUG Leu 17	UCG Ser 7	UAG Ter –	UGG Trp 10	G
C	CUU 65	CCU 41	CAU His 18	CGU 7	U
	CUC Leu 167	CCC Pro 119	CAC His 79	CGC Arg 25	C
	CUA 276	CCA 52	CAA Gln 81	CGA 29	A
	CUG 45	CCG 7	CAG Gln 9	CGG 2	G
A	AUU Ile 125	ACU 51	AAU Asn 33	AGU Ser 14	U
	AUC Ile 196	ACC Thr 155	AAC Asn 130	AGC Ser 39	C
	AUA 166	ACA 133	AAA Lys 85	AGA Ter –	A
	AUG Met 40	ACG 10	AAG Lys 10	AGG Ter –	G
G	GUU 30	GCU 43	GAU Asp 15	GGU 24	U
	GUC Val 49	GCC Ala 124	GAC Asp 55	GGC Gly 88	C
	GUA 71	GCA 80	GAA Glu 64	GGA 67	A
	GUG 18	GCG 8	GAG Glu 24	GGG 34	G

Fig. 4. The human mitochondrial genetic code showing the predicted total number of codons used in the map shown in Fig. 2. The boxed codons show how the genetic code is read by the mt-tRNAs. The 4 codon/1 amino acid boxes are each read by single tRNAs with U in the first position of the antiocodon. The non-family or split boxes with two codons for one amino acid are read by tRNAs having the correct G:U wobble anticodons. Because of the uncertainty in the number of methionine tRNAs and their codon response as discussed in the text, these are not boxed

pairs at the base of the anticodon loop and the unique CC before the anticodon sequence CAU (see Wrede et al. 1979). This tRNA should thus correspond to an initiator tRNA that recognises the codon AUG. Two possible candidates for a tRNA[Met] specific for AUG/A have been located in the human sequence, but one of these structures is not preserved in the bovine sequence and the other is in a region where we have no bovine data. Thus at this stage we are unable to say how many methionine tRNAs are specified by mammalian mt or what their coding properties are.

Different Decoding Mechanism

The number of tRNAs (22 - 23) are not sufficient to decode all the mt codons using the rules of G:U wobble (Crick 1966) where in this genetic system at least 30 tRNAs would be required. All codons except AGA and AGG are used and no tRNAs are imported from the cytoplasm (Aujame and Freeman 1979; Roe and Chen, unpublished). A mechanism using only 22 - 23 tRNAs to decode all the mt codons is suggested by the observation that

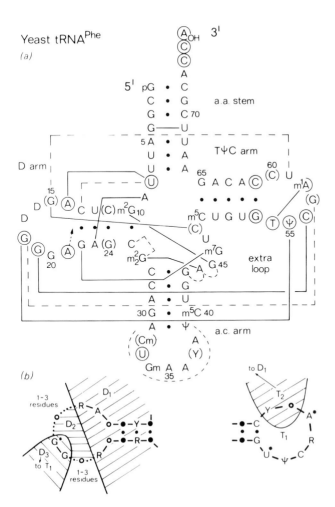

Yeast tRNA^Phe (a)

Fig. 5. The general structure of tRNA using yeast tRNA^Phe as an example showing the circled invariant bases and bracketed semi-invariant bases and their tertiary interactions (Ladner et al. 1975)

for the genetic code boxes containing four codons for one amino acid (family boxes) there is only one tRNA specific for each of these boxes (Barrell et al. 1980). These eight tRNAs all have first position U in the anticodon as judged by the DNA sequence, the state of modification not beeing known for these tRNAs. This tRNA using the G:U wobble rules would only decode the bottom two codons in these boxes, i.e. those ending in A and G. No tRNAs with anticodons corresponding to the upper two codons in these boxes, i.e. those ending in U and C, are found in the mitochondria. Thus these tRNAs must decode all four codons in these boxes either by the first U in the anticodon recognising all four bases in the third position of the codon (U:N wobble) or by a "two out of three" mechanism as proposed by Lagerkvist (1978). A similar situation has been found in yeast (Bonitz et al. 1980) and in *Neurospora* (Heckman et al. 1980). The other genetic code boxes, i.e. those containing two sets of two codons for different amino acids, are read normally by their tRNAs having the predicted G:U wobble anticodons (Fig. 4). However, there must be a discrimination mechanism whereby the tRNAs with first position U for the bottom two codons in the non-family boxes do not read the top two codons in these boxes, as is proposed for the

22

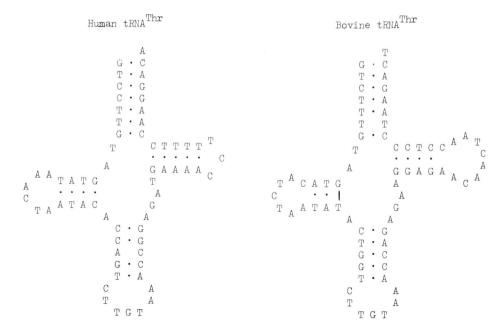

Fig. 6. Human and bovine tRNA^{Thr}

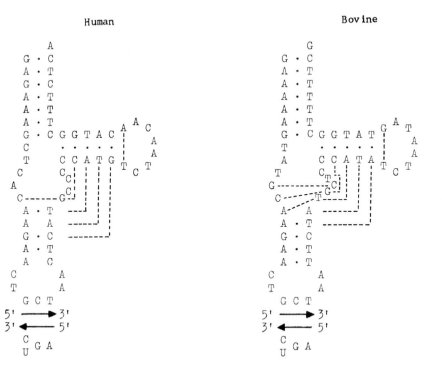

Fig. 7. Human and bovine tRNA^{Ser}

family boxes, and thus cause the wrong amino acids to be inserted. In *Neurospora* Heckman et al. (1980) find that the tRNA specific for the four codons in the family boxes has an unmodified U in the first position of the anticodon, whereas the U in the first position of the tRNAs specific for the bottom two codons of the non-family boxes is modified. This may serve to prevent these tRNAs from recognising the top two codons for a different amino acid in these boxes. It is likely that a similar pattern of modification will be found for the mammalian mt tRNAs.

Different tRNA Structures

Mammalian mt tRNA structures are different from all other known structures. Over 200 tRNA sequences have been compiled by Sprinzl et al. (1979), and all the chain elongation tRNAs have the same constant features in the "D-loop" and "TψC-loop shown in Figure 5. Only one human mt tRNA, a leucine tRNA specific for UUA/G, shows these constant features. The rest of these tRNAs vary in the sequence and size of these loops. For example, the constant seven bases in the TψC-loop vary from three to nine bases. This variability of sequence and loop size is also found when the same tRNA species is compared between human and bovine, as shown in the example for tRNA^{Thr} in Figure 6.

In normal tRNAs the constant bases in these two loops interact in the tertiary structure model for yeast tRNA^{Phe} (Ladner et al. 1975; Kim et al. 1974). Because of the tremendous variation in these regions of mammalian mt tRNAs it has not been possible to predict the homologous interactions, if they exist, in these tRNAs.

The most unusual tRNA structure found is that of the mammalian mt tRNA^{Ser} specific for the codons AGU/C shown in Figure 7. This unusual structure, which lacks the "D-loop" and "D-arm" has been shown to be a tRNA by sequence analysis of the bovine tRNA as well as that of the human and bovine gene sequences. It is post-transcriptionally modified by the addition of CCA$_{OH}$ and at least one minor base has been identified in the tRNA sequence. As predicted by the anticodon, this tRNA will accept serine when isolated mitochondria are incubated in the presence of ^{3}H-serine (M.H.L. de Bruijn, unpublished results). How this tRNA folds up into a stable tertiary structure is at present a mystery.

Summary

The unusual features discussed above present a picture of a genetic system which has been isolated for a considerable time from the mainstream of prokaryotic and eukaryotic evolution. Whether these features represent a more primitive system or a very highly evolved specialist system is difficult to tell. An attractive theory of mitochondrial evolution is that they evolved from primitive bacteria that entered into a symbiotic relationship with the cell at an early stage in the evolution of eukaryotes. Unlike the homologous sequences found in wheat mt ribosomal RNAs with bacterial rRNA, a similar comparison with the mammalian mt rRNAs has not revealed any recent common ancestor for the mammalian mt genetic system with that of any other system (Eperon et al. 1980).

References

Aloni Y, Attardi G (1971) Symmetrical in vivo transcription of mitochondrial DNA in HeLa cells. Proc Natl Acad Sci USA 68:1757-1761

Angerer L, Davidson N, Murphy W, Lynch D, Attardi G (1976) An electron microscope study of the relative positions of the 4S and ribosomal RNA genes in HeLa cell mitochondrial DNA. Cell 9:81-90

Attardi G, Cantatore P, Ching E, Crews S, Gelfland R, Merkel C, Ojala D (1979) The organisation of the genes in the human mitochondrial genome and their mode of transcription. In: Cummings D, Borst P, Dawid J, Weissman S, Fox CF (eds) Proc ICN-UCLA Symp on Extrachromosomal DNA. Academic Press, New York, pp 443-469

Aujame L, Freeman KB (1979) Mammalian mitochondrial transfer RNAs: chromatographic properties, size and origin. Nucleic Acids Res 6:455-469

Barrell BG, Bankier AT, Drouin J (1979) A different genetic code in human mitochondria. Nature (London) 282:189-194

Barrell BG, Anderson S, Bankier AT, de Bruijn MHL, Chen E, Coulson AR, Drouin J, Eperon IC, Nierlich DP, Roe BA, Sanger F, Schreier PH, Smith AJH, Staden R, Young IG (1980) Different pattern of codon recognition by mammalian mitochondrial tRNAs. Proc Natl Acad Sci USA 77:3164-3166

Bonitz SG, Berlani R, Coruzzi G, Li M, Nobrega FG, Nobrega MP, Thalenfeld BE, Tzagoloff A, Macino G (1980) Assembly of the mitochondrial membrane system: Decoding rules of yeast mitochondria. Proc Natl Acad Sci USA 77:3167-3170

Borst P (1977) Structure and function of mitochondrial DNA. TIBS 2:31-34

Crews S, Ojala D, Posakony J, Nishiguchi J, Attardi G (1979) Nucleotide sequence of a region of human mitochondrial DNA containing the precisely identified origin of replication. Nature (London) 277:192-198

Crick FHC (1966) Codon — Anticodon pairing: The wobble hypothesis. J Mol Biol 19:548-555

Drouin J (1980) Human mitochondrial DNA: Cloning in E. coli. J Mol Biol 140:15-34

Dubin DT, Jones TH, Cleaves GR (1974) An unmethylated "3S" RNA in hamster mitochondria: A 5S RNA equivalent? Biochem Biophys Res Commun 56:401-406

Eperon IC, Anderson S, Nierlich DP (1980) The distinctive sequence of human mitochondrial ribosomal RNA genes. Nature (London) in press

Fox TD (1979) Five TGA stop codons occur within the translated sequence of the yeast mitochondrial gene for cytochrome C oxidase subunit II. Proc Natl Acad Sci USA 76:6534-6538

Gronenborn B, Messing J (1978) Methylation of single-stranded DNA in vitro introduces new restriction endonuclease cleavage sites. Nature (London) 272:375-377

Hare JF, Ching E, Attardi G (1980) Isolation, subunit composition, and site of synthesis of human cytochrome c oxidase. Biochemistry 19:2023-2030

Heckman JE, Sarnoff J, Alzner-De Weerd B, Yyn S, RajBhandary UL (1980) Novel features in the genetic code and codon reading patterns in Neurospora crassa mitochondria based on sequence of six mitochondrial tRNAs. Proc Natl Acad Sci USA 77:3159-3163

Hensgens LAM, Grivell LA, Borst P, Bos JL (1979) Nucleotide sequence of the mitochondrial structural gene for subunit 9 of yeast ATPase complex. Proc Natl Acad Sci USA 76:1663-1667

Kim SH, Suddath FL, Quigley GJ, McPherson A, Sussman JL, Wang AHJ, Seeman NC, Rich A (1974) Science 185:435-440

Kozak M (1978) How do eucaryotic ribosomes select initiation regions in messenger RNA. Cell 15:1109-1123

Ladner JE, Jack A, Robertus JD, Brown RS, Rhodes D, Clark BFC, Klug A (1975) Structure of yeast phenylalanine transfer RNA at 2.5 Å resolution. Proc Natl Acad Sci USA 72:4414-4418

Lagerkvist U (1978) "Two out of three": An alternative method for codon reading. Proc Natl Acad Sci USA 75:1759-1762

Li M, Tzagoloff A (1979) Assembly of the mitochondrial membrane system: Sequences of yeast mitochondrial valine and an unusual threonine tRNA gene. Cell 18:47-53

Macino G, Tzagoloff A (1979) Assembly of the mitochondrial membrane system: Partial sequence of a mitochondrial ATPase gene in Saccharomyces cerevisiae. Proc Natl Acad Sci USA 76:131-135

Macino G, Coruzzi G, Nobrega FG, Li M, Tzagoloff A (1979) Use of the UGA terminator as a tryptophan codon in yeast mitochondria. Proc Natl Acad Sci USA 76:3784-3785

Martens PA, Clayton DA (1979) Mechanism of mitochondrial DNA replication in mouse L-cells: Localization and sequence of the light-strand origin of replication. J Mol Biol 135:327-351

Murphy W, Attardi B, Tu C, Attardi G (1975) Evidence for complete symmetrical transcription in vivo of mitochondrial DNA in HeLa cells. J Mol Biol 99:809-814

Ojala D, Attardi G (1977) A detailed physical map of HeLa cell mitochondrial DNA and its alignment with the positions of known genetic markers. Plasmid 1:78-105

Ojala D, Montoya J, Attardi G (1980) The putative mRNA for subunit II of cytochrome c oxidase in human mitochondria starts directly at the translation initiator codon. Submitted to Nature (London)

Robberson D, Aloni Y, Attardi G, Davidson N (1972) Expression of the mitochondrial genome in HeLa cells. VIII The relative position of ribosomal RNA genes in mitochondrial DNA. J Mol Biol 64:313-317

Sanger F, Coulson AR (1978) The use of thin acrylamide gels for DNA sequencing. FEBS Lett 87:107-110

Sanger F, Nicklen S, Coulson AR (1977) DNA sequencing with chain-terminating inhibitors. Proc Natl Acad Sci USA 74:5463-5467

Sanger F, Coulson AR, Barrell BG, Smith AJH, Roe BA (1980) Cloning in single-stranded bacteriophage as an aid to rapid DNA sequencing. J Mol Biol 143:161-178

Schreier PH, Cortese R (1979) A fast and simple method for sequencing DNA cloned in the single-stranded bacteriophage M13. J Mol Biol 129:169-172

Sebald W, Wachter E (1978) In: Schäfer G, Klingenberger M (eds) 29th Mosbacher Colloq. Energy conservation in biological membranes. Springer, Berlin Heidelberg New York, p 228

Smith AJH (1979) The use of exonuclease III for preparing single-stranded DNA for use as a template in the chain terminator sequencing method. Nucleic Acids Res 6:831-848

Sprinzl M, Grüter F, Gauss DH (1979) Compilation of tRNA sequences. Nucleic Acids Res Spec Suppl r1-r19

Staden R (1977) Sequence data handling by computer. Nucleic Acids Res 4:4037-4051

Staden R (1978) Further procedures for sequence analysis by computer. Nucleic Acids Res 5:1013-1015

Staden R (1979) A strategy of DNA sequencing employing computer programs. Nucleic Acids Res 6:2601-1610

Staden R (1980a) A computer program to search for tRNA genes. Nucleic Acids Res 8:817-825

Staden R (1980b) A new computer method for the storage and manipulation of DNA gel reading data. Nucleic Acids Res in press

Steffens GJ, Buse G (1979) Studies on cytochrome c oxidase, IV Primary structure and function of subunit II. Hoppe-Seyler's Z Physiol Chem 360:613-619

Walz A, Pirrotta V, Ineichen K (1976) λ repressor regulates the switch between Pr and Prm promoters. Nature (London) 262:665-669

Wrede P, Woo NH, Rich A (1979) Initiator tRNAs have a unique anticodon loop conformation. Proc Natl Acad Sci USA 76:3289-3293

The Optional Introns in Yeast Mitochondrial DNA[1]

P. Borst[2]

Introduction

Research on mitochondrial biogenesis has concentrated to a large extent on yeast in recent years. The reasons for this are obvious: yeast is easy to grow and easy to label with radioactive precursors; mitochondrial biogenesis can be manipulated in yeast by glucose repression or anaerobiosis; and finally, mitochondrial genetics is highly developed in yeast, allowing a concerted genetic and biochemical attack on the problems involved in mitochondrial biogenesis. In view of the preceding article on yeast, I shall concentrate here on two topics. In the main part of my article I shall discuss our present knowledge of the nature of the large insertions/deletions in yeast mtDNA, discovered by Sanders et al. (1977) in our lab. In the last part I shall briefly compare some of the common and uncommon features of mitochondrial biogenesis in fungi, protozoa, animal cells, and plants.

The "Sanders" Inserts

Intact yeast mtDNA molecules were first found in 1968 by Hollenberg et al. (1970) in electron micrographs of lysed yeast mitochondria. Unfortunately, up till today nobody has succeeded in isolating a sizeable fraction of these 25-µm circles intact. I attribute this to the high nuclease content of mitochondrial preparations, which hit the DNA in the brief moment between mitochondrial lysis and denaturation of the nucleases, but other explanations have not been excluded. Final proof that the 25-µm circles are the only type of mtDNA present in yeast came from the restriction enzyme analysis by Sanders et al. (1975, 1976, 1977), which showed that also the broken DNA molecules in these mtDNA preparations are derived from the same class of 25-µm circles.

When Sanders and Heyting analyzed the mtDNAs from a whole series of yeast strains with restriction enzymes, they found that these could be grouped in four classes (Sanders et al. 1977), which differ by major insertions and deletions of 1 to 3 kb. Representatives of three of these are shown in Figure 1.

The first question that one can ask about these insertions is: "Are they comparable to the Insertion Sequences found in bacterial DNA and in the nuclear DNAs of some eukaryotes"? The hallmark of bacterial In-

---

[1]Abbreviations: kb, kilo-base pair(s); bp, base pair(s); kd, kilo-dalton(s); rRNA, ribosomal RNA; mRNA, messenger RNA; DBM-paper, diazobenzyloxymethyl-cellulose paper; $I_{\alpha\beta}$, intron between coding regions α and β

[2]Section for Medical Enzymology and Molecular Biology, Laboratory of Biochemistry, University of Amsterdam, Jan Swammerdam Institute, P.O. Box 60.000, 1005 GA Amsterdam, The Netherlands

28

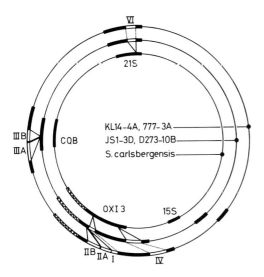

Fig. 1. The major insertions/deletions in yeast mtDNA that give rise to the strain differences. The *circles* represent the mitochondrial genomes of *Saccharomyces cerevisiae* KL14-4A, 777-3A, JS1-3D and D273-10B and *S. carlsbergensis* NCYC 74. The mitochondrial loci *cob* and *oxi*-3 and the genes for the 21S and 15S rRNAs are shown as *closed bars*. The *cross-hatched region* on the *oxi*-3 bar indicates that many *oxi*-3 transcripts have short regions of homology with the adjacent *oli*-2 region (Van Ommen et al. 1979). The *open bars* represent the various large insertions responsible for size differences of mitochondrial loci in different yeast strains. The nomenclature of these is according to Sanders et al. (1977), apart from insertions II and III which have since been found to consist of closely adjacent smaller insertions, termed IIA, IIB, IIIA and IIIB (Van Ommen et al. 1979). (From Van Ommen et al. 1980)

sertion Sequences is their mobility. There are usually multiple copies per genome and they move around. We have checked this by DNA-DNA hybridization for the mtDNA insertions I, II (Hensgens, L.A.M., unpubl.), III (Grivell, L.A. et al., unpubl.), IV (Borst et al. 1977) and VI (Heyting et al. 1979; Bos et al. 1980). In all these cases labeled fragments from the insert hybridize only with the homologous insert. It follows that there is only one copy of these inserts per genome and that this copy is at a fixed position. The Sanders insertions/deletions are, therefore, not insertion sequences in the usual sense of the word. Moreover, the genetic information in these inserts, if any, is not present in strains that lack these inserts, so they cannot carry information that is essential for mitochondrial biogenesis in all *Saccharomyces* strains.

The main reason for studying the Sanders inserts in more detail was their position on the mitochondrial map. As shown in Figure 1 each of these inserts is found in a region known to contain a mitochondrial gene. This provided the incentive for further experiments.

Insert VI is an Intervening Sequence in the 21S rRNA Gene

A more detailed restriction enzyme map of the region in which the gene for 21S ribosomal RNA (rRNA) is located, showed that this RNA hybridized to DNA fragments that were not contiguous on the physical map and suggested that insert VI was actually within the rRNA gene (Fig. 2).

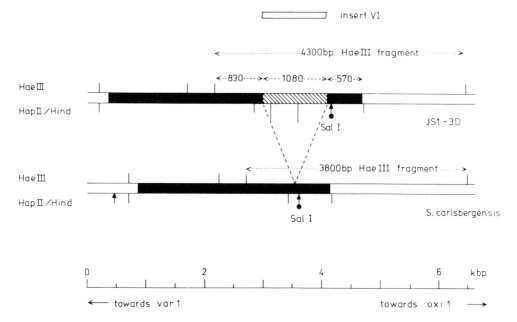

Fig. 2. Physical maps of the 21S rRNA gene in *S. cerevisiae* strain JS1-3D and in *S. carlsbergensis*. The *large bars* are restriction maps of the region of the 21S rRNA gene (cf. Fig. 1) in yeast mtDNA. Recognition sites for the restriction endonucleases HaeIII, HapII, HindII+III and SalI are given. The *black bars* give the position of mature rRNA sequences, the *hatched part* is the intron in the rRNA gene. The transcription is from *left* to *right* in this and other figures. (From Bos et al. 1980)

In that same period Jeffreys and Flavell (1977) discovered in our lab that the gene for β-globin in rabbits contains an intervening sequence. It was rather obvious that the same might be the case with the gene for rRNA in yeast mitochondria and this was proven by electron microscopy of DNA-rRNA hybrids (Bos et al. 1978a). We then asked whether strains that lack insert VI have no intervening sequence at all, or a small one which cannot be detected by restriction enzyme analysis or electron microscopy. This question was analyzed by S₁ nuclease digestion of homologous hybrids between rRNA and a DNA fragment from a strain that lacks insert VI. S₁ nuclease will cut a mismatched region in the hybrid and this will happen even if the DNA loop consists of a few nucleotides. No mismatch was found in this way and we therefore conclude that strains without insert VI also lack an intervening sequence at that position in the rRNA gene (Bos et al. 1980).

Split genes in nuclear DNA are completely transcribed and the intron sequence is removed from the precursor RNA by splicing. A similar mechanism has been found for the rRNA gene in strains that contain Insertion VI (Bos et al. 1980). An interesting point is that the excised insert is relatively stable and present at higher concentration than other intermediates in splicing (Bos et al. 1980). Another important point is that splicing occurs exactly at the borders of insert VI and hence that the sequence of mature rRNA in the region of the insert is the same whether the rRNA comes from a strain with or without insert VI (see Bos et al. 1980).

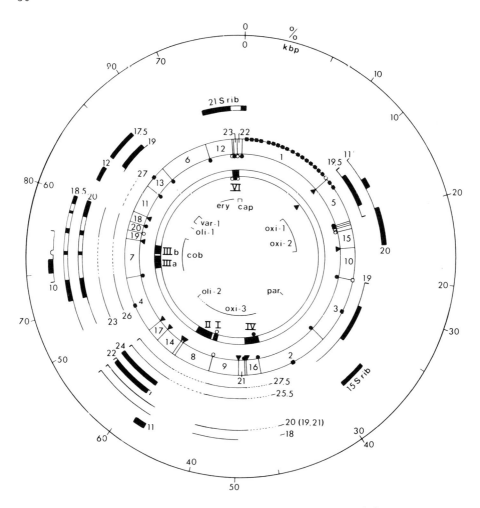

Fig. 3. Detailed transcription map of the mtDNA of *S. cerevisiae* KL14-4A. The positions of the genetic loci are shown inside the *inner ring*. The *major ring* represents the restriction map of the mtDNA with the fragments numbered in order of decreasing size. Restriction sites for EcoRI (▼), HindII (●) and HindIII (○) are indicated inside this major ring. The *inner ring* shows the positions of the large insertions in *S. cerevisiae* KL14-4A not present in *S. carlsbergensis* (■). The approximate positions of 4S RNA genes is given on the *outer circle of the major ring* (■). The *open squares* represent methionine-tRNA genes. The positions of the transcripts are shown outside the map, the *thin lines* indicating uncertainty in the exact location and the *heavy bars* their approximate length. *Openings in the RNA bars* correspond to the positions of intervening sequences in DNA. Homology with inserted regions is indicated by *dotted lines*. (From Van Ommen et al. 1979)

Analysis of the transcripts of the rRNA gene in petite mutants by Rabinowitz and co-workers has shown the presence of RNAs that extend far beyond the 3'-end of the gene (Synenki et al. 1979; Morimoto et al. 1979). J.P. Schouten, K.A. Osinga and H.F. Tabak (pers. comm.) have also obtained evidence for a longer precursor in wild-type yeast mitochondria, even in *Saccharomyces carlsbergensis*, a strain not containing

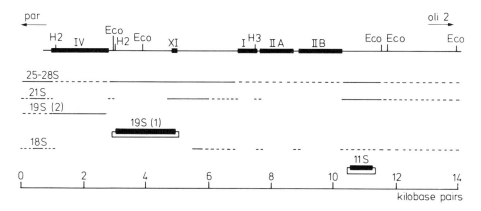

Fig. 4. Preliminary transcription map of *oxi*-3. The figure shows a portion of the restriction map of *S. cerevisiae* strain KL14-4A. *Left* to *right* in this figure is clock-wise in Figs. 1 and 3. The *upper line* presents the restriction map with recognition sites for endonucleases HhaI, HindII (H2), HindIII (H3), EcoRI, and BamHI. Each transcript is indicated by its sedimentation coefficient and the approximate extent of homology with the DNA is indicated by (-), whereas (...) denotes uncertainty in the exact localization. (Adapted from Van Ommen et al. 1980)

insert VI. Synthesis of mature rRNA may therefore involve more processing steps than the splicing required to remove insert VI.

Insertions I, II and IV, *oxi*-3 and the Discovery of Circular RNAs in Mitochondria

If insert VI is an intervening sequence in the rRNA gene, the obvious possibility is that the other inserts are intervening sequences in other genes. The first evidence for this came from a detailed analysis of mitochondrial transcripts by Van Ommen, Groot and Grivell (Van Ommen and Groot 1977, 1979) in our lab. As shown in Figure 3, there are two regions where a complex series of overlapping transcripts can be found: the *oxi*-3 region which contains inserts I, II and IV and the *cob* region, where insert III is located.

The *oxi*-3 region (Fig. 4) probably contains the structural gene for subunit I of cytochrome c oxidase. Physical mapping of the defects in *oxi*-3 mutants by Grivell and Moorman (1977) established that *oxi*-3 includes inserts I, II and IV and in strain KL14-4A is at least 11,000 bp, i.e., nearly 10 times the minimal size required to encode subunit I (molecular weight about 40,000). Using a coupled transcription-translation system they then showed that DNA fragments from both sides of the *oxi*-3 region gave rise to antigenic determinants of cytochrome c oxidase in the in vitro system, whereas the large HindII+III fragment from the middle of the region (map units 3 - 7.5 in Fig. 4) did not. They proposed (Grivell and Moorman 1977) that the *oxi*-3 gene is split and this was the first suggestion that structural genes for mitochondrial proteins may be split as well.

Further analysis of the transcripts in the *oxi*-3 region has confirmed the presence of spliced transcripts (Fig. 4) and shown that insert II consists of two separate inserts — IIA and IIB. We do not know yet which of these transcripts contains the messenger RNA (mRNA) activity

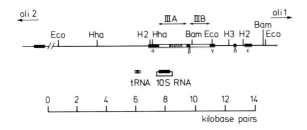

Fig. 5. The *cob* region in mtDNA from *S. cerevisiae* strain KL14-4A. Sites for BamHI (Bam), EcoRI (Eco; O), HhaI (Hha), HindII (H2) and HindIII (H3) are shown and transcription runs from *left to right*. Coding sequences for 18S *cob* mRNA, as localized by electron microscopy (Grivell et al. 1979), are shown as *thickened bars*. They are designated α, β, γ, δ, ε and correspond in position to genetically identified *box* loci 4/5, 8, 1, 2-1 and 6, respectively (Slonimski et al. 1978; Grivell et al. 1979). Additional sequences, coding for approximately 400 - 800 nucleotides of the 5' portion of the 18S mRNA are located leftwards of the HindIII site in α, but these have not yet been mapped accurately. They are certainly not present between the HindII and the HhaI site and probably not located before the next EcoRI site (Grivell, L.A., pers. comm.). The *hatched region* in insert IIIA represents the reading frame that is linked to α by the removal of 10S RNA (see text). (Modified from Van Ommen et al. 1980)

for subunit I of oxidase and, although it seems likely that the inserts in this region will be introns, this remains to be proven.

An interesting side product of our analysis of the transcripts of this region was the discovery of mitochondrial RNA molecules which behave as covalently-closed, single-stranded circles (Grivell et al. 1979; Arnberg et al. 1980). The 11S circular RNA (see Fig. 4) is one of the most abundant transcripts of the whole *oxi*-3 region and is present in all strains, including *S. carlsbergensis*. A second type of circles (19S; see Fig. 4), may arise from longer RNA molecules, which form tailed (not covalently-closed) circles by rather stable internal base pairing. By DNA-RNA hybridization the circular part of these neck-ties has been shown to have the same sequence as the 19S circles (Arnberg et al. 1980). Circles could obviously arise as a side product of splicing if both products of the splicing operation are ligated. The base pairing observed in the 26S neck-ties is not a pre-requisite for splicing in mitochondria, however, because such extensive base pairing is not possible in the precursor to 21S rRNA (Bos et al. 1980).

Insert III is an Intron in the Structural Gene for Cytochrome b

I have mentioned already that insert III lies in the *cob* region and, in fact, it is now well established that it is an intron in the gene for apocytochrome b. We first established by hybrid-arrested translation that mRNA activity for apocytochrome b is contained in an 18S RNA (Grivell et al. 1979). We then made hybrids of this 18S RNA with complementary DNA and mapped the position of the gene by electron microscopy (Bos et al. 1978b; Grivell et al. 1979). The electron microscopy has been complemented by hybridization studies and partial sequence analysis (Grivell, L.A., De Haan, M. and Boer, P.H., unpubl.). The results presented in Figure 5 show a number of interesting points:

1. Insert III is split into two segments separated by a brief coding region.

2. There are two additional introns downstream that are present in all strains.

3. The sequence that codes for the 5'-terminal 400 nucleotides of the 18S RNA has not yet been located. It is certainly not within 3.8 kb of the α-region and probably not within 6 kb. This implies that there is at least one additional large intron upstream of α. This intron contains the gene for glutamyl-tRNA and this is the first example of an intron containing an indispensable gene.

4. The whole b gene is transcribed into a large precursor which is spliced via a complex series of steps to yield mature 18S RNA (Van Ommen et al. 1979, 1980; Van Ommen and Groot 1977; Grivell et al. 1979; Halbreich et al. 1980; Church et al. 1979; Haid et al. 1980). An early step in this splicing event is the removal of part of the α-β intron to yield a 650 nucleotide RNA (10S) present only in yeast strains that contain insert III. Halbreich et al. (1980) have recently found circular RNA molecules in petite mutants that retain part of the *cob* region and have suggested that these may be derived from insert III. Following their report we have demonstrated circular forms of the 10S RNA in wild-type mitochondria by electron microscopy of hybrids between 10S RNA and complementary DNA (Arnberg, A.C. and Grivell, L.A., unpubl.). These molecules were overlooked in our previous work because of the excess of 11S *oxi*-3 circles present in wild-type mtRNA.

5. Mutants in which the first part of the α-β intron is spliced out, but not the remainder, synthesize a 42 kd protein which can be precipitated with an antibody against cytochrome b (Kreike et al. 1979; Solioz and Schatz 1979). Since the coding segment α contains information only for the first 15 kd of apocytochrome b (these are approximations from mobility in gels), the α-β intron must contain an open reading frame for 27 kd (about 800 bp), which is linked to α by the excision of the 10S RNA (see Fig. 5). This intronic reading frame has also been found by Jacq et al. (1980) by sequence analysis of the α-β intron.

For further information about the genetics and sequence of the gene for apocytochrome b, the reader is referred to the contributions of Tzagoloff in this volume.

Promotors in Yeast mtDNA

Let us now briefly consider what we know about the control of mitochondrial RNA synthesis in yeast mitochondria. Some of the map regions studied yield very long transcripts. In the *cob* region the primary transcript must be at least 13,000 nucleotides, as deduced from the minimal distance of exonic regions on the DNA. In the *oxi*-3 region, transcripts up to at least 15,000 nucleotides have been observed by hybridizing RNA blotted onto DBM-paper with specific DNA probes (Hensgens, L.A.M., pers. comm.). Although we do not know whether these are obligatory precursors to functional mRNAs or side products, the former possibility seems the more likely. Transcripts longer than the 4400 nucleotides, minimally required to transcribe the 21S rRNA and the intron contained in it, have also been found.

At least three regions of the genome are, therefore, transcribed into very long RNAs. Moreover, nearly all RNAs analyzed thus far are transcribed from the same DNA strand; this includes the *cob*, *oli*-1 and *oxi*-1 transcripts, both rRNAs and several tRNAs (Grivell et al. 1979). So far only one tRNA gene has been located on the opposite strand (Tzagoloff et al. 1979). This suggests the possibility that yeast mtDNA is transcribed into two giant precursor RNAs, one complementary to each strand, followed by processing of these RNAs into mature rRNAs, tRNAs, and mRNAs. This would make RNA synthesis in yeast mitochondria analogous to RNA synthesis in human mitochondria where complete transcription of both strands has been rigorously proven (Murphy et al. 1975). Such a mechanism would allow the stoicheiometric synthesis of the two rRNAs, notwithstanding the fact that the genes for these RNAs are more than 25,000 bp apart on yeast DNA (Sanders et al. 1975, 1977). Moreover, processing of such a large precursor might require correct folding of the RNA, involving long-range interactions. This could explain why no complete tRNAs are made in petites containing the tRNA region (between *cap*-1 and *oxi*-1) unless also a DNA segment from the *par*-1 region is present (Levens et al. 1979). (Of course, the *par*-1 region could also code for a discrete RNA required for tRNA processing.)

Linnane et al. (1980) have recently found in RNA preparations from yeast mitochondria duplex nucleic acid molecules that they interpret as 25-µm duplex RNA circles, containing a full-length duplex copy of both strands of mtDNA. The evidence for this interpretation is far from complete in my opinion and I still see other problems with the concept that transcription in yeast mitochondria is organized in a similar fashion as in human mitochondria:

1. In pulse-labeling experiments Groot et al. (1974) found only a minimal amount of self-complementary RNA in yeast mitochondria.

2. Using a purified complex of yeast mtDNA and its homologous RNA polymerase, Levens et al. (1979) have found a preferential synthesis of rRNAs. Similar, though quantitatively less striking, results were obtained with isolated polymerase added to purified DNA.

Clearly, none of these results is conclusive. Processing of some segments of the precursor RNAs could be very rapid and the RNA polymerase preparations could still contain processing enzymes or give spurious results for other reasons. The matter remains unresolved in my opinion.

The Role of Optional Introns in mtDNA

Before going into the possible functions of introns in mtDNA, I have to re-emphasize the fact that some introns contain open reading frames. As mentioned before, this was first found in the *cob* region, where some mutants — which splice out the 10S RNA but not the remainder of the α-β intron (see Fig. 5) — accumulate a 42 kd protein, which precipitates with antibodies against cytochrome b (Kreike et al. 1979). The possibility that this 42 kd protein is also made in wild-type mitochondria is supported by the splicing pathway of *cob* RNAs. The removal of the intron(s) in the leader sequence and the 10S RNA are early events in processing, whereas the removal of the remainder of $I_{\alpha\beta}$ is a late event and even a holding step in processing, resulting in a relatively high concentration of the 22S precursor RNA which codes for the 42 kd fusion protein (Van Ommen et al. 1980).

More recently an open reading frame of 705 nucleotides has been found in the intron of the 21S rRNA gene by Dujon (Dujon 1980). There is no direct proof that the protein corresponding to this reading frame is made in vivo, but the presence of detectable amounts of excised intron RNA would in principle allow its synthesis.

What could be the function of these optional intron proteins, only present in some strains but not in others? Three possibilities can be envisaged:

1. The splicing of the intron in which it is located (Kreike et al.

1979; Jacq et al. 1980; Van Ommen et al. 1980).

2. A function in the polarity of recombination (Dujon 1980).

3. A regulatory function which does not provide a clear advantage under laboratory conditions.

Strong evidence for the first possibility has recently been obtained by Jacq et al. (1980). Mutants in the *box*-3 locus of the *cob* region (equivalent to the second half of the α-β intron in Fig. 5), are defective in splicing and accumulate long *cob* transcripts, which contain $I_{\beta\gamma}$, $I_{\gamma\delta}$, $I_{\delta\varepsilon}$, and part of $I_{\alpha\beta}$. In two of these mutants Jacq et al. found either a double amino acid substitution or a stop introduced into the C-terminal part of the fusion protein.

Although this hypothesis provides an elegant explanation for many (though not all[1]) observations in the *cob* region it does not readily explain the open reading frame in the intron in the 21S rRNA. As pointed out by Dujon (1980), Faye et al. (1979) have presented evidence that normal 21S RNA can be formed in petites, which we now know to contain the intron, and this observation has been confirmed using sequence-specific hybridization probes by J.P. Schouten, K.A. Osinga and H.F. Tabak (pers. comm.). This proves that correct splicing can occur without the intron protein, since petite mutants are unable to synthesize any mitochondrial protein at all. It is possible, however, that in this case splicing is only accelerated by the intron protein, because the extra intron also happens to be recognized by the splicing enzyme that takes out the other introns present in all strains. This enzyme is presumably imported. This is very much an ad hoc explanation and the possibility remains that different intron proteins have different types of functions.

Even if all mitochondrial introns were found to code for splicing enzymes designed to remove the very intronic sequences from which they arise, this does not explain why mtDNA contains introns at all. Figure 6 summarizes some of the ideas put forward to explain the presence of

Why Split Genes?

1. Regulation of gene expression in complex genomes (Scherrer and Marcaud 1968).

2. Splicing is required for the exit of RNAs from the nucleus (Hamer and Leder 1979).

3. Limiting homology in duplicated genes (Tiemeier et al. 1978)

4. Splicing is required for the formation of a stable RNA (Gruss et al. 1979).

5. Alternative splicing of large precursor RNAs to yield different mRNAs (multiple choice genes).

6. Facilitation of the reassortment of gene segments to yield new genes (Gilbert 1978).

Fig. 6

---

[1]Halbreich et al. (1980) have presented hybridization experiments with mitochondrial RNAs from petite mutants containing most of the *cob* region. These experiments seem to show longer RNAs from which not only 10S RNA, but also the rest of $I_{\alpha\beta}$ and $I_{\beta\gamma}$ have been removed. This could mean that the removal of these introns can also occur in the absence of the 42 kd fusion protein.

introns in nuclear DNA. Although the mitochondrial genome is not complex, it is possible that gene expression is mainly controlled by RNA processing and that splicing is used to make this control more effective. Although I do not like this idea, it is readily testable and this should be done. The next three ideas are clearly irrelevant to our problem: there are no known gene duplications in yeast mtDNA (sub 3) and splicing cannot be required to yield a stable mature RNA (sub 4), as shown by the existence of the same gene in different strains with and without introns. Alternative splicing of large precursor RNAs to yield different mRNAs (sub 5), as observed with some small DNA viruses, may occur in the *cob* region, as discussed, but it would seem difficult to account for the introns in nuclear tRNA genes or in mitochondrial rRNA genes on this basis. Finally, Gilbert's suggestion (sub 6) does not apply to mitochondria where the gene complement seems to remain remarkably stable throughout evolution.

Nevertheless, I think that possibilities 5 and 6 still provide the most reasonable basis to explain the presence of introns in mtDNA. As also pointed out by Gilbert (1978), introns increase the length of genes and, therefore, increase the possibility for genetic recombination within genes. Likewise, splicing increases the chance of creating new versions of a gene by occasional mutations resulting in missplicing. In this way introns also increase the possibility of creating new alleles of a given gene (rather than creating new genes). This flexibility might be especially useful in the genes for rRNA and apocytochrome b, as the products of these genes are the targets for various inhibitors that yeast may encounter in nature. Why the gene for subunit I of cytochrome c oxidase should also benefit from this flexibility is not clear, however.

If alternative splicing plays a role in creating allelic diversity, it will not be so easy to determine where the protein-coding sequences in a gene are, just from DNA sequence analysis alone. It may thus be necessary to sequence the mRNA (or its complementary DNA) to define precisely which sequences end up in mRNA.

Finally, the possibility should not be overlooked that (some) introns may not serve any function at all (see Borst 1979). Once a splicing system is operative, there may not be a very strong selection pressure against introns if they are efficiently removed at the RNA level. If the DNA replication machinery has a preference for a certain size of DNA (as seems likely on theoretical grounds: Borst et al. 1976), there may even be a selective pressure to retain extra segments of DNA without genetic function.

In summary, it is clear that we cannot say yet why mtDNA should contain (optional) introns. Yeast mitochondria provide a nice test system, however, to study the possible functions of introns by following them through evolution and manipulating them by mutagenesis.

Common and Uncommon Features of mtDNA Through Nature

Whereas chloroplast DNAs from plants and unicellular algae are circular DNA molecules with only limited variation in size and base composition, the diversity in mtDNAs is much greater. In Table 1 I have listed a number of examples that illustrate this point. From the preceding papers

Table 1. Size and structure of some mitochondrial DNAs (Borst 1977)

Species	Structure	m.w. $(\times 10^{-6})$
Animals (from flatworm to man)	Circular	9 – 12
Higher plants	Circular	70
Fungi		
Baker's yeast (*Saccharomyces*)	Circular	49
Kluyveromyces	Circular	22
Protozoa		
Acanthamoeba	Circular	27
Malarial parasite (*Plasmodium*)	Circular	18
Paramecium	Linear	27
Tetrahymena	Linear	30–36
Kinetoplastidae	Circle network	2000 – 20,000
Trypanosoma brucei	Mini-circle	0.6
	Maxi-circle	13

it should be clear that the gene complement of human and yeast mtDNAs do not differ much, notwithstanding a five-fold difference in size. Yeast only spreads its genes more widely with long introns, with long leader sequences on some mRNAs and AT-rich blocks between genes (see Borst and Grivell 1978). It is, therefore, not surprising that another yeast, *Kluyveromyces lactis*, can manage with only 22 x 10^6 daltons mtDNA (Groot and Van Harten-Loosbroek 1980).

mtDNA in higher plants is even larger than that of yeast. This DNA may code for more proteins, because the spectrum of proteins made in isolated plant mitochondria is much more complex than that found in other mitochondria (Forde et al. 1978). It is of interest that the mitochondrial ribosome of plants is the only one among mitochondrial ribosomes thus far that contains a 5S RNA in the large ribosomal sub-unit (see Leaver 1975). It could be that plant mitochondria are more closely related to the (hypothetical) prokaryotic ancestor of present-day mitochondria than most other mitochondria. Plant mtDNA may, therefore, turn out to contain genes that are present in the nucleus in yeast or animals.

Not all mtDNAs are circular and, in the case of *Paramecium* and *Tetrahymena* mtDNAs, there is strong evidence that the linearity is not an isolation artifact. *Tetrahymena* mtDNA has other odd features. It contains a (sub-) terminal duplication-inversion in which the gene for the large rRNA

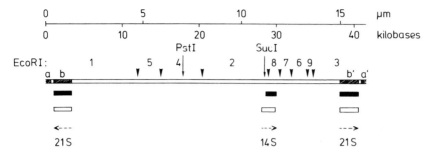

Fig. 7. The physical map of mtDNA from *Tetrahymena pyriformis* strain ST. The clea
vage sites for restriction endonucleases EcoRI, PstI and SacI are shown. Regions *a*
and *b* are the two parts of the terminal duplication-inversion in this DNA, *b'* is
complementary to *b*, *a'* to *a*. The figure shows the position of the two rRNA genes
(for the 21S and 14S rRNAs, respectively) determined by hybridization experiments
with restriction fragments (■) and by electron microscopy of DNA-RNA hybrids (□).
--- indicates the direction of transcription. (From Goldbach et al. 1978a)

is located (Goldbach et al. 1978a,b); there are thus two copies of
this gene and only one for the small rRNA (see Fig. 7). A second odd
feature is the presence of "frayed" ends (Goldbach et al. 1977, 1979).
We think that these features are related to the peculiar problems involved in the correct duplication of linear DNAs. The analogy with the
linear DNA from herpes viruses is noteworthy.

Finally, the most unusual mtDNA in nature is also found in protozoa
and this is the kinetoplast DNA of trypanosomes. The mtDNA (usually
called kinetoplast DNA) of these organisms consists of enormous networks of catenated circles, containing two components, mini-circles
and maxi-circles. In the two representatives of this family that we
have studied, the mini-circles are heterogeneous in sequence and not
transcribed. Their function is not known. The maxi-circles are homogeneous in sequence and transcribed and their genetic complexity is
in the same range as that of mtDNAs in other organisms (see Borst and
Hoeijmakers 1979a,b). Why trypanosomes have chosen to organize their
mtDNA in such a complex and unusual fashion is unclear. One may expect, however, that other aspects of mitochondrial biogenesis in trypanosomes will turn out to be unusual as well. One example is summarized
in Figure 8. Although I am convinced that these 9S and 12S RNAs are
micro-rRNAs, much smaller still than the rRNAs of animal mitochondria,
attempts to isolate a ribosome containing them has failed thus far.
The sequence analysis of the genes coding for these RNAs will show
whether there is any homology with other rRNAs[1].

---

[1]Since this talk was given, the sequence of the 9S and 12S RNAs has been determined
and found to have significant homology with the rRNAs of *Escherichia coli* and human
mitochondria (I.C. Eperon, Medical Research Council, Cambridge, UK, pers. comm.)

Do Trypanosomes Have Micro-rRNAs in Their Mitochondria?

1. The major RNA species in trypanosome mitochondria are (see Simpson and Simpson 1978):

    12 S = 1000 nucleotides
     9 S = 500 nucleotides
     4 S

2. The 9S and 12S RNAs are present in a 1:1 ratio (Hoeijmakers, J.H.J., unpubl. results).

3. RNA blot gels show that 12S RNA is the largest major RNA complementary to kinetoplast DNA (Hoeijmakers, J.H.J., unpubl. results).

4. The genes for the 9S and 12S RNAs cover less than 1900 bp on the maxi-circle (Borst and Hoeijmakers 1979b; Hoeijmakers and Borst 1978).

5. Size and sequence of 9S and 12S RNAs are conserved in evolution (Borst and Hoeijmakers 1979b; Simpson and Simpson 1978).

6. The 9S and 12S RNAs are not retained on oligo(dT)-cellulose in contrast to the minor RNAs that hybridize to kinetoplast DNA (Hoeijmakers, J.H.J., unpubl. results).

Fig. 8

These examples should suffice to show that the world of mitochondrial biogenesis does not end with HeLa cells, yeast and a little *Neurospora*. Although mitochondrial biogenesis is probably similar in outline throughout nature, the details of the process are rather different in trypanosomes, plants, and man. It looks as if in the friendly cytoplasm of the eukaryotic host, selection pressures are less uniform and less harsh than in the outside world. Moreover, with such a limited set of genes, the mitochondrial genetic system may survive accidents that are lethal for free-living organisms. This is in my opinion the most plausible explanation for the altered genetic code in animal and yeast mitochondria and not their descendance from a prokaryotic ancestor with a different code. I am sure that the field of mitochondrial biogenesis will have other surprises for us in store and that the subject will not be closed with the complete sequence of human and yeast mtDNAs.

Acknowledgments. In our work on mitochondrial biogenesis we have greatly benefited from yeast mutants kindly made available to us by Dr R.J. Schweyen (Genetisches Institut der Universität München, Germany), Dr P.P. Slonimski (Centre de Génétique Moléculaire du C.N.R.S., Gif-sur-Yvette, France) and Dr A. Tzagoloff (Fairchild Center for Life Sciences, Columbia University, New York, N.Y., USA). I am indebted to Dr A.C. Arnberg and Dr E.F.J. Van Bruggen (Department of Biochemistry, The University, Groningen, The Netherlands) for allowing me to quote unpublished results obtained in collaborative projects. I thank students and colleagues at the Amsterdam laboratory, especially Dr L.A. Grivell, for providing me with their recent results and comments on the manuscript. This work was supported in part by a grant from The Netherlands Foundation for Chemical Research (SON) with financial aid from The Netherlands Organization for the Advancement of Pure Research (ZWO).

References

Arnberg AC, Van Ommen G-JB, Grivell LA, Van Bruggen EFJ, Borst P (1980) Cell 19: 313-319
Borst P (1977) In: Brinkley BR, Porter KR (eds) International cell biology 1976-1977. Rockefeller University Press, New York, pp 237-244

Borst P (1979) In: Engberg J, Klenow H, Leick V (eds) Specific eukaryotic genes: Structural organization and function. Munksgaard, Copenhagen, pp 244-255

Borst P, Grivell LA (1978) Cell 15:705-723

Borst P, Hoeijmakers JHJ (1979a) Plasmid 2.20-40

Borst P, Hoeijmakers JHJ (1979b) In: Cummings DJ, Borst P, Dawid IB, Weissman SM, Fox CF (eds) Extrachromosomal DNA: ICN-UCLA Symp Mol Cell Biol, vol 15. Academic Press, London New York, pp 515-531

Borst P, Heyting C, Sanders JPM (1976) In: Bücher Th, Neupert W, Sebald W, Werner S (eds) Genetics and biogenesis of chloroplasts and mitochondria. North-Holland, Amsterdam, pp 525-533

Borst P, Bos JL, Grivell LA, Groot GSP, Heyting C, Moorman AFM, Sanders JPM, Talen JL, Van Kreijl CF, Van Ommen GJB (1977) In: Bandlow W, Schweyen RJ, Wolf K, Kaudewitz F (eds) Mitochondria 1977: Genetics and biogenesis of mitochondria. De Gruyter, Berlin, pp 213-254

Bos JL, Heyting C, Borst P, Arnberg AC, Van Bruggen EFJ (1978a) Nature (London) 275:336-338

Bos JL, Van Kreijl CF, Ploegaert FH, Mol JNM, Borst P (1978b) Nucleic Acids Res 5:4563-4578

Bos JL, Osinga KA, Van der Horst G, Hecht NB, Tabak HF, Van Ommen GJB, Borst P (1980) Cell 20:207-214

Church GM, Slonimski PP, Gilbert W (1979) Cell 18:1209-1215

Dujon B (1980) Cell 20:185-197

Faye G, Dennebouy N, Kujawa C, Jacq C (1979) Mol Gen Genet 168:101-109

Forde BG, Oliver RJC, Leaver CJ (1978) Proc Natl Acad Sci USA 75:3841-3845

Gilbert W (1978) Nature (London) 271:501-502

Goldbach RW, Arnberg AC, Van Bruggen EFJ, Defize J, Borst P (1977) Biochim Biophys Acta 477:37-50

Goldbach RW, Bollen-De Boer JE, Van Bruggen EJF, Borst P (1978a) Biochim Biophys Acta 521:187-197

Goldbach RW, Borst P, Bollen-De Boer JE, Van Bruggen EFJ (1978b) Biochim Biophys Acta 521:169-186

Goldbach RW, Bollen-De Boer JE, Van Bruggen EFJ, Borst P (1979) Biochim Biophys Acta 562:400-417

Grivell LA, Moorman AFM (1977) In: Bandlow W, Schweyen RJ, Wolf K, Kaudewitz F (eds) Mitochondria 1977: Genetics and biogenesis of mitochondria. De Gruyter, Berlin, pp 371-384

Grivell LA, Arnberg AC, Boer PH, Borst P, Bos JL, Van Bruggen EFJ, Groot GSP, Hecht NB, Hensgens LAM, Van Ommen GJB, Tabak HF (1979) In: Cummings DJ, Borst P, Dawid IB, Weissman SM, Fox CF (eds) Extrachromsomal DNA: ICN-UCLA Symp Mol Cell Biol, vol 15. Academic Press, London New York, pp 305-324

Groot GSP, Van Harten-Loosbroek N (1980) Curr Gen 1:133-135

Groot GSP, Flavell RA, Van Ommen GJB, Grivell LA (1974) Nature (London) 252:167-169

Gruss P, Lai C-J, Dhar R, Khoury G (1979) Proc Natl Acad Sci USA 76:4317-4321

Haid A, Grosch G, Schmelzer C, Schweyen RJ, Kaudewitz F (1980) Curr Gen 1:155-161

Halbreich A, Pajot P, Foucher M, Grandchamp C, Slonimski PP (1980) Cell 19:321-329

Hamer DH, Leder P (1979) Cell 18:1299-1302

Heyting C, Meijlink FCPW, Verbeet MPh, Sanders JPM, Bos JL, Borst P (1979) Mol Gen Genet 168:231-246

Hoeijmakers JHJ, Borst P (1978) Biochim Biophys Acta 521:407-411

Hollenberg CP, Borst P, Van Bruggen EFJ (1970) Biochim Biophys Acta 209:1-15

Jacq C, Lazowska J, Slonimski PP (1980) CR Acad Sci Paris Ser D 290:1-4

Jeffreys AJ, Flavell RA (1977) Cell 12:1097-1108

Kreike J, Bechmann H, Van Hemert FJ, Schweyen RJ, Boer PH, Kaudewitz F, Groot GSP (1979) Eur J Biochem 101:607-617

Leaver CJ (1975) In: Harborne JB, VanSumere CF (eds) The chemistry and biochemistry of plant proteins, Phytochem Soc Symp, ser 11. Academic Press, London New York, pp 137-165

Levens D, Edwards J, Locker J, Lustig A, Merten S, Morimoto R, Synenki R, Rabinowitz M (1979) In: Cummings DJ, Borst P, Dawid IB, Weissman SM, Fox CF (eds) Extrachromosomal DNA: ICN-UCLA Symp Mol Cell Biol, vol 15. Academic Press, London New York, pp 287-304

Linnane AW, Marzuki S, Nagley P, Roberts H, Beilharz M, Choo WM, Cobon GS, Murphy
  M, Orian JM (1980) In: Davies DR, Hopwood DA (eds) The plant genome: Proc 4th
  John Innes Symp and 2nd Int Haploid Conf. John Innes Charity, Norwich, in press
Morimoto R, Locker J, Synenki RM, Rabinowitz M (1979) J Biol Chem 254:12461-12470
Murphy WI, Attardi B, Tu C, Attardi G (1975) J Mol Biol 99:809-814
Sanders JPM, Heyting C, Borst P (1975) Biochem Biophys Res Commun 65:699-707
Sanders JPM, Heyting C, DiFranco A, Borst P, Slonimski PP (1976) In: Saccone C,
  Kroon AM (eds) The genetic function of mitochondrial DNA. North-Holland, Amsterdam,
  pp 259-272
Sanders JPM, Heyting C, Verbeet MPh, Meijlink FCPW, Borst P (1977) Mol Gen Genet
  157:239-261
Scherrer K, Marcaud L (1968) J Cell Physiol Suppl 1 72:181-212
Simpson L, Simpson AM (1978) Cell 14:169-178
Slonimski PP, Claisse ML, Foucher M, Jacq C, Kochko A, Lamouroux A, Pajot P, Perrodin
  G, Spyridakis A, Wambier-Kluppel ML (1978) In: Bacila M, Horecker BL, Stoppani
  AOM (eds) Biochemistry and genetics of yeast: Pure and applied aspects. Academic
  Press, London New York, pp 391-401
Solioz M, Schatz G (1979) J Biol Chem 254:9331-9334
Synenki RM, Merten S, Christiansen T, Locker J, Rabinowitz M (1979) Abstr Cold
  Spring Harbor Lab Meet Mol Biol Yeast. Cold Spring Harbor, NJ, p 70
Tiemeier DC, Tilghman SM, Polsky FI, Seidman JG, Leder A, Edgell MH, Leder P (1978)
  Cell 14:237-245
Tzagoloff A, Macino G, Nobrega MP, Li M (1979) In: Cummings DJ, Borst P, Dawid IB,
  Weissman SM, Fox CF (eds) Extrachromosomal DNA: ICN-UCLA Symp Mol Cell Biol,
  vol 15. Academic Press, London New York, pp 339-355
Van Ommen GJB, Groot GSP (1977) In: Bandlow W, Schweyen RJ, Wolf K, Kaudewitz F
  (eds) Mitochondria 1977: Genetics and biogenesis of mitochondria. De Gruyter,
  Berlin, pp 415-424
Van Ommen GJB, Groot GSP, Grivell LA (1979) Cell 18:511-523
Van Ommen GJB, Boer PH, Groot GSP, De Haan M, Roosendaal E, Grivell LA, Haid A,
  Schweyen RJ (1980) Cell 20:173-183

Note added in proof: Further experiments have located the missing leader sequence
of the *cob* 18S mRNA (page 33 and legend to Fig. 5) adjacent to the α exon. These
sequences only hybridize at temperatures close to the t_m of the hybrid, but not under the standard conditions used previously. (Amsterdam, October 22nd, 1980)

Synthesis and Assembly of Mitochondrial Membrane Proteins

S. Werner[1]

Introduction

The progress made in a certain scientific area does not proceed at a constant rate, but is subjected to large alterations. It appears to me that a remarkable step forward has been made in the field of mitochondrial biogenesis during the last two years, although this is not comparable to the period of "dramatic moves" which occurred in the early seventies. The aim of this chapter is not to give an extensive review on the numerous publications which have emerged recently, is not to include a volume of detailed facts, and is not to enter into the diversities found in the various organisms investigated. (This information is supplied thoroughly by the proceedings of recent symposia.) Rather, an attempt has been made to survey a selected number of experimental works of exemplary character and subject them to a critical appraisal.

The methodology used for the investigation of synthesis and assembly of mitochondrial membrane proteins (including techniques for the preparation of membrane proteins, isotope labeling, immunoprecipitation, surface labeling of proteins, use of mutants with impaired mitochondrial assembly, in vitro translation systems, small-scale amino acid sequencing, and the application of physical methods, such as X-ray diffraction, optical polarization spectroscopy and high resolution electron microscopy) has not changed very much in the last years; instead, it has been a continuous improvement and refinement of existing know-how which has brought some exiting new insights.

In the following, the potentials of these methods will become evident when tracing the route of mitochondrial membrane proteins from their site of translation on the ribosomal machinery to their final points of deposition within the membrane. Throughout, emphasis has been put on the biogenetic aspects of this process.

A discussion of the role of the lipid entity for the assembly of functional membrane units is beyond the scope of this chapter.

General

The essential structural feature of the mitochondrion is the double membrane system. The inner layer of the mitochondrial membrane is folded, obviously enlarging the surface area, and forms the so-called cristae (see Fig. 1). There is still some disagreement about the existence of contact sites between the two membranes — junctions which

[1] Institut für Physiologische Chemie, Physikalische Biochemie und Zellbiologie der Universität München, FRG

44

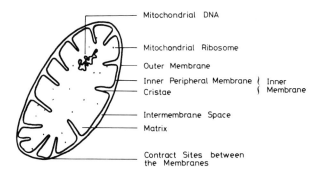

Mitochondrial DNA

Mitochondrial Ribosome

Outer Membrane

Inner Peripheral Membrane ⎱ Inner
Cristae   ⎰ Membrane

Intermembrane Space

Matrix

Contract Sites between
the Membranes

Fig. 1. A schematic representation of the typical
morphological features of
a mitochondrion. Size and
shape of the organelles as
well as the differentiation
of the cristae vary greatly
among various organisms and
during different develop-
mental stages

could be involved in the transport of proteins into the organelle. The
outer membrane comprises only a small percentage of the total mitochondrial mass. 80% - 95% of the total membrane-bound mitochondrial protein is associated with the inner membrane [1]. The protein fractions
of the two membrane systems can be subdivided into two classes: intrinsic and extrinsic proteins, based upon their ability to be solubilized in aqueous media. Since the extrinsic proteins are thought to
bind to the surface of membranes by predominantly electrostatic forces,
it is sometimes difficult to distinguish between genuine membrane-
associated proteins and merely adsorbed proteins.

Outer Membrane

The outer envelope of the organelle is made up of about 60% lipid and
40% protein [2] and thus is one of the most lipid-rich membranes known.
The protein entity is composed of a rather heterogeneous spectrum of
components including enzymes such as acyl-CoA synthetase, amine oxidase,
NADH-cytochrome b-5 reductase, cytochrome b-5, glycerophosphate acyltransferase, kynurenine 3-monooxygenase, lysophosphatidate acyltransferase, phospholipase A_2 [3], a nonheme iron sulfur protein [4] and
another integral membrane protein responsible for the selective binding of hexokinase [5]. It is a well-established fact that all proteins
comprising the outer membrane are encoded by nuclear genes and translated on cytoplasmic ribosomes [6, 7]. Very little, however, is known
about the structure of these components. Only the positioning of some
membrane enzymes has been investigated more extensively. In these
studies, the accessibility of outer membrane enzymes to proteases and
antibodies under various conditions have been examined to indicate
whether they are located on the surface of or within the outer mitochondrial membrane (e.g., [8 - 11]). It has been found that the rotenone-
sensitive NADH-cytochrome c reductase is located on the outer surface
of the membrane, whereas both fatty acyl-CoA synthetase and glycerol-
phosphate acyltransferase lie on the inner surface of the outer mitochondrial membrane [8, 10]. The classical "marker enzyme" for the outer
membrane, monoamine-oxidase, is of special interest because of its
binary functional nature (β-phenetylamine oxidizing activity and 5-
hydroxytryptamine oxidizing activity) which depends on the lipid environment [12]. This enzyme probably exists only in a single molecular
form [13]. The distribution of the two monoamine-oxidizing activities
detectable on the opposing sides of the outer membrane could suggest
that the enzyme is situated at both sides of the membrane or could
indicate a transmembraneous enzyme displaying a different activity on
each side. The latter proposal is probably invalidated by the fact
that there is only one FAD molecule per molecule monoamine oxidase.

The enzyme seems to be composed of two identical subunits with a molecular weight about 62,000 each [13, 14]. The distribution of a single protein species on both sides of a cellular membrane is novel in that all proteins currently investigated are either transmembraneous or are situated on only one side of the membrane [15].

Inner Membrane

The inner envelope of mitochondria belongs to one of the most intensively studied biological membrane systems. This is primarily due to the fact that it contains the machinery for oxidative phosphorylation. The membrane is relatively rich in protein, as are energy-transducing membranes in general [16]. It is composed of approximately 75% protein and 25% lipid by weight [17]. About 60% - 70% of the protein moiety accounts for the fraction of intrinsic proteins, judged from its extraction behavior [18].

A large number of integral proteins of the inner membrane [3] have been characterized mainly with respect to functional aspects, including various enzymes (e.g., ferrochelatase and glycerol 3-phosphate dehydrogenase) and diverse ion or metabolite translocators. This is in contrast to the components of the respiratory chain and the ATP-synthesizing system (they constitute about 30% - 40% of the total protein of the membrane [19]) which have been isolated as well-defined protein complexes. Thus, the following sections of this chapter will deal mainly with those energy-transducing multi-protein complexes of the membrane.

Translation of Mitochondrial Proteins at Two Different Sites Within the Cell

Although mitochondria are capable of synthesizing their own proteins, the overwhelming portion (about 90%) is actually of extramitochondrial origin [20 - 22]. These proteins are specified by nuclear genes and synthesized on extramitochondrial ribosomes, which differ characteristically from their counterparts found within the organelle (for review see [23]. In order to identify specifically the products of extra- and intramitochondrial protein synthesis, two rather simple approaches are generally used:

1. Labeling of growing cells with radioactive amino acids in the presence of specific inhibitors of cytoplasmic (cycloheximide) or mitochondrial (chloramphenicol) protein synthesis.

2. Incorporation of radioactive amino acids by isolated mitochondria.

The first approach is mainly suited for the investigation of lower eukaryotic cells or for tissue cultures of weakly differentiated mammalian cells (e.g., HeLa cells), but is difficult to apply to poorly growing or nongrowing cell types such as hepatocytes. Moreover, both approaches may be unsatisfactory since there is an interdependence between cytosolic and mitochondrial protein synthesis (see [24]). Therefore, for in vivo experiments, highly sophisticated labeling techniques — including a transitory incubation of the cells with inhibitors of the protein synthesis — have been elaborated to compensate, at least partly, for these difficulties [25]. Using both approaches, the following three essential results were obtained:

1. All mitochondrial translation products are associated with the inner membrane fraction.

46

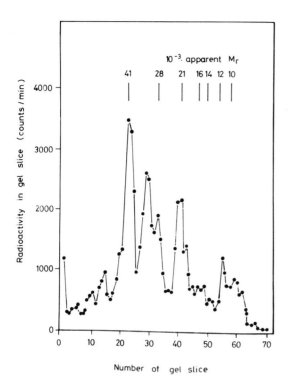

Fig. 2. Mitochondrial translation products in *Neurospora crassa*. Cells were labeled with $[^3H]$-leucine in the presence of cycloheximide and the mitochondrial protein was analyzed by dodecylsulfate gel electrophoresis. The gel was calibrated with subunits of cytochrome oxidase

2. Only a small number of protein species (about 10) ranging in molecular weight from less than 10,000 to about 60,000 are synthesized on mitochondrial ribosomes (see also Fig. 2).

3. Mitochondrial translation products alone seem not to form any function membrane units but are always closely associated with nuclear gene products.

In Table 1, the mitochondrially synthesized proteins identified to date are listed with their relation to functional complexes. A more direct *biochemical* approach for the identification of mitochondrial gene products is their translation in a cell-free system (e.g., prepared from *E. coli*) programmed either by a mitochondrial messenger RNA fraction or by isolated mtDNA using a coupled transcription-translation system [38 - 40]. Actually, these studies did not succeed to a sufficient degree: the products formed were predominantly of low molecular weight and it has been difficult to align them to known mitochondrial polypeptides.

The reasons for the untranslatability of mitochondrial mRNA's in heterologous systems are unclear. The unexpected use of extraordinary base codons in the mitochondrion specifying for the amino acids tryptophan, threonine, and methionine [41] may offer a possible explanation.

Two major differences between the mitochondrial translation products and the cytoplasmically synthesized components of the inner membrane have been found.

1. The mitochondrially made polypeptides represent the most hydrophobic proteins of the membrane. Thus, the products solubilize only in

Table 1. Mitochondrial translation products in *Neurospora crassa* and yeast

Mitochondrial complexes	Mitochondrial products	Molecular weights of polypeptides[a]	References
Cytochrome oxidase	The three largest subunits (1, 2 and 3)	41,000 (42,000) 28,000 (34,500) 21,000 (20,000)	26 - 29
Cytochrome bc_1 complex	Cytochrome b	30,000 (32,000)	30, 31
ATPase complex	Two to three subunits	19,000 (22,000) 11,000 (12,000) (7,500)	32 - 34
Ribosome	Polypeptide associated with the small ribosomal subunit	52,000 (42,000 - 47,000)	35 - 37

[a]Molecular weights for yeast polypeptides are given in parentheses

the presence of powerful ionic detergents and tend to form aggregates. Their content of nonpolar amino acids is usually 10% - 20% higher than that of intrinsic membrane proteins made on cytoplasmic ribosomes [42].

2. The initiator tRNA for mitochondrial translation is fmet-tRNA^{met} as in prokaryotes, in contrast to the met-tRNA^{met} used by the cytosolic ribosome (e.g., [43 - 46]). The formyl group is retained in the nascent polypeptide chain because the mitochondria (at least of yeast) appear to be deficient in deformylase [47]. This characteristic feature may be very helpful for the identification of putative precursor proteins synthesized on mitochondrial ribosomes (see below).

Precursors of Mitochondrial Membrane Proteins

Pulse and pulse-chase experiments with radioactive amino acids carried out in vivo with *Neurospora* and yeast cells have shown that cytoplasmically synthesized mitochondrial membrane proteins can first be detected in the cytosol and subsequently in mitochondria. Such a delayed appearance of membrane proteins in the organelle was shown for cytochrome c, and the ADP/ATP translocator protein [48], for several subunits of the ATPase [49, 50], and for the small subunits of cytochrome oxidase [25]. The polypeptides were isolated from the postribosomal supernatant after cell fractionation by immunoprecipitation with specific antisera. All these isolated precursor proteins, however, exhibited an apparent molecular weight identical with that of the corresponding authentic products prepared from mitochondrial membranes. Recently, a surprising result has been obtained when these polypeptides were translated either in a reticulocyte or a wheat germ system programmed with polyadenylated RNA from yeast and *Neurospora*, respectively. The immunological analysis of the products formed revealed that larger precursors of several cytoplasmically made subunits of the ATPase [51, 52], and of the cytochrome bc_1-complex [53] had been synthesized. These larger polypeptides show an extension, presumably N-terminal, of 20 - 60 amino acids when compared with the "authentic" membrane proteins (Table 2). The identification of the translated polypeptide precursors was not only based on their immunological cross-reactivity, but also on proteolytic fingerprint patterns [51, 53] and partial sequence

48

Table 2. Precursors to cytoplasmically made membrane proteins of mitochondria
translated by in vitro systems

Polypeptide	Source of mRNA	Translation system[a]	Molecular weight		Reference
			Precursor	Mature form	
F_1 ATPase α subunit	Yeast	RC	64,000	58,000	51
F_1 ATPase β subunit	Yeast	RC	56,000	54,000	51
F_1 ATPase γ subunit	Yeast	RC	40,000	34,000	51
ATPase proteolipid subunit	Neurospora crassa	WG	12,000	8,000	52
bc_1 complex subunit V	Yeast	RC	27,000	25,000	53
Cytochrome c_1	Yeast	RC	37,000	31,000	53

[a]Reticulocyte system (RC), wheat germ system (WG)

analysis [52]. Moreover, it has been demonstrated that the in vitro
synthesized products were cleaved by the addition of isolated mitochondria to the cell-free system and hence the mature forms of polypeptides
could be isolated from the mitochondrial fraction [50].

In subsequent studies, the existence of larger precursors of cytoplasmically made membrane proteins were established also by in vivo experiments. In contrast to earlier approaches subjecting the pulse-
labeled cells first to a fractionation procedure, the yeast spheroplasts were lysed immediately in a boiling dodecylsulfate solution
[51]. The immunological analysis of the solubilized yeast proteins revealed that these larger precursors were synthesized in vivo as well.
Furthermore, most of the larger precursors made in vivo disappeared
if the pulse-labeled cell protein was subsequently chased. In their
place, the mature polypeptide subunits were found [51, 53].

Is it a general feature of the mitochondrial membrane proteins translated on cytoplasmic ribosomes that they are initially made in the
form of extended precursors? The answer seems to be no. Pulse-labeling
experiments in vivo, as well as the cell-free synthesis of mitochondrial proteins both in homologous and heterologous systems, have failed
to demonstrate precursors of higher apparent molecular weights in the
cases of cytochrome c and of the ADP/ATP carrier protein [48, 54 - 56].
In order to confirm that these two proteins (one is an extrinsic membrane protein, the other a hydrophobic intrinsic membrane protein) are
not synthesized as larger precursors, the synthesis of Neurospora proteins in the reticulocyte lysate was carried out in the presence of
formyl-(^{35}S)methionyl-transfer RNA [55, 50]. The analysis of the immuno-precipitated polypeptides starting with formyl-methionine indicated that the primary translation products of both protein species
do not carry additional sequences.

A rather interesting finding has been reported concerning the four
cytoplasmically made subunits of cytochrome oxidase [57, 58]. The authors have isolated a 55,000 molecular weight polypeptide from the
postribosomal supernatant of yeast cells. This component translated
on cycloheximide-sensitive ribosomes exhibited a precursor-product relationship to the cytochrome oxidase subunits IV to VII and it moreover

cross-reacted with antibodies raised against the four enzyme subunits of cytoplasmic origin. The authors suggest that this polypeptide — exhibiting a molecular weight larger by 8000 daltons than the combined molecular weight of the oxidase subunits IV to VII — represents a "polyprotein" precursor to all four cytoplasmically translated enzyme components. This large precursor has to be cleaved into the appropriate subunits before they are assembled with the mitochondrially synthesized subunits to form a functional oxidase complex. These spectacular findings clash with the results obtained very recently from a cell-free translation system programmed with yeast RNA (Schatz pers. comm.). Two extended precursor polypeptides corresponding to the individual subunits V and VI of the oxidase were identified. The "pre-pieces" are in the range of 2000 and 6000 daltons. Moreover, it has been shown that both of these individual precursor polypeptides have been translated as separate chains on cytoplasmic ribosomes. This result is in accordance with the observation made from subunits of other oligomeric complexes (see above) namely, that if extended precursors are found, they are synthesized as discrete precursor polypeptides and not as polyproteins.

The occurrence of extended precursor polypeptides seems not to be restricted to cytoplasmically made membrane proteins. Recently, a larger precursor for the mitochondrially synthesized subunit 1 of cytochrome oxidase has been reported [59]. The precursor protein detected in a cytoplasmic mutant of *Neurospora crassa* has an apparent molecular weight of about 45,000 and can be converted in vivo into the mature enzyme subunit (apparent molecular weight 41,000), as judged by its labeling kinetics. Amino acid sequencing of the precursor isolated immunologically revealed that the precursor protein starts with formyl-methionine, whereas the mature form displays an open sequence (the first 15 amino acids of the N-terminus are known) beginning with serine (Machleidt and Werner). Furthermore, the existence of a N-terminally extended precursor for another mitochondrial synthesized protein — subunit 2 of cytochrome oxidase — has also been predicted [60]. (This might be relevant only for lower eukaryotic cells.) If one compares the N-terminal amino acid sequence of the beef heart subunit [61] with that of the corresponding *Neurospora* polypeptide [60], not only was a striking homology found (12 out of 38 residue positions were identical), but the arrangement of the N-terminus may indicate the existence of an extended precursor protein for the fungal polypeptide. This prediction has been strongly supported by data obtained from base sequencing of the corresponding mitochondrial DNA fragment from yeast [62]. In Figure 3, a convincing "assembly" of these three lines of information has been attempted. Very recently, this suggestion of a larger precursor of subunit 2 of cytochrome oxidase seems to have been confirmed by the identification of a polypeptide isolated from yeast mitochondria treated with low concentrations of the drug aurintricarboxylic acid [63]. This particular polypeptide, displaying a molecular weight approximately 1500 daltons larger than the mature subunit, accumulates in the organelle under the specified conditions and shows the characteristics for a precursor to this enzyme subunit.

What is the biological significance of these pre-pieces observed for several precursors of membrane proteins? Concerning the proteins made inside the mitochondria and which need not be transported across membranes, one is tempted to speculate that the extending portion may be involved in the assembly process and/or in the orientation of an oligomeric protein complex within the membrane. On the other hand, a possible role of those mitochondrial pre-proteins synthesized in the cytosol may be closely linked to the mechanism enabling their import into the organelle.

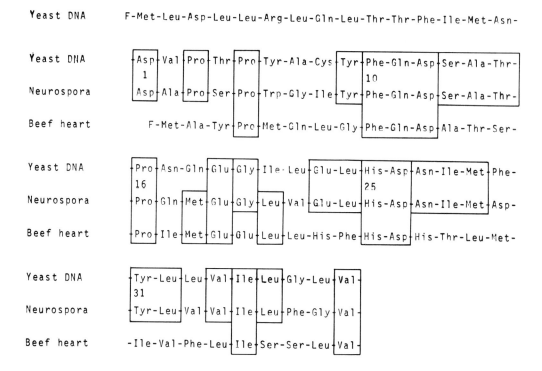

Fig. 3. N-terminal amino acid sequences of subunit 2 of cytochrome oxidase. The data is obtained from the base sequences of the corresponding DNA fragment from yeast [62], and from amino acid sequencing of the enzyme subunit isolated from *Neurospora* [60] and beef heart [61]

Transfer of Mitochondrial Membrane Proteins from the Cytosol into the Organelle

Most of the proteins of mitochondria are synthesized on cytosolic ribosomes and must be transferred into these organelles. The mechanism by which these proteins find their way into the mitochondrion is still not explained. It has been suggested that a special subclass of 80 S cytoplasmic ribosomes bound to the outer mitochondrial membrane "inject" those proteins destined for import directly into the organelle in the manner of a vectorial co-translational process [64, 65]. This hypothesis is most probably incorrect, since kinetic studies in vivo and in vitro have shown that discrete pools of extramitochondrial precursors of mitochondrial membrane proteins do exist and that those precursors are then translocated by a posttranslational mechanism [48, 50]. Since an accumulation of larger precursors for some of the membrane proteins in the cytosol has been observed, it is unlikely that these polypeptides detected in the cytoplasma could simply by explained as material selectively eluted from mitochondria during the preparation procedure. Finally, it has been demonstrated that the transfer of these precursors into the organelle occurs also in the absence of protein synthesis [50] — a feature not indicative of a cotranslational process.

The intriguing question remains, how do these completely translated precursors pass into the membrane compartment of the mitochondrion?

For those precursors synthesized in the form of an extended polypeptide chain, their translocation is accompanied by a processing of the precursors to their mature forms as they are obtained from isolated membrane units, e.g., ATPase synthetase or cytochrome bc_1-complex. Such a processing of larger precursors independent from concommitant protein synthesis has been shown both in vivo and in vitro [51, 53]. Because this unidirectional transmembrane movement appears to be linked closely to a proteolytic processing of the precursors, the underlying mechanism for the transfer has been termed "vectorial processing" [50]. The driving force for this process could be the (irreversible) cleavage of the pre-pieces of these precursors. No information is available to date about the involvement of the extended portion of the polypeptides in the recognition of specific "binding sites" of the membrane (receptors?). Also unanswered is the question about the localization and specificity of the required "proteases".

Interestingly enough, it has been demonstrated that the depletion of the mitochondrial matrix ATP in intact yeast spheroplasts prevents the processing of extended precursors of membrane components to their mature forms [66]. This interpretation is based on experiments employing simultaneously specific inhibitors for both oxidative phosphorylation and ATP transport [66]. Such energy-dependent processing of cytoplasmically made precursors, however, seems to apply only to the precursors of integral membrane proteins and not to those mitochondrial components located in the intermembrane space, such as cytochrome c peroxidase.

On the other hand, the in vivo studies may not be suitable to distinguish if energy is necessary for the binding of the precursor to the membrane, for the processing of the precursor or for the proper insertion of the mature polypeptides into the membrane (following or during the proteolytic cleavage), or even for all these processes. An attempt to approach this problem has been made recently using an in vitro translation system (reticulocyte lysate) supplemented with isolated mitochondria (Neupert, pers. comm.). In this case, however, the cytoplasmically translated precursor investigated (ADP/ATP translocator) is not synthesized as an extended polypeptide. Protection against the action of proteases was taken as a criterion for the insertion of this precursor into the mitochondrial membrane. It has been found that, although the precursor became firmly bound to the mitochondrion in the presence of the uncoupler carbonyl cyanide m-chlorophenyl-hydrazone (CCCP), the protein still remained sensitive to the proteolytic enzyme employed. In contrast, using mitochondria with a functioning ATP-generating system, the precursor seems to be integrated into the mitochondrial membrane, since the polypeptide could no longer be attacked by the added protease.

Nevertheless, it remains unclear whether there is a fundamental difference between the mechanism(s) governing the translocation of precursors made in the form of extended polypeptides and of precursors without additional sequences. It is pertinent that integral membrane proteins facing the cytoplasmic side of the inner mitochondrial envelope do not require pre-pieces for their insertion? Can the extended portion of larger precursor polypeptides be viewed as a means of activating a protein for assembly?

Assembly of Mitochondrial Membrane Protein

Several general models have been suggested to approach the problem of membrane assembly. One, an adaptation of the signal hypothesis [67, 68] emphasizes the role of catalysis for membrane assembly and involves a specific protein transport channel [15]. It is proposed that the force of polypeptide chain elongation drives the protein into the membrane. The function of the pre-sequence would consist in the recognition of a specific membrane "pore". In contrast, the theory, termed the "Membrane Triggered Folding Hypothesis" [69], assumes that integral membrane proteins are capable of more than one functional conformation and that the thermodynamics of protein folding governs membrane assembly with little intervention of catalysis. This means that the information for the (spontaneous) assembly process lies in the structure of the protein itself. The pre-sequences of precursors, in this case, would have the function of modifying the folding of the protein. A membrane-triggered conformational change can be accompanied by proteolysis, but this is not obligatory. Also, the attachment of a prosthetic group — such as heme — to a polypeptide could easily account for a subsequent change in the protein folding. A somewhat similar hypothesis was presented recently, termed the "Direct Transfer Model" [70]. The transfer of proteins across or into membranes is here explained by the energetics of partitioning of the amino acid side chains between water and lipid. The rather complex data available on mitochondrial membrane proteins seems to considerably favor the latter two models. Amino acid sequencing of the precursors involved and the study of the action of pre-peptide proteases in both in vivo and in vitro systems should provide further valuable information in this area.

The unique feature of the mitochondrial membrane assembly is its dependency on two different sources of protein synthesis. This implies a coordinated delivery of the components of cytoplasmic and mitochondrial origin in proper amounts. A number of years ago, discrete pools of unassembled but integrated subunits of oligomeric complexes (e.g., cytochrome oxidase [71] were detected in the membrane. It has also been shown that translation products may accumulate to a certain extent in the membrane if protein synthesis of either the cytoplasmic or the mitochondrial compartment is switched off by means of specific inhibitors [25]. In a similar way, a large number of nuclear and mitochondrial mutants of yeast cells [24, 72 – 74] and of *Neurospora* [75 – 81], both deficient in the respiratory chain, have been investigated in order to study the effect of the various translation products on the assembly process of oligomeric membrane complexes, such as cytochrome oxidase and ATPase. It was found that these mutations can affect production, integration, and assembly of mitochondrial proteins at many different levels. The obviously very complex mechanisms for the coordination and mutual regulation of the proteins are still poorly understood. Although it is well established that nuclear mutations may control the transcription and/or translation of the mitochondrial system, it is not clear whether also the reverse is true. Some recent results from *Neurospora* likewise seem to support the latter idea (Bertrand pers. comm.).

The difficulty involved in interpreting the individual observations may be documented by just one example. There has been considerable indirect evidence that the cytoplasmically made subunits of mitochondrial membrane complexes stimulate the synthesis of their mitochondrially made "partner subunits" (for review see [24]). This clue was strongly supported by in vitro experiments to test the protein synthesis of isolated mitochondria. The rate of synthesis in this system was greatly stimulated by the addition of a dialysed cytosolic supernatant containing precursors of cytoplasmically made membrane components [82]. Very recently, evidence has been obtained (Schatz, pers. comm.) that this stimulatory effect of the supernatant is not due to cytoplasmically made precursor proteins, as suggested previously, but is caused by the presence of nondialysable GMP, which is converted to GDP under assay conditions. Supplementing the in vitro system with the nucleotide GDP alone has essentially the same effect on the rate of mitochondrial protein synthesis.

A particularly interesting problem is the mode in which multi-protein membrane complexes consisting of polypeptide subunits from either translation site are assembled from the individual pools of precursors already present in the inner membrane. What is the time sequence of the assembly process leading to functional units? A direct approach to isolate assembly intermediates for cytochrome oxidase has been attempted with *Neurospora* wild-type cells [83]. Unfortunately, only small intermediates (consisting of two to three subunits) or rather large ones (six to seven subunits) were identified, and no sequence of the process could be established. The data obtained, however, indicates that the largest mitochondrially made component (subunit 1) has a key role in the assembly process of this protein complex. The view that subunit 1 of the enzyme serves as a "crystallization point" for the oxidase assembly is supported by investigations of cytochrome oxidase-deficient mutants either lacking or having altered specifically this subunit [76, 80]. Many other indirect studies using cells conditionally deficient in cytochrome oxidase, such as certain temperature-sensitive mutants [78], copper-depleated cells [75], mutants defective in heme synthesis [84], anaerobically grown yeast cells [28, 85, 86], germinating spores [87] and in vitro complementation of mitochondrial and cytosolic products [88], did not supply additional substantial information on how the polypeptide components interact to form a functional unit. In a similar way, the very elegant in vitro reconstitution experiments performed with subunits of the F_1ATPase-complex (e.g., see [89]) contribute predominantly information about functional aspects of the enzyme, but can tell little about the formation of the complex in situ.

Structure of Membrane Polypeptides and Architecture of Supramolecular Protein Complexes

Complete amino acid sequences of several subunits of cytochrome oxidase [61, 90 - 93] and the so-called "proteolipid" from the ATPase complex [94] have been published. Undoubtedly, cytochrome oxidase is one of the best-characterized components of the mitochondrial membrane on a structural basis. Nevertheless, there is still considerable uncertainty as to the number and the size of its subunits. Whereas the yeast enzyme seems to contain seven [24, 95] and the oxidase in *Neurospora* most probably eight polypeptides [96], the number of subunits constituting the cytochrome complex in mammalian cells may be as many as twelve [97, 98]. There is, however, substantial agreement on the contribution of the mitochondrial translation system: the three largest subunits have been designated with consistency. Almost all subunits reportedly bind the prosthetic groups of the enzyme — heme a and copper — depending on the procedure used for the dissociation of the complex (for review see [99]). Concerning this problem, the resolution of the primary structure of particularly three subunits isolated from the beef heart oxidase has led to most interesting insights. The amino acid sequence data of the mitochondrially made subunit II indicates that this polypeptide is involved in the binding of copper, since striking sequence homologies between this oxidase subunit and the well-known copper proteins of the azurin/plastocyanin family were found [100]. Another case of sequence homology was observed with the cytoplasmically synthesized subunits VII [100] and V [90]. They show remarkable similarity to sequences of bacterial cytochromes and to the β-chain of hemoglobin, respectively. Thus, these polypeptides are possible candidates for the binding sites of heme a.

The main component of the membrane factor, F_o, of the ATPase complex — a very hydrophobic polypeptide with a molecular weight of about 8000 daltons, mostly referred to as "proteolipid" — has been isolated from mitochondria of different organisms and the complete sequences are now known [94]. It turned out that this protein, present most likely in a hexameric form within the membrane [101], has two extremely hydrophobic stretches each consisting of about 25 amino acid residues. Moreover, a glutamic acid residue located in the middle of the second hydrophobic segment has been identified as the reaction site with the inhibitor dicyclohexylcarbodiimide (DCCD) [94]. The striking clustering of both hydrophobic and hydrophilic amino acid side chains implies a specific arrangement of the polypeptide chain in the membrane. It has been speculated that the two hydrophobic segments traverse the lipid bilayer of the membrane, whereas the polar regions are either exposed to the water phase or are in contact with other subunits of the complex [94]. This interpretation could be helpful towards understanding the mechanism of the ATPase proton channel which seems to consist predominantly of this hydrophobic subunit (e.g., see [102]).

Elucidation of the three-dimensional arrangement of the ATPase complex in the membrane is facilitated by the capability of dissociating the complex gradually into components and reconstituting it with restoration of both structural and functional properties. The juxtapositions of the individual subunits have been explored by chemical cross-linking (see [102]). Although the exact stoichiometry of the individual components is not yet established, more or less detailed ideas about the morphology of the complex have been suggested [102].

Even more difficult to interpret are the results obtained with the cytochrome oxidase complex. Specific antibodies [103], iodination and reaction with membrane-impermeable protein-modifying compounds [104 - 106] have been used to identify those polypeptide subunits available from one or the other side of the membrane. Cross-linking studies of the isolated enzyme have provided data on the proximity of subunits within the complex [107]. In addition, cross-linking of cytochrome c to the oxidase has localized the binding site for this substrate to subunit II in the beef heart enzyme [108, 109], but to the subunit III in the yeast enzyme [110]. Finally, experiments in which fluorescence resonance energy transfer has been applied to determine the distances between the heme group of cytochrome c and the nearest heme of the oxidase gave a rough idea about the location of this prosthetic group within the membrane [101]. Most of these results are summarized in Figure 4, depicting a rather hypothetical topographical model of the oxidase. It is consistent with the conclusions reached by other approaches (see below) proving that the enzyme spans the inner mitochondrial membrane and is asymetrically arranged across this barrier.

Crystals of cytochrome oxidase suitable for high-resolution X-ray crystallography are not yet available. It has, however, been possible to prepare "crystalline membranes" containing the oxidase which have been examined by electron microscopy. The optical diffraction patterns of the micrographs were analyzed [112]. The results provide further evidence for the asymetrical geometry of the oxidase molecule. The observed crystalline arrays were composed of two membrane layers which form top and bottom of a collapsed vesicle. Each cytochrome oxidase molecule protruded into solution on the inside surface of the artificial vesicle, but less than half as far on the outside surface. It is not known, however, whether the inside surface of the vesicles corresponds to the inside surface of the mitochondrial inner membrane. In this particular crystalline form, the molecules occurred as dimers, possibly reflecting the state of the unit in its natural environment.

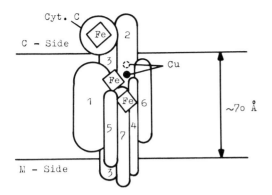

Fig. 4. A hypothetical model of cyto-chrome oxidase

Another successful attempt to generate two-dimensional crystals of a mitochondrial protein was reported recently for the ubiquionine: cyto-chrome reductase [113], a multiprotein enzyme contributing about 10% of the inner mitochondrial membrane [114]. In this case, the "crystal-lization" process was achieved for the first time with a highly puri-fied preparation of the mitochondrial protein. The enzyme was isolated from *Neurospora* in the form of a dimer (molecular weight about 550,000). The monomeric unit comprises two b cytochromes, one cytochrome c_1 and six other subunits without known prosthetic groups [115]. To obtain membrane crystals from the isolated enzyme, the detergent bound to the protein was gradually replaced by phospholipids [113]. High reso-lution electron microscopy revealed that the vesicles obtained con-sisted of dimeric reductase complexes arranged in highly ordered ar-rays, in such a way that the alternate protein molecules are oriented in an up-and-down manner. This is unlike the membrane crystals reported for cytochrome oxidase [112] in which the protein is arranged unidi-rectionally. The dimensions of one dimeric cytochrome c reductase mo-lecule in projection are ∿90 Å by 70 Å. This is only about half the Stokes diameter of the (dimer) detergent solubilized enzyme prepared initially from the mitochondria. This inconsistency between the di-mensions of the crystallized enzyme in projection and its apparent hydrodynamic diameter in solution, indicates that cytochrome c reduc-tase is a rod-like enzyme spanning the inner membrane [113]. By anal-yzing various projections of the membrane crystals, a three-dimensional image of the protein complex has been reconstructed (K. Leonard and H. Weiss, pers. comm.). The model for the enzyme designated according to this information will be very helpful to afford a clearer under-standing of how a functional multi-protein unit is assembled within the mitochondrial membrane.

Prospects

A broad spectrum of efficient methods for investigating the biogenesis of mitochondrial membranes is now available. Although the interplay between the mitochondrial and cytoplasmic systems is still poorly un-derstood, at least the question about the number of products contri-buted by the mitochondrion should be solved soon. How different the answer might be given in detail for various organisms has been docu-mented by the proteolipid component of the ATPase complex, a polypep-tide which is made on cytoplasmic ribosomes in *Neurospora*, but is a mitochondrial translation product in yeast [94].

The mechanism by which proteins synthesized in the cytosol are transported into the organelle could be explored successfully by the use of various homologous and heterologous in vitro systems and by extending the studies to a larger variety of protein species. The role of transient "pre-pieces" for the translocation and/or integration process of the imported polypeptides will become more clear once the amino acid sequences of the elongated portions of the polypeptide precursors are available. In analogy, similar insights might be gained from the determination of pre-sequences of mitochondrial products. The elaboration of a suitable in vitro system for the translation of mitochondrial mRNA would be of considerable advantage in this context. The localization and characterization of the mitochondrial proteases involved in the processing of "pre-pieces" supplement the set of experiments.

Currently, we have little information on how multi-subunit complexes of the membrane become assembled. To approach this problem, we might profit from the numerous nuclear and mitochondrial mutants from various micro-organisms displaying impaired mitochondrial assembly. The goal is to establish the time sequence for the formation of these oligomeric structures.

Finally, revealing the micro-architecture of multi-protein membrane units will depend on the progress in several fields: structural analysis of purified membrane polypeptides, cross-linking and surface labeling studies, controlled dissociation and reconstitution experiments, and the production of crystals from isolated membrane complexes meeting the requirements for high resolution electron microscopy and X-ray diffraction studies.

References

1. Wainio WW (1970) The mammalian mitochondrial respiratory chain. Academic Press, London New York, p 499
2. Hallermayer G, Neupert W (1974) Hoppe-Seyler's Z Physiol Chem 355:279-288
3. Altmann PL, Katz DD (eds) (1976) Biological handbooks I. Cell biology: Bethesda, Maryland. Fed Am Soc Exp Biol, p 179
4. Backström D, Hoffström I, Gustafsson I, Ehrenberg A (1973) Biochem Biophys Res Commun 53:596-602
5. Felgner PL, Messer JL, Wilson JE (1979) J Biol Chem 254:4946-4949
6. Neupert W, Brdiczka D, Bücher Th (1967) Biochem Biophys Res Commun 27:488-493
7. Beatti DS, Basford RE, Koritz SB (1967) Biochemistry 6:3099-3106
8. Kuylenstierna B, Nicholls DG, Hoomoller S, Ernster L (1970) Eur J Biochem 12: 419-426
9. Mayer RJ, Hubscher G (1971) Biochem J 124:491-500
10. Nimmo HG (1979) FEBS Lett 101:262-264
11. Russell SM, Davey I, Mayer RJ (1979) Biochem J 181:7-14
12. Tipton KF, Houslay MD, Mantle TJ (1976) Ciba Found Symp New Series 39:5-33
13. Minamiura N, Yasunobu KT (1978) Arch Biochem Biophys 189:481-489
14. Salach JI (1978) In: Fleischer S, Packer L (eds) Methods in enzymology, vol 53. Academic Press, London New York, pp 495-501
15. Rothman JE, Lenard K (1977) Science 195:743-753
16. DePierre JW, Ernster L (1977) Annu Rev Biochem 46:201-262
17. Crane FL, Sun FF (1972) In: King TE, Klingenberg M (eds) Electron and coupled energy transfer in biological systems, vol IB. M Dekker, New York, pp 477-587
18. Harmon HJ, Hall JD, Crane FL (1974) Biochem Biophys Acta 344:119-155
19. Ernster L (1975) FEBS Symp 35:257-285
20. Sebald W, Hofstötter T, Hacker D, Bücher Th (1969) FEBS Lett 2:177-180
21. Henson DP, Perlman P, Weber CN, Mahler HR (1968) Biochemistry 7:4445-4454
22. Groot GSP, Rouslin W, Schatz G (1972) J Biol Chem 247:1735-1742

23. Buetow BE, Wood WM (1978) In: Roodyn DB (ed) Subcellular biochemistry, vol V. Plenum Publ Corp, New York, pp 1-85
24. Schatz G, Mason TL (1974) Annu Rev Biochem 43:51-87
25. Sebald W, Werner S, Weiss H (1979) In: Fleischer S, Packer L (eds) Methods in enzymology, vol 56. Academic Press, London New York, pp 50-58
26. Weiss H, Sebald W, Bücher Th (1971) Eur J Biochem 22:19-26
27. Sebald W, Machleidt W, Otto J (1977) Eur J Biochem 38:311-324
28. Mason TL, Schatz G (1973) J Biol Chem 248:1355-1360
29. Rubin MS, Tzagoloff A (1973) J Biol Chem 248:4275-4279
30. Weiss H, Ziganke B (1976) Eur J Biochem 41:63-71
31. Katan NB, Pool L, Groot GSP (1976) Eur J Biochem 65:95-105
32. Ebner E, Mason TL, Schatz G (1973) J Biol Chem 248:5369-5378
33. Jackl G, Sebald W (1975) Eur J Biochem 54:97-106
34. Tzagoloff A, Meagher P (1972) J Biol Chem 247:594-603
35. La Polla RJ, Lambowitz AM (1977) J Mol Biol 116:189-205
36. Butow RA, Vincent RD, Strausberg RL, Zanders E, Perlman PS (1977) In: Bandlow W, Schweyen RJ, Wolf K, Kaudewitz F (eds) Mitochondria 1977, genetics and biogenesis of mitochondria. De Gruyter, Berlin, p 317
37. Perlman PS, Douglas MG, Strausberg RL, Butow RA (1977) J Mol Biol 115:675-694
38. Padmanaban G, Hendler F, Patzer J, Ryan R, Rabinowitz M (1975) Proc Natl Acad Sci USA 72:4293-4297
39. Scragg AH, Thomas BY (1975) Eur J Biochem 56:183-192
40. Moormann AFM (1978) Thesis Univ Amsterdam
41. Hall BD (1979) Nature (London) 282:129-130
42. Capaldi RA (1978) In: Fleischer S, Hatefi Y, Maclennan DH, Tzagoloff A (eds) Molecular biology of membranes. Plenum Press, New York, pp 103-119
43. Smith AE, Marker KA (1968) J Mol Biol 38:241-243
44. Galper JB, Darnell JE (1969) Biochem Biophys Res Commun 34:205-214
45. Sala F, Küntzel H (1970) Eur J Biochem 15:280-286
46. Bianchetti R, Lucchini G, Sartirana ML (1971) Biochem Biophys Res Commun 42:97-102
47. Feldman F, Mahler RH (1974) J Biol Chem 249:3702-3709
48. Hallermayer G, Zimmermann R, Neupert W (1977) Eur J Biochem 81:523-532
49. Sebald W (1977) Biochim Biophys Acta 463:1-27
50. Schatz G (1979) FEBS Lett 103:203-211
51. Maccecchini ML, Rudin Y, Blobel G, Schatz G (1979) Proc Natl Acad Sci USA 76:343-347
52. Michel R, Wachter E, Sebald W (1979) FEBS Lett 101:373-376
53. Coté C, Solioz M, Schatz G (1979) J Biol Chem 254:1437-1439
54. Harmey M, Hallermayer G, Korb H, Neupert W (1977) Eur J Biochem 81:533-544
55. Zimmermann R, Paluch U, Sprinzl M, Neupert W (1979) Eur J Biochem 99:247-252
56. Zimmermann R, Paluch U, Neupert W (1979) FEBS Lett 108:141-146
57. Poyton RO, McKemmie E (1979) J Biol Chem 254:6763-6771
58. Poyton RO, McKemmie E (1979) J Biol Chem 254:6772-6780
59. Werner S, Bertrand H (1979) Eur J Biochem 99:463-470
60. Machleidt W, Werner S (1979) FEBS Lett 107:327-330
61. Steffens GJ, Buse G (1979) Hoppe-Seyler's Z Physiol Chem 360:613-619
62. Coruzzi G, Tzagoloff A (1979) J Biol Chem 254:9324-9330
63. Sevarino KA, Poyton RO (1980) Proc Natl Acad Sci USA 77:141-146
64. Kellems RE, Butow RA (1972) J Biol Chem 247:8043-8050
65. Bennett WF, Gutierrez-Hartmann A, Butow PA (1976) In: Bücher Th, Neupert W, Sebald W, Werner S (eds) Genetics and biogenesis of chloroplasts and mitochondria. North-Holland, Amsterdam, pp 801-806
66. Nelson N, Schatz G (1979) Proc Natl Acad Sci USA 76:4365-4369
67. Blobel G, Dobberstein B (1975) J Cell Biol 67:835-851
68. Blobel G, Dobberstein B (1975) J Cell Biol 67:852-862
69. Wicker W (1979) Annu Rev Biochem 48:23-45
70. von Heijne G, Blomberg G (1979) Eur J Biochem 97:175-181
71. Weiss H, Schwab AJ, Werner S (1975) In: Tzagoloff (ed) Membrane biogenesis. Plenum Press, New York, pp 125-153

73. Eccleshall TR, Needleman RB, Storm EM, Buchferer B, Marmur J (1978) Nature (London) 273, 67-70
74. Cabral F, Solioz M, Rudin Y, Schatz G, Clavilier L, Slonimski P (1978) J Biol Chem 253:297-304
75. Werner S, Schwab AJ, Neupert W (1974) Eur J Biochem 49:607-617
76. Nargang FE, Bertrand H, Werner S (1978) J Biol Chem 253:6364-6369
77. Nargang FE, Bertrand H (1978) Mol Gen Genet 166:15-23
78. Nargang FE, Bertrand H, Werner S (1979) Eur J Biochem 102:297-307
79. Bertrand H, Werner S (1977) Eur J Biochem 79:599-606
80. Bertrand H, Werner S (1979) Eur J Biochem 98:9-18
81. Sebald W, Sebald-Althaus M, Wachter E (1977) In: Bandlow W, Schweyen RJ, Wolf K, Kaudewitz F (eds) Mitochondrial 1977, genetics and biogenesis of mitochondria. Walter de Gruyter, Berlin, pp 433-440
82. Poyton R, Kavanagh J (1976) Proc Natl Acad Sci USA 73:3947-3951
83. Werner S, Neuner-Wild M (1976) Bücher Th, Neupert W, Sebald W, Werner S (eds) Genetics and biogenesis of chloroplasts and mitochondria. North Holland, Amsterdam, pp 199-206
84. Saltzgaber-Müller J, Schatz G (1978) J Biol Chem 253:305-310
85. Groot GSP, Poyton RO (1975) Nature (London) 255:238
86. Woodrow G, Schatz G (1979) J Biol Chem 254:6088-6093
87. Josephson M, Brambl R (1980) Biochem Biophys Acta 606:125-137
88. Chandrasekaran K, Dharmalingam K, Jayaraman J (1980) Eur J Biochem 103:471-480
89. Kagawa Y (1978) In: Kotani M (ed) Advances in biophysics, vol 10. Japan Scientic Societis Press and University Park Press, Tokyo, pp 209-247
90. Tanaka M, Haniu M, Yasunobu KT, Yu CA, Yu L, Wei YH, King TE (1977) Biochem Biophys Res Commun 76:1014-1019
91. Buse G, Steffens GJ (1978) Hoppe-Seyler's Z Physiol Chem 359:1005-1009
92. Sacher R, Steffens GJ, Buse G (1979) Hoppe-Seyler's Z Physiol Chem 360:1385-1392
93. Steffens GCM, Steffens GJ, Buse G (1979) Hoppe-Seyler's Z Physiol Chem 360:1641-1650
94. Sebald W, Hoppe I, Wachter E (1979) In: Quagliariello E, Palmieri F, Papa S, Klingenberg M (eds) Function and molecular aspects of biomembrane transport. Elsevier/North Holland, Biomedical Press, Amsterdam New York, pp 63-74
95. Poyton KO, McKemmie E, George-Nascimiento C (1978) J Biol Chem 253:6303-6306
96. Werner S (1977) Eur J Biochem 79:103-110
97. Buse G, Steffens GJ, Steffens GCM, Sacher R (1978) In: Button PL, Leigh L, Scarper A (eds) Frontier of biological energetics, vol II. Academic Press, New York, pp 799-807
98. Merle P, Kadenbach B (1980) Eur J Biochem in press
99. Malmström GB (1979) Biochim Biophys Acta 549:281-303
100. Buse G, Steffens GI, Steffens GCM (1978) Hoppe-Seyler's Z Physiol Chem 359:1011-1013
101. Sebald W, Graf I, Lukins HB (1979) Eur J Biochem 93:587-599
102. Senior AE (1979) In: Capaldi RA (ed) Membrane proteins in energy transduction. Marcel Dekker Inc, New York, pp 233-278
103. Chan SHP, Tracey RP (1978) Eur J Biochem 89:595-605
104. Eytan GD, Schatz G (1975) J Biol Chem 250:767-774
105. Eytan GD, Carroll RC, Schatz G, Racker E (1975) J Biol Chem 250:8598-8603
106. Ludwig B, Downer NW, Capaldi RA (1979) Biochemistry 18:1401-1407
107. Briggs MM, Capaldi RA (1977) Biochemistry 16:73-77
108. Briggs MM, Capaldi RA (1977) Biochem Biophys Res Commun 80:553-559
109. Bisson R, Gutweniger H, Montecúcco C, Colonna R, Zanotti A, Azzi A (1977) FEBS Lett 81:147-150
110. Birchmeier W, Kohler CE, Schatz G (1976) Proc Natl Acad Sci USA 73:4334-4338
111. Dockter ME, Steinemann A, Schatz G (1978) J Biol Chem 253:311-317
112. Henderson R, Capaldi RA, Leigh JS (1977) J Mol Biol 112:631-648
113. Wingfield P, Arad T, Leonard K, Weiss H (1979) Nature (London) 280:696-697
114. Rieske JS (1976) Biochem Biophys Acta 456:195-247
115. Weiss H, Korb H (1979) Eur J Biochem 99:139-149

Primary Structure of Mitochondrial Membrane Proteins: Evolutionary, Genetic and Functional Aspects

G. Buse[1]

Introduction

When early life was confronted with a lack of usable free energy in
an environment running short of reduced substrate (Broda 1975), cyclic
photophosphorylation (Arnon 1959) provided an effective possibility
of energy conservation linked to solar energy. With the bacterial ancestors of chloroplasts and mitochondria this reaction cycle probably
developed to photosynthesis and respiration. However, it was not prokaryotes which made the most successful use of these achievements but
the eukaryotic cell with its superior genetic capabilities.

Together with obtaining insight into these functions, we will unravel
the history of these events.

As far as the mitochondrion is concerned, the most recent success in
this field comes from the biochemical analysis of both types of informative macromolecules, nucleic acids, and proteins. Here we consider those biochemical structures whose genetic informatin at least to
the most significant extent, remained encoded in the mitochondrion:
the electron transfer complex III (ubiquinone-cytochrome-c-oxidoreductase) and complex IV (cytochrome-c-O_2-oxidoreductase, EC 1.9.3.1) and
the linked ATP synthase, complex V (ATPase or ATP phosphorhydrolase,
EC 3.6.1.3) of the inner mitochondrial membrane.

Facts and Findings

The Inner Mitochondrial Membrane

Figure 1 comprises the main reaction centers of the terminal respiratory chain and the ATP synthase of the inner mitochondrial membrane.
Neither reaction mechanisms nor the stoichiometries occurring at these
sites are presently well understood. Thus the data included in Figure
1 should be taken as exemplifying the basic principles only. The genetic system of yeast mitochondria synthesizes 7-9 proteins (Borst and
Grivell 1978; Tzagoloff et al. 1979). These are:

    1 protein component of complex III (bc$_1$), the cytochrome b subunit.
    3 components of complex IV, the polypeptides I, II and III.
    2-4 components (in yeast) of complex V, including the protonophoric
    subunit IX.

One further protein seems to be present which functions in the mitochondrial biosynthetic apparatus (Lambowitz et al. 1976). Thus most
of the proteins originating from the mitochondrion are constituents
of the respiratory electron transfer chain or its energy conserving

---

[1]RWTH Aachen, Abteilung für Physiologische Chemie, 5100 Aachen, FRG

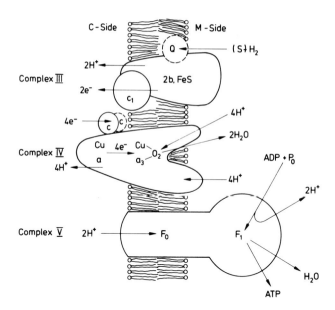

Fig. 1. Schematic drawing of mitochondrial electron transfer chain (complexes III and IV) and ATP synthase (complex V). For reaction stoichiometries at these sites see the series of papers in Frontiers of Biological Energetics (1978) V 1,2, Dutton et al. (eds), North Holland Publ. Co., Amsterdam. C,M Side cycloplasmic, matrix side of the inner mitochondrial membrane; Q ubiquinone; a, a_3, b, c, c_1 cytochromes; Cu copper proteins; F_0, F_1 segments of ATP synthase

ATP synthesizing system. Integrated in multisubunit complexes they constitute a considerable part of the inner mitochondrial membrane. This membrane contains additional enzyme complexes (I and II) and group transducing proteins for the metabolic adaptation between the organelle and the cell such as the ADP-ATP carrier protein (Klingenberg et al. 1975). Together with a number of other indications (Lehninger 1965) this gives the impression of a symbiotic organism which has been reduced to those main functions which once gave rise to the symbiosis and was provided with additional systems for the adaptation to the host.

The fact that the complexes III, IV and V are membrane-integrated makes them difficult objects for biochemical analysis. Thus several decades have passed since the discovery of the respiratory cytochromes (Warburg 1924; Keilin 1925) and first ideas about mechanism of energy conservation (Slater 1949; Mitchell 1961) until now preparative and analytical techniques allow the purification and analysis of constitutive components as first steps towards a detailed molecular description, which was already years ago achieved for several of the smaller, soluble auxiliary transport proteins, for instance cytochrome c and myoglobin/hemoglobin. Considerable misunderstanding results even today from the oversimplifying identification of bands found in SDS gels with subunits of the enzymes, which blurs the complexity of the inner membrane and obscures the important criterion of mitochondrial or cytoplasmic origin of the components. Furthermore the question whether the set of polypeptides constituting the complexes obtained from the most investigated sources, yeast, Neurospora, and mammalian tissues are identical in number and stoichiometry cannot finally be answered today. While in the recent past more weight was put to their identity now some ideas of difference are emerging (Kadenbach 1980).

Complex III

The ubiquinone-cytochrome-c-oxidoreductase is assumed to contain cytochrome b and cytochrome c_1 in a ratio of $2:1$ (Weiss 1978), furthermore

a nonheme iron protein is integrated (Rieske 1976). About 10 protein
components can be obtained with SDS-gel techniques (Capaldi et al.
1977). The apparent molecular weights range between 8,000 and 50,000
daltons, stoichiometries of 1 : 1 and 2 : 1 respectively have been found
(Weiss 1978). Only cytochrome b has been shown to be coded in the
yeast mitochondrion (Ross and Schatz 1976). This protein penetrates
in the membrane. The protein primary structure of this component is
presently being deduced from the sequence of the yeast mtDNA gene.
However, a difference exists between its gene structure and the apparent molecular weight (30,000 daltons) obtained with the polypeptide.
This difference has raised the question whether introns exist in the
gene (Mahler et al. 1978). Though there has been some protein chemical
work with the components of complex III, no primary structure analysis
has been completed.

Complex IV

Our recent analyses of cytochrome-c-oxidase from beef heart have shown
the existence of 12 polypeptides in a complex containing two heme a
and two copper atoms, the functional monomer. Dependent on the preparation the enzyme retains between 5% to 25% phospholipids mostly cardiolipin and phosphatidyl-ethanolamine (Steffens and Buse 1976). The protein components seems to be present in a 1 : 1 stoichiometric ratio,
(Merle and Kadenbach 1980) though this has not been finally established.
In the enzymes obtained from yeast and *Neurospora* and probably also in
the beef heart enzyme, the three largest polypeptides originate from
mitochondrial genes (Cabral et al. 1978). The number of cytoplasmically
synthesized components, however, may be different compared with fungi
enzymes. The sum of the molecular weights of the 12 polypeptides of the
beef heart oxidase is about 180,000 and thus agrees very well with the
10 - 12 nMol heme a/mg protein found normally in the preparations of
the beef heart oxidase (Steffens and Buse 1976). Table 1 lists the 12
polypeptides of the oxidase, gives their apparent or correct molecular

Table 1. Protein components of beef heart cytochrome c-oxidase

Polypeptide	Mol.weight	Synthesis	N-Terminal Sequence
			1 2 3 4
I	36,000[a]	mit.	Formyl-Met-Phe-Ile-Asn-
II	26,021	mit.	Formyl-Met-Ala-Tyr-Pro-
III	22,000	mit.	(Met)-Thr-His-Gln-
IV	17,153	cyt.	Ala-His-Gly-Ser-
V	12,436	cyt.	Ser-His-Gly-Ser-
VIa	14,000[a]	cyt.	Ala-Ser-Gly-Gly-
VIb	11,000[a]	cyt.	Ala-Ser-Ala-Ala-
VIc	11,000[a]	cyt.	Ser-Thr-Ala-Leu-
VII	10,026	cyt.	Acetyl-Ala-Glu-Asp-Ile-
VIIIa	5,541	cyt.	Ser-His-Tyr-Glu-
VIIIb	6,000[a]	cyt.	Ile-Thr-Ala-Lys-
VIIIc	6,000[a]	cyt.	Phe-Glu-Asn-Arg-

[a]apparent molecular weight

weights, their biosynthetic origin and classifies them chemically from their N-terminal sequences. The roman numerals refer to the fractions obtained in the preparative and analytical SDS gel separation proce- dures (Steffens and Buse 1976; Buse and Steffens 1976, 1978). We have now completely sequenced four of these components (II, IV, VII and VIIIa) and have partial information on the other polypeptides. Poly- peptide V has been sequenced by Tanaka et al. (1977). Up to now no com- plete protein structures from oxidase subunits obtained from other species have been published.

The shortest polypeptides of the bovine oxidase are found in the frac- tion VIII. There seem to be three peptide chains of similar construc- tion as found with VIIIa (Buse and Steffens 1978) (see Table 2). This chain has only 47 residues and its structure is clearly divided into three domains. The first — from residue 1 - 20 — is hydrophilic, the

Table 2. Primary structures of cytochrome c-oxidase (hydrophobic sequences blockface typed):
a) Amino acid sequence of polypeptide VIIIa of the beef heart enzyme
b) Amino acid sequence of polypeptide IV of the beef heart enzyme

<table>
<tr><td colspan="2"></td><td>20</td></tr>
<tr><td colspan="3">Ser-His-Tyr-Glu-Glu-Gly-Pro-Gly-Lys-Asn-Ile-Pro-Phe-Ser-Val-Glu-Asn-Lys-Trp-Arg-</td></tr>
<tr><td colspan="2"></td><td>40</td></tr>
<tr><td colspan="3">Leu-Leu-Ala-Met-Met-Thr-Leu-Phe-Phe-Gly-Ser-Gly-Phe-Ala-Ala-Pro-Phe-Phe-Ile-Val-</td></tr>
<tr><td>47</td><td></td><td></td></tr>
<tr><td colspan="2">Arg-His-Gln-Leu-Leu-Lys-Lys</td><td>a</td></tr>
</table>

<table>
<tr><td colspan="2"></td><td>20</td></tr>
<tr><td colspan="3">Ala-His-Gly-Ser-Val-Val-Lys-Ser-Glu-Asp-Tyr-Ala-Leu-Pro-Ser-Tyr-Val-Asp-Arg-Arg-</td></tr>
<tr><td colspan="2"></td><td>40</td></tr>
<tr><td colspan="3">Asp-Tyr-Pro-Leu-Pro-Asp-Val-Ala-His-Val-Lys-Asn-Leu-Ser-Ala-Ser-Gln-Lys-Ala-Leu-</td></tr>
<tr><td colspan="2"></td><td>60</td></tr>
<tr><td colspan="3">Lys-Glu-Lys-Glu-Lys-Ala-Ser-Trp-Ser-Ser-Leu-Ser-Ile-Asp-Glu-Lys-Val-Glu-Leu-Tyr-</td></tr>
<tr><td colspan="2"></td><td>80</td></tr>
<tr><td colspan="3">Arg-Leu-Lys-Phe-Lys-Glu-Ser-Phe-Ala-Glu-Met-Asn-Arg-Ser-Thr-Asn-Glu-Trp-Lys-Thr-</td></tr>
<tr><td colspan="2"></td><td>100</td></tr>
<tr><td colspan="3">Val-Val-Gly-Ala-Ala-Met-Phe-Phe-Ile-Gly-Phe-Thr-Ala-Leu-Leu-Leu-Ile-Trp-Glu-Lys-</td></tr>
<tr><td colspan="2"></td><td>120</td></tr>
<tr><td colspan="3">His-Tyr-Val-Tyr-Gly-Pro-Ile-Pro-His-Thr-Phe-Glu-Glu-Glu-Trp-Val-Ala-Lys-Gln-Thr-</td></tr>
<tr><td colspan="2"></td><td>140</td></tr>
<tr><td colspan="3">Lys-Arg-Met-Leu-Asp-Met-Lys-Val-Ala-Pro-Ile-Gln-Gly-Phe-Ser-Ala-Lys-Trp-Asp-Tyr-</td></tr>
<tr><td>147</td><td></td><td></td></tr>
<tr><td colspan="2">Asp-Lys-Asn-Glu-Trp-Lys-Lys</td><td>b</td></tr>
</table>

second domain from 21 - 40 is hydrophobic with only a few OH-groups as possibly hydrogen-bonding hydrophiles. A third short domain is again hydrophilic with mostly cationic residues and a characteristic -Lys-Lys sequence at the C-terminus. The chemical character suggests that these chains are not enzymatic, i.e., copper or heme-binding subunits, but anchor peptides which span the membrane and contact the hydrophobic parts. Furthermore they may provide specific ionic groups (two ba- sic or one basic one acidic) of the adjacent lysine, arginine, and glutamic acid residues to ionic groups of cardiolipin (two acidic) or phosphatidyl-ethanolamine (one acidic one basic) and with the rest of the hydrophilic N- and C-terminal parts bind to other more extrinsic

63

polypeptides of the complex. It should be mentioned that a series of
short polypeptides with analog structures has been found as coat proteins of the filamentous bacteriophages, for instance the fd and Pf1
phage coat proteins with 50 and 46 residues respectively (Nakashima
and Konigsberg 1974; Nakashima et al. 1975). This structural analogy
may correspond to a functional analogy, i.e., giving access to a "bacterial" membrane. It should also be mentioned that these polypeptides
(fraction VIII) can be extracted from the oxidase together with the
phospholipids for instances with chloroform/methanol.

We may shortly discuss the cytoplasmic polypeptides IV, V, and VII.
The function of these components is unknown. Polypeptides V (Tanaka
et al. 1977) and VII (Steffens et al. 1979a,b) do not contain a hydrophobic segment and for the latter case have been shown to be accessible
for chemical probes and trypsin digestion with the active enzyme. Component IV (Sacher et al. 1979a,b), the largest cytoplasmically synthesized chain of the oxidase, is rather hydrophilic but again contains a sequence of 18 hydrophobic residues with two adjacent ionic
groups and a tryptophan residue (Table 2b). IV has been located by
chemical labeling at the inner (m) side of the mitochondrial inner
membrane (Ludwig et al. 1979). As in VIIIa the terminal -Lys-Lys may
indicate the site-specific cleavage from a cytoplasmic precursor. Certain segments of the sequence of IV show a faint similarity to heme
proteins of the cytochrome c type. A decision of the question of such
a relationship — and thus the identification of the function of this
chain — is not yet possible from these data and may have been obscured
by the vaste time of separate evolution. It has been shown that reacting the oxidase with trypsin in the presence of cholate or isoelectric
focusing in urea (Freedman et al. 1979), leads to an active copper and
heme-binding oxidase containing subunits I and II, and IV as the only
cytoplasmic component.

Together with its location at the matrix side of the membrane, these
data indicate an important contribution to the O_2-reducing function
of complex IV.

The largest protein which we have sequenced so far is the mitochondrial
subunit II (Steffens and Buse 1979) (Table 3). It has been shown to
penetrate the inner mitochondrial membrane (Ludwig et al. 1979) and
corresponding to this the character of a membrane protein can again be
recognized by the presence of two hydrophobic segments of like structure as found in IV and VIII ranging from residues 27 to 47 and 63 to
81. The amino acid sequence of this mitochondrial component shows a
distinct homology to copper proteins of the bacterial azurin and chloroplast plastocyanin type (Table 4). This includes the identification
of invariant residues (His 102, Cys 106, His 204, Met 207) which from
X-ray data have been shown to build up the copper complex in the blue
oxidases (Coleman et al. 1978; Adman et al. 1978). Thus this mitochondrial protein is identified as one of the functional subunits of the
oxidase. The characteristic hydrophobic sequences of this membrane
protein have no correspondence in the structure of the soluble bacterial and chloroplast copper proteins. It has recently been discussed
whether this protein does not only bind copper but also heme a (Winter
et al. 1980) and thus is the copper-heme subunit which has been postulated from ESR data (Palmer et al. 1976) to react with O_2. However,
the sidedness and the fact that this subunit can be cross-linked with
cytochrome c (Bisson et al. 1977) are in favor of an electron transfer
function, oxidizing cytochrome c.

64

Table 3. Primary structures of cytochrome c-oxidase (hydrophobic sequences in blockface type):
B) Amino acid sequence of polypeptide II of the beef heart enzyme
H) Amino acid sequence of polypeptide II of the human placenta enzyme deduced from mtDNA cox I gene (Barrell et al. 1979)

```
                                                                    20
B  f-Met-Ala-Tyr-Pro-Met-Gln-Leu-Gly-Phe-Gln-Asp-Ala-Thr-Ser-Pro-Ile-Met-Glu-Glu-Leu-
H  f-Met-Ala-His-Ala-Ala-Gln-Val-Gly-Leu-Gln-Asp-Ala-Thr-Ser-Pro-Ile-Met-Glu-Glu-Leu-
                                                                              40
   Leu-His-Phe-His-Asp-His-THR-LEU-MET-ILE-VAL-PHE-LEU-ILE-SER-SER-LEU-VAL-LEU-TYR-
   Ile-Thr-Phe-His-Asp-His-ALA-LEU-MET-ILE-ILE-PHE-LEU-ILE-CYS-PHE-LEU-VAL-LEU-TYR-
                                                                              60
   ILE-ILE-SER-LEU-MET-LEU-Thr-Thr-Lys-Leu-Thr-His-Thr-Ser-Thr-Met-Asp-Ala-Gln-Glu-
   ALA-LEU-PHE-LEU-THR-LEU-Thr-Thr-Lys-Leu-Thr-Asn-Thr-Asn-Ile-Ser-Asp-Ala-Gln-Glu-
                                                                              80
   Val-Glu-THR-ILE-TRP-THR-ILE-LEU-PRO-ALA-ILE-ILE-LEU-ILE-LEU-ILE-ALA-LEU-PRO-SER-
   Met-Glu-THR-VAL-TRP-THR-ILE-LEU-PRO-ALA-ILE-ILE-LEU-VAL-LEU-ILE-ALA-LEU-PRO-SER-
                                ...                                           100
   LEU-Arg-Ile-Leu-Tyr-Met-Met-Asp-Glu-Ile-Asn-Asn-Pro-Ser-Leu-Thr-Val-Lys-Thr-Met-
   LEU-Arg-Ile-Leu-Tyr-MET-Thr-Asp-Glu-Val-Asn-Asp-Pro-Ser-Leu-Thr-Ile-Lys-Ser-Ile-
                                                                              120
   Gly-His-Gln-Trp-Tyr-Trp-Ser-Tyr-Glu-Tyr-Thr-Asp-Tyr-Glu-Asp-Leu-Ser-Phe-Asp-Ser-
   Gly-His-Gln-Trp-Tyr-Trp-Thr-Tyr-Glu-Tyr-Thr-Asp-Tyr-Gly-Gly-Leu-Ile-Phe-Asn-Ser-
                                ...                                           140
   Tyr-Met-Ile-Pro-Thr-Ser-Glu-Leu-Lys-Pro-Gly-Glu-Leu-Arg-Leu-Leu-Glu-Val-Asp-Asn-
   Tyr-MET-Leu-Pro-Pro-Leu-Phe-Leu-Glu-Pro-Gly-Asp-Leu-Arg-Leu-Leu-Asp-Val-Asp-Asn-
                                                                              160
   Arg-Val-Val-Leu-Pro-Met-Glu-Met-Thr-Ile-Arg-Met-Leu-Val-Ser-Ser-Glu-Asp-Val-Leu-
   Arg-Val-Val-Leu-Pro-Ile-Glu-Ala-Pro-Ile-Arg-MET-MET-Ile-Thr-Ser-Gln-Asp-Val-Leu-
                                                                              180
   His-Ser-Trp-Ala-Val-Pro-Ser-Leu-Gly-Leu-Lys-Thr-Asp-Ala-Ile-Pro-Gly-Arg-Leu-Asn-
   His-Ser-Trp-Ala-Val-Pro-Thr-Leu-Gly-Leu-Lys-Thr-Asp-Ala-Ile-Pro-Gly-Arg-Leu-Asn-
                                ...                                           200
   Gln-Thr-Thr-Leu-Met-Ser-Ser-Arg-Pro-Gly-Leu-Tyr-Tyr-Gly-Gln-Cys-Ser-Glu-Ile-Cys-
   Gln-Thr-Thr-Phe-Thr-Ala-Thr-Arg-Pro-Gly-Val-Tyr-Tyr-Gly-Gln-Cys-Ser-Glu-Ile-Cys-
                                                                              220
   Gly-Ser-Asn-His-Ser-Phe-Met-Pro-Ile-Val-Leu-Glu-Leu-Val-Pro-Leu-Lys-Tyr-Phe-Glu-
   Gly-Ala-Asn-His-Ser-Phe-Met-Pro-Ile-Val-Leu-Glu-Leu-Ile-Pro-Leu-Lys-Ile-Phe-Glu-'
                        227
   Lys-Trp-Ser-Ala-Ser-Met-Leu
   Met-Gly-Pro-Val-Phe-Thr-Leu
```

Met, translated from AUA; Trp, translated from UGA

Table 4. Alignment of azurin *(Pseudomonas fluorescens)*, plastocyanin (French bean) and cytochrome c-oxidase subunit II

```
Azurin          H-Ala-Glu-Cys-Lys-Val-Asp-Val-Asp-Ser-Thr-Asp-Gln  -  Met-Ser-Phe-Asn-Thr-
                                                         10
Polypeptide II  -Met-Asp-Ala-Gln-Glu-Val-Glu-Thr-Ile-Trp-Thr-Ile-Leu-Pro-Ala-Ile-Ile-Leu-Ile-
                          56       60                              70
Plastocyanin    H-Leu-Glu-Val-Leu-Leu-Gly-Ser-Gly-Asp-Gly-Ser-Leu-Val-Phe-Val-Pro-
                                                         10


-Lys-Glu-Ile-Thr-Ile-Asp-Lys-Ser-Cys-Lys  -  Thr-Phe-Thr-Val-Asn-Leu-Thr-His-Ser-Gly-Ser-Leu-
          20                                      30                                  Ser-Leu
-Leu  -  Ile-Ala-Leu-Pro-Ser  -  Leu-Arg-Ile-Leu-Tyr-Met-Met-Asp-Glu-Ile-Asn-Asn-Pro-Ser-Leu-
               80                                             90
-Ser-Glu-Phe-Ser-Val-Pro-Ser-Gly-Glu-Lys-Ile-Val-Phe-Lys  -  -  -  -  Asn-Asn-Ala-Gly-Phe-
          20                                                          31


-Pro-Lys-Asn-Val-Met-Gly-His-Asn-Trp-Val-Leu-Ser-Lys  -  Ser-Ala-Asp-Met-Ala-Gly-Ile-Ala-Thr-
   40            46            30                         Ser-Ala                        60
-Thr-Val-Lys-Thr-Met-Gly-His-Gln-Trp-Tyr-Trp-Ser-Tyr-Glu-Tyr-Thr-Asp-Tyr-Glu-Asp-Leu-Ser-Phe-
                 100   102                        110
-Pro  -  -  -  -  -  His-Asn-Val-Ala-Phe  -  -  -  -  -  -  -  -  -  -  -  -  -  Asp-Glu-
   36                 37            40


-Asp-Gly-Met-Ala-Ala-Gly-Ile-Asp-Lys-Asp-Tyr-Leu-Lys-Pro-Gly-Asp-Ser-Arg-Val-Ile-Ala  -  His-
                                70                                         80
-Asp-Ser-Tyr-Met  -  -  Ile-Pro-Thr-Ser-Glu-Leu-Lys-Pro-Gly-Glu-Leu-Arg-Leu-Leu-Glu-Val-Asp-
    120                                      130
-Asp-Glu-Ile-Pro-Ala-Gly-Val-Asp-Ala-Val-Lys-Ile-Ser-Met-Pro-Glu-Glu-Glu-Leu-Leu-Asn  -  -
                              50                                  60


-Thr-Lys-Ile-Ile-Gly-Ser-Gly-Glu-Lys-Asp-Ser-Val-Thr-Phe-Asp-Val-Ser-Lys  -  -  -  -  -
                      90                              100
-Asn-Arg-Val-Val-Leu-Pro-Met-Glu-Met  -  Thr-Ile-Arg-Met-Leu-Val-Ser-Ser-Glu-Asp-Val-Leu-His-
 140                                              150                              160
-  -  -  -  Ala-Pro-Gly-Glu-Thr-Tyr-Val-Val-Thr  -  -  -  -  -  -  -  -  -  -  -  -
                      70


-  -  -  -  -  -  -  -  -  -  -  -  -  -  -  -  -  -  -  -  -  -  -  -  -  -  -
-Ser-Trp-Ala-Val-Pro-Ser-Leu-Gly-Leu-Lys-Thr-Asp-Ala-Ile-Pro-Gly-Arg-Leu-Asn-Gln-Thr-Thr-Leu-
                              170                                    180
-  -  -  -  -  -  -  -  -  -  -  -  -  -  -  -  -  -  -  -  -  -  -  -  -  -  -


-  Leu-Thr-Ala-Gly-Glu-Ser-Tyr-Glu-Phe-Phe-Cys-Ser  -  Phe-Pro-Gly  -  -  His-Asn  -  Ser-
                          110      112                                  117           Ser
-Met-Ser-Ser-Arg-Pro-Gly-Leu-Tyr-Tyr-Gly-Gln-Cys-Ser-Glu-Ile-Cys-Gly-Ser-Asn-His  -  -  Ser-
                 190              196              200              204
-  Leu-Asp-Thr-Lys-Gly-Thr-Tyr-Ser-Phe-Tyr-Cys-Ser  -  -  Pro  -  -  -  His-Gln-Gly-Ala-
                 80              84                                   87           90


-Met-Met-Lys-Gly-Ala-Val-Val-Leu-Lys-OH
    121                  128
-Phe-Met-Pro-Ile-Val-Leu-Glu-Leu-Val-Pro-Leu-Lys-Tyr-Phe-Glu-Lys-Trp-Ser-Ala-Ser-Met-Leu-OH
    207         210                          220                              227
-Gly-Met-Val-Gly-Lys-Val-Thr-Val-Asn-OH
    92                  99
```

• Point I mutations of subunit II to azurin or plastocyanin

Complex V

In the case of complex V the reversible ATPase (ATP-phosphohydrolase
EC 3.6.1.3) or ATP synthase up to 12 different polypeptides have been
separated (Tzagoloff and Meagher 1971). Most protein chemical work
has been done with the F_O 8000 daltons polypeptide (Sebald et al. 1979).
This component can be extracted and purified from the complex with
chloroform/methanol and has therefore been named "proteolipid" (Cattell
et al. 1971). It is assumed to be the protonophoric subunit of the F_O
fragment (Hoppe et al. 1980). Complete primary structures of this subunit have been obtained by Sebald and his coworkers from a variety of
organelles including mitochondria from *Neurospora*, yeast and bovine
heart, chloroplasts from spinach, and bacteria (*E. coli* and the thermophile PS-3) (Table 5).

Surprisingly some of these sequences do not start with N-terminal
formylmethionine and thus do not immediately indicate their mitochondrial origin or are even synthesized in the cytoplasma (Sebald et al.
1976). Though exemplifying the entire world of living organisms, all
the sequences show remarkable homology not only concerning single invariant residues but also the general constancy of two hydrophobic
segments. The first hydrophobic segment of about 25 residues length
is constructed in a manner similar to the seven hydrophobic sequences
in bacteriorhodopsin which are said to cross the membrane as helical
rods (Ovchinnikov et al. 1979). The most significant character is
the presence of one glutamyl or aspartyl residue invariably included

Table 5. Primary structures of the ATPase (ATP synthase) proteolipid subunit IX,
after Sebald et al. (1979). Mitochondrial sequences from *Neurospora crassa*, bovine
heart, *Saccharomyces cerevisiae*, chloroplast sequence from spinach, bacterial sequences from *E. coli* and the thermophile PS-3

```
                          10                                      20
N. cr. Mit.  Tyr-Ser-Ser-Glu-Ile-Ala-Gln-Ala-Met-Val-Glu-Val-Ser-Lys-Asn-Leu-Gly-Met-Gly-Ser-Ala-Ala-Ile-Gly-Leu-
Bovine Mit.                          Asp-Ile-Asp-Thr-Ala-Ala-Lys-Phe-Ile-Gly-Ala-Gly-Ala-Ala-Thr-Val-Gly-Val-
S. cer.Mit.              f-Met-Gln-Leu-Val-Leu-Ala-Ala-Lys-Tyr-Ile-Gly-Ala-Gly-Ile-Ser-Thr-Ile-Gly-Leu-
Spin. Chl.            f-Met-Asn-Pro-Leu-Ile-Ala-Ala-Ala-Ser-Val-Ile-Ala-Ala-Gly-Leu-Ala-Val-Gly-Leu-Ala-Ser-
E. coli              f-Met-Glu-Asn-Leu-Asn-Met-Asp-Leu-Leu-Tyr-Met-Ala-Ala-Ala-Val-Met-Met-Gly-Leu-Ala-Ala-
PS-3                     f-Met-Ser-Leu-Gly-Val-Leu-Ala-Ala-Ala-Ile-Ala-Val-Gly-Leu-Gly-Ala-

                     30                              40                          50
N. cr. Mit.  Thr-Gly-Ala-Gly-Ile-Gly-Ile-Gly-Leu-Val-Phe-Ala-Ala-Leu-Leu-Asn-Gly-Val-Ala-Arg-Asn-Pro-Ala-Leu-Arg-
Bovine Mit.  Ala-Gly-Ser-Gly-Ala-Gly-Ile-Gly-Thr-Val-Phe-Gly-Ser-Leu-Ile-Ile-Gly-Tyr-Ala-Arg-Asn-Pro-Ser-Leu-Lys-
S. cer.Mit.  Leu-Gly-Ala-Gly-Ile-Gly-Ile-Gly-Leu-Val-Phe-Ala-Ala-Leu-Ile-Asn-Gly-Val-Ser-Arg-Asn-Pro-Ser-Ile-Lys-
Spin. Chl.   Ile-Gly-Pro-Gly-Val-Gly-Gln-Gly-Thr-Ala-Ala-Gly-Gln-Ala-Val-Gly-Ile-Ala-Arg-Gln-Pro-Glu-Ala-Glu-
E. coli      Ile-Gly-Ala-Ala-Ile-Gly-Ile-Gly-Ile-Leu-Gly-Gly-Lys-Phe-Leu-Gln-Gly-Ala-Ala-Arg-Gln-Pro-Asp-Leu-Ile-
PS-3         Leu-Gly-Ala-Gly-Ile-Gly-Asn-Gly-Leu-Ile-Val-Ser-Arg-Thr-Ile-Glu-Gly-Ile-Ala-Arg-Gln-Pro-Glu-Leu-Arg-

                           60                          70
N. cr. Mit.  Gly-Gln-Leu-Phe-Ser-Tyr-Ala-Ile-Leu-Gly-Phe-Ala-Phe-Val-Glu-Ala-Ile-Gly-Leu-Phe-Asp-Leu-Met-Val-Ala-
Bovine Mit.  Gln-Gln-Leu-Phe-Ser-Tyr-Ala-Ile-Leu-Gly-Phe-Ala-Leu-Ser-Glu-Ala-Met-Gly-Leu-Phe-Cys-Leu-Met-Val-Ala-
S. cer.Mit.  Asp-Thr-Val-Phe-Pro-Met-Ala-Ile-Leu-Gly-Phe-Ala-Leu-Ser-Glu-Ala-Thr-Gly-Leu-Phe-Cys-Leu-Met-Val-Ser-
Spin. Chl.   Gly-Lys-Ile-Arg-Gly-Thr-Leu-Leu-Leu-Ser-Leu-Ala-Phe-Met-Glu-Ala-Leu-Thr-Ile-Tyr-Gly-Leu-Val-Val-Ala-
E. coli      Pro-Leu-Leu-Arg-Thr-Gln-Phe-Phe-Ile-Val-Met-Gly-Leu-Val-Asp-Ala-Ile-Pro-Met-Ile-Ala-Val-Gly-Leu-Gly-
PS-3         Pro-Val-Leu-Gln-Thr-Thr-Met-Phe-Ile-Gly-Val-Ala-Leu-Val-Glu-Ala-Leu-Pro-Ile-Ile-Gly-Val-Val-Phe-Ser-

                  80
N.cr. Mit.   Leu-Met-Ala-Lys-Phe-Thr
Bovine Mit.  Phe-Leu-Ile-Leu-Phe-Ala-Met
S. cer.Mit.  Phe-Leu-Leu-Leu-Phe-Gly-Val
Spin. Chl.   Leu-Ala-Leu-Leu-Phe-Ala-Asn-Pro-Phe-Val
E. coli      Leu-Tyr-Val-Met-Phe-Ala-Val-Ala
PS-3         Phe-Ile-Tyr-Leu-Gly-Arg
```

PRIMARY STRUCTURES OF ATPase PROTEOLIPID.

in the second hydrophobic segment of the proteolipid. This residue
specifically reacts with dicyclohexylcarbodiimide (DCCD). Already if
occurring with only one out of six of these subunits present in the
ATPase complex (Kagawa 1978) this reaction completely inhibits the
protonophoric function which according to current ideas on the mechanisms of energy conservation (Mitchell 1978) is essential for ATP
synthesis, coupled to proton pumping across bacterial, chloroplasts,
and mitochondrial membranes.

Conclusions and Hypotheses

Though our present knowledge about the protein chemistry of the respiratory membrane is rather incomplete and can even in this state not
be completely referred to here, we are now able to draw some conclusions and arrive at some hypotheses from the facts that have accumulated.

Primordial Electron Transfer

Component analysis of bacterial, chloroplast (Blankenship and Parson
1978), and mitochondrial electron transfer has now revealed the existence of members of the cytochrome b (b), cytochrome c (f, c_2, c-551)
and copper protein (azurins, plastocyanins, stellacyanins and cytochrome oxidase subunit II) families in the low potential segments of
these chains. At the high potential side a ferredoxin-like protein
(FeS) has been suggested (Evans and Buchanan 1965; Tagawa et al. 1963).
As far as "mitochondrial" cytochrome c are concerned one must consider
whether a change of the genetic coding from the organelle to the eukaryotic host has occurred, or whether there exists a relationship of
one of the mitochondrial heme proteins integrated in the complexes III
or IV to this protein family. The similarity of the copper subunit II
of the oxidase to plastocyanines or stellacyanins is not closer than
to the bacterial azurins. These data thus clearly point to the origin
of mitochondria from bacterial ancestors. They allow us to constitute
the scheme of a primordial electron transport chain (Fig. 2) which
may have occurred more than $3 \cdot 10^9$ years ago (Baltscheffsky 1978).

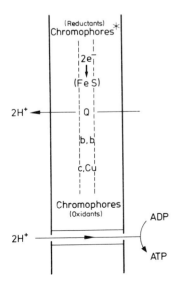

Fig. 2. Hypothetical scheme of a primordial cycle
of electron transport and energy conservation.
FeS nonheme iron protein; *Q* quinone; *b,c* cytochromes;
Cu copper protein

Cyclic Photophosphorylation

The most probable function which this system may have had at its beginning is that of cyclic photophosphorylation (Broda 1975), providing
metabolic energy for different synthetic needs in a reducing milieu
running short of some of the necessary components (photerger). Already
at that time the light-induced electron transport was used for energy
conservation by proton pumping across a membrane and reversed proton
flux-dependent ATP synthesis. The most common vehicle used for the
purpose of proton pumping was of the quinone type (Q) and is found as
ubiquinone or menaquinone in present-day photophosphorylating bacteria
or as plastoquinone in photosynthetic electron transport of chloroplasts
(Blankenship and Parson 1978) and as ubiquinone in the mitochondrial
respiratory chain. These functions are completed with the ATP synthesizing enzyme, the protonophoric membrane-penetrating subunit of which
has been analyzed as a common member from bacteria and both organelles,
showing an obvious relationship in all these cases. Again no close
similarity exists between the chloroplasts and mitochondrial sequences,
indicating their bacterial origin.

The Bacterial Heritage of Mitochondria

The components that have been named so far apparently are those that
represent the old bacterial heritage in present-day mitochondria. The
essence of the symbiosis may be seen in the opening of the cyclic process to a linear array allowing for the adaptation to different re-
ductants and oxidants. While in present-day chloroplasts the system
is seen at work between the photosystem II (PS 680) and I (PS 700), it
has been adapted to the reduced substrate and O_2 in mitochondria. It
is in line with the logic of this process that the latter possibility
as the symbiosis itself was only achieved in a genetically capable
organism, the eukaryotic cell. In this connection several cases of
transfer of genetic information from the organelle to the nucleus must
be assumed (Borst and Grivell 1978) as in the case of the *Neurospora*
ATP synthase proteolipid (Jackl and Sebald 1975).

Origin of O_2-Reductive Capability

After the present spectroscopic evidence (Palmer et al. 1976) the
need for the reduction of O_2 to 2 H_2O was the development of a heme/
copper complex providing the possibility of transferring 4 electrons
to O_2 and reacting on the matrix site of the mitochondrial membrane
to bring the consumption of 4 protons in accord with the proton gradient across this membrane. The component IV of the oxidase which probably is synthesized as part of a cytoplasmic precursor (Schmelzer and
Heinrich 1980) in this respect may be considered as a significant contribution of the eukaryotic cell. However, the O_2-reducing capacity
probably developed in bacteria and in the beginning may have had an
O_2 depoisoning function.

The "Universal" Code

The recently achieved possibility of comparing amino acid sequences
of mitochondrial proteins as the ATP-synthase proteolipid and the cytochrome-c-oxidase subunit II with their corresponding mitochondrial DNA
gene sequences (Coruzzi and Tzagoloff 1979; Fox 1979; Barrell et al.
1979; Hensgens et al. 1979), surprisingly revealed the existence of
an altered genetic code in these organelles. Besides the restricted

use of synonymous triplets found especially in yeast, one of the established differences is the UGA coding for tryptophan. Furthermore an additional AUA coding for methionine must be concluded from the comparison of the human placenta subunit II gene with two transcribed AUG (methionine) codons and the beef heart oxidase subunit II with 16 methionine residues. In this connection it is important to notice that the two AUG triplets transcribed from the placenta subunit II gene are N-terminal formyl methionine and methionine 204, found to be invariant in our copper protein alignment. The same applies to the invariant tryptophan, which at least in the human placenta gene remains to be coded solely by UGG. The code changes are thus probably more recent then the occurrance of the symbiosis and the main gene transfers from the mitochondrion to the nucleus, respectively. Despite the simple fact that the genetic code today is not universal, the only convincing explanation is that these changes are derived from one common code in the ancestry of bacteria, organelles, and the eukaryotic cell.

Acknowledgment. This work was supported by the Sonderforschungsbereich 160 "Biologische Membranen" of the Deutsche Forschungsgemeinschaft.

References

Adman ET, Stenkamp RE, Sieker LC, Jensen LH (1978) J Mol Biol 123:35-47

Arnon DJ (1959) Nature (London) 184:10

Baltscheffsky H (1978) In: Schäfer G, Klingenberg M (eds) Energy conservation in biological membranes, 29th Colloquium-Mosbach. Springer, Berlin Heidelberg New York, pp 3-18

Barrell BG, Bankier AT, Drouin J (1979) Nature (London) 282:189-194

Bisson R, Gutweniger H, Montecucco R, Colonna A, Zanotti A, Azzi A (1977) FEBS Lett 81:147-150

Blankenship RE, Parson WW (1978) Annu Rev Biochem 47:635-653

Borst P, Grivell LA (1978) Cell 15:705-723

Broda E (1975) The evolution of the bioenergetic processes. Pergamon Press, New York, pp 53 ff

Buse G, Steffens G (1976) In: Bücher et al. (eds) Genetics and biogenesis of chloroplasts and mitochondria. North Holland Publ Co, Amsterdam, pp 189-196

Buse G, Steffens GJ (1978) Hoppe-Seyler's Z Physiol Chem 359:1005-1009

Cabral F, Solioz M, Rudin Y, Schatz G, Clavilier L, Slonimski PP (1978) J Biol Chem 253:297-304

Capaldi RA, Bell RL, Branchek T (1977) Bioch Biophys Res Commun 74:425-433

Cattell KJ, Lindop CR, Knight J, Beechey RB (1971) Biochem J 125:169-177

Coleman PM, Freeman HC, Guss JM, Murata MA, Noris VA, Ramshaw JAM, Venkatappa MP (1978) Nature (London) 272:319-324

Coruzzi G, Tzagoloff A (1979) J Biol Chem 254:9324-9330

Evans MCW, Buchanan BBC (1965) Proc Natl Acad Sci USA 53:1420

Freedman JA, Tracy RP, Chan SHP (1979) J Biol Chem 254:4305-4308

Fox T (1979) Proc Natl Acad Sci USA 76:6534-6538

Hensgens LAM, Grivell LA, Borst P, Bos JL (1979) Proc Natl Acad Sci USA 76:1663-1667

Jackl G, Sebald W (1975) Eur J Biochem 54:97-106

Kadenbach B (1980) personal communication

Kagawa Y (1978) Biochem Biophys Acta 505:45-93

Keilin D (1925) Proc R Soc London Ser B 98:312-339

Klingenberg M, Aquila H, Reccio P, Buchanan BB, Eiermann W, Hachenberg H (1975) In: Quagliariello E et al (eds) Electron transfer chains and oxidative phosphorylation. North Holland Publ Co, Amsterdam, p 431

Lambowitz AM, Chua NH, Luck DJL (1976) J Mol Biol 107:223-253

Lehninger AL (1965) The mitochondrion. Benjamin Menlo Park, California

Ludwig B, Downer NW, Capaldi R (1979) Biochemistry 18:1401-1407

Mahler HR, Hanson DK, Miller DH (1978) In: Dutton L et al. (eds) Frontiers of biological energetics, vol I. Academic Press, London New York, pp 135-145

Merle P, Kadenbach B (1980) Eur J Biochem 105:499-507

Mitchell P (1961) Nature (London) 191:144-148

Mitchell P (1978) In: Dutton L et al (eds) Frontiers of biological energetics, vol I. Academic Press, London New York, pp 3-11

Nakashima Y, Konigsberg W (1974) J Mol Biol 88:598-600

Nakashima Y, Wiseman RL, Konigsberg W, Marvin DA (1975) Nature (London) 253:68-71

Ovchinnikov YA, Abdulaev NG, Feigina MY, Kiselev AW, Lobanov NA (1979) FEBS Lett 100:219-224

Palmer G, Babcock GT, Vickery LE (1976) Proc Natl Acad Sci USA 73:2206-2210

Rieske JS (1976) Biochem Biophys Acta 456:195-247

Ross E, Schatz G (1976) J Biol Chem 251:1997-2004

Sacher R, Buse G, Steffens GJ (1979a) Hoppe-Seyler's Z Physiol Chem 360:1377-1383

Sacher R, Steffens GJ, Buse G (1979b) Hoppe-Seyler's Z Physiol Chem 360:1385-1392

Schmelzer E, Heinrich PC (1980) J Biol Chem in press

Sebald W, Graf T, Wild G (1976) In: Bücher Th et al (eds) Genetics and biogenesis of chloroplasts and mitochondria. North Holland Publ Co, Amsterdam, pp 167-174

Sebald W, Hoppe J, Wachter E (1979) In: Quagliariello E, Palmieri F, Papa S, Klingenberg M (eds) Proc Int Symp Function Mol Aspects Biomembr Transport, Selva di Fasano (Italy). Elsevier/North-Holland Biomedical Press, Amsterdam New York, pp 63-74

Slater EC (1949) Biochem J 45:14

Steffens G, Buse G (1976) Hoppe-Seyler's Z Physiol Chem 357:1125-1137

Steffens GJ, Buse G (1979) Hoppe-Seyler's Z Physiol Chem 360:613-619

Steffens GCM, Steffens GJ, Buse G, Witte L, Nau H (1979a) Hoppe-Seyler's Z Physiol Chem 360:1633-1640

Steffens GCM, Steffens GJ, Buse G (1979b) Hoppe-Seyler's Z Physiol Chem 360:1641-1650

Tagawa K, Tjujimoto JY, Arnon DJ (1963) Proc Natl Acad Sci USA 49:567

Tanaka M, Haniu M, Yasunobu KT, Yu CA, Yu L, Wei YH, King TE (1977) Biochem Biophys Res Commun 76:1014-1019

Tzagoloff A, Meagher P (1971) J Biol Chem 246:7328-7336

Tzagoloff A, Macino G, Sebald W (1979) Annu Rev Biochem 48:419-441

Warburg O (1924) Biochem Z 152:479-494

Weiss H (1978) In: Schäfer G, Klingenberg M (eds) Energy conservation in biological membranes, 29th Mosbach Colloquium. Springer, Berlin Heidelberg New York, pp 314-321

Winter DB, Foulke FG, Mason HS (1980) In: Int Symp Oxidases and Related Redox Systems ISOX III Albany. Pergamon Press, New York, in press

Early Biological Evolution Derived from Chemical Structures

M. O. Dayhoff and R. M. Schwartz[1]

Introduction

Chemists have known for many years that enzymes from one organism would function with those from another in the test tube but that the chemical structures of homologous enzymes from different species were usually not identical. Powerful sequencing methods have permitted the elucidation of the details of these differing covalent structures. It is now clear that many protein and nucleic acid sequences are "living fossils" in the sense that their structures have been dynamically conserved by evolution over billions of years: recognizably related forms are found in eukaryotes and prokaryotes and are believed to be of common evolutionary origin, having evolved by many small changes in the sequences (Bryson and Vogel 1965; Dayhoff et al. 1976). By comparison of sequences in many contemporary species it has been possible to infer the evolutionary connections of the species. This history provides a powerful organizing concept in biology, a kind of interpolation scheme that can be used to predict how different each species is from the others. The history of eukaryotes, particularly that of the animals, is now well understood. Schemes derived from morphological and fossil data and from sequences are in agreement. However, the connections of bacteria have remained obscure because the morphological data are inadequate. Recently, enough sequence information from prokaryotes has become available to permit us to construct a biologically comprehensive phylogenetic tree (Schwartz and Dayhoff 1978a,b; Dayhoff and Schwartz 1978). This tree was derived from the covalent chemical structures of the sequences by objective mathematical methods, a line of reasoning that is independent of the derivations from biological traits, metabolic capacities, or fossil evidence.

Evolutionary history is conveniently represented by a tree. Each point on the tree corresponds to a time, a biological species, and one or more macromolecular sequences within the species. There is one point that corresponds to the earliest time and to the ancestral organism and sequence. We will call this point the base of the tree. Time advances on all branches emanating from the base. The topology of the branches gives the relative order of events. During evolution, sequences ·in different species have gradually and independently accumulated changes yielding the sequences found today in the various organisms represented at the ends of the branches. In the trees in this paper we have drawn the lengths of the branches proportional to the amount of evolutionary change in the sequences.

Our working hypothesis is that, for those sequences that perform basic metabolic functions, genetic transfer between major types of bacteria followed by permanent acceptance in evolution is very infrequent. All

---

[1]Department of Physiology and Biophysics, Georgetown University Medical School, and National Biomedical Research Foundation, Washington, D.C. 20007, USA

such basic sequences within an organism share the same phylogenetic history, and any one of them can be used to infer its course.

We will first discuss the method used, then we will investigate the trees derived from different kinds of molecules, and finally, we will combine the evidence from these trees to show the overall picture of evolution.

Methods

The mechanisms of evolution acting on the protein sequences are sufficiently well understood that the usefulness of any proposed method for reconstructing phylogeny can be tested. Sequences connected by a known history of divergence and amount of evolutionary change can be constructed by simulating the evolutionary change using random numbers. We have described a computer model (Dayhoff and Schwartz 1979) that is simple, yet adequate for this purpose. By considering many groups of simulated sequences, each group related by the same evolutionary history but generated with different random numbers, a measure of the accuracy of a given method can be assessed and different methods can be compared.

We have found the least-squares matrix method most reliable for distantly related sequences (Dayhoff 1976, 1979); therefore, we used it for constructing the trees in this paper. In this method, sequences are aligned to reflect the genetic changes accepted in evolution and a matrix of percentage difference between them is calculated. These percentages are corrected for inferred superimposed mutations to yield the "observed" matrix of accepted point mutations per 100 residues (PAMs). A double minimization is required. For a predetermined order of connection or topology of the sequences, the computer program MATTOP determines the branch lengths that give a weighted least-squares fit of the reconstructed matrix to the observed matrix. The results for all of the topologies must be compared. We have evaluated many variations of criteria to infer the true evolutionary history and found that the topology that gives the smallest sum of the absolute values of the branch lengths provides the best estimate. The least-squares matrix method makes no assumptions about rates of change on the various branches in a tree and says nothing about the location of the base of the tree. It is not valid to assume equality of rates, especially for the major prokaryote groups treated here, for which even in the same kind of sequences there has frequently been twice as much, and sometimes ten times as much, change in one line as in another during the same time interval.

The occasional occurrence of a gene duplication that persists in evolution provides us with evidence to identify the base of a tree and to infer the temporal order of divergences of the species. Suppose that two related proteins, A and B, both occur in a number of species. (These proteins may be distinguishable because they perform slightly different functions in the species, or because of gross structural differences, or perhaps only because of their segregation on a tree due to point mutations.) A typical topology is shown in Figure 1a. The most parsimonious description of the history is that there was a single gene duplication somewhere on the line between nodes 1 and 2 and that both of these nodes are identified with the same species that diverged to give the chicken and mammalian lines. The same phylogenetic history is given by each half of the tree, as summarized in Figure 1b, where the corresponding nodes and branch tips from the upper tree are superimposed. In the lower tree, each species appears only once, but the

Placing the root using a gene duplication

Evolutionary Topology of Proteins

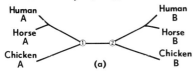

(a)

Phylogenetic Tree of Species

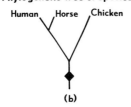

(b)

Fig. 1. (a) A typical topology obtained from two related kinds of sequences, A and B, from three species. The computer program finds the best order of connection of the sequences and the branch lengths (without regard to the angles between the branches). The A and B sequences segregate. (b) The corresponding phylogenetic history. The branch lengths for the two kinds of sequences were averaged and the gene duplication was presumed to be halfway between nodes 1 and 2. It is represented by a diamond on the figure

order of connections is derived from more than one protein. The branch lengths are average values given by the two kinds of proteins. Where rates of change are similar, the approximate congruence of the subtrees provides further support for this history.

Confusion of interpretation of a tree arises only when a duplication has occurred that cannot be identified from the available data because only one of the forms is known for some species and only the other is known for the rest. In this case an earlier gene duplication may be misinterpreted as a species divergence, giving the impression that the species are more distant than they really are. As data accumulate, the significant gene duplications will be identified because both forms will usually persist in some species.

Results and Discussion

Three and one-half years ago there was sufficient sequence information for us to make a comprehensive evolutionary tree, including the position of the base, derived objectively from sequence data (Schwartz and Dayhoff 1978a,b; Dayhoff and Schwartz 1978). This tree, based on sequences of the c-type cytochromes, ferredoxins, and 5S ribosomal RNA molecules, is consistent with newer sequence data incorporated here.

Bacterial Ferredoxins

The bacterial ferredoxins, whose active centers are 4Fe-4S clusters, are found in a broad spectrum of organisms and participate in such fundamental biochemical processes as reductive carboxylation, nitrogen fixation, and dissimilatory (energy-producing) sulfate reduction (Rao and Hall 1977). Sequences have been elucidated by a number of workers, including Bruschi (1979) and Hase et al. (1978a). See Schwartz and Dayhoff (1979b) for other references. The tree of Figure 2 is derived from these sequences. A gene-doubling shared by all of these sequences makes it possible to deduce the point of earliest time in this tree and, by inference, provides a time orientation in the other trees presented here. Because all the sequences show evidence of gene-doubling, this

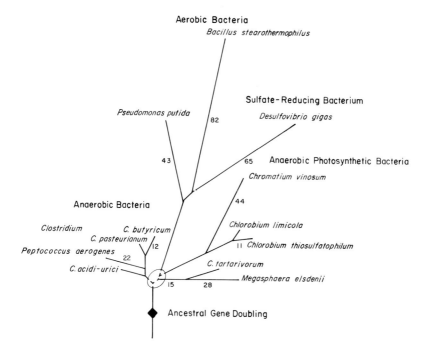

Aerobic Bacteria
Bacillus stearothermophilus

Sulfate-Reducing Bacterium
Desulfovibrio gigas

Pseudomonas putida

82

43

65   Anaerobic Photosynthetic Bacteria

Chromatium vinosum

44

Anaerobic Bacteria

Clostridium   *C. butyricum*
             C. pasteurianum
Peptococcus aerogenes
                        12
             22
C. acidi-urici

Chlorobium limicola

11 *Chlorobium thiosulfatophilum*

C. tartarivorum
                Megasphaera elsdenii
15   28

Ancestral Gene Doubling

Fig. 2. Evolutionary tree based on bacterial ferredoxins. Branch lengths are drawn proportional to the amount of evolutionary change. Selected values in PAMs are shown. The order of divergence of the branches leading to the pseudomonads, the anaerobic photosynthetic bacteria, and the clostridia was not clearly resolved, as indicated by the dotted lines. (Figure taken with permission from Dayhoff and Schwartz 1980)

event must have occurred prior to the species divergences. The organisms located near the base (*Clostridium, Megasphaera,* and *Peptococcus)* are all anaerobic, heterotrophic bacteria whose sequences are still very similar to the extremely ancient protein that duplicated. It has long been thought that, of the extant bacteria, these species most closely reflect the metabolic capacities of the earliest organisms (Broda 1975). In Figure 2, *Chlorobium* and *Chromatium* are pictured as having diverged very early from an ancestral heterotrophic bacterium, although the point of divergence is not clearly resolved. *Chromatium* and *Chlorobium* are anaerobic bacteria capable of photosynthesis using H_2S as an exogenous electron donor. The *Bacillus stearothermophilus, Desulfovibrio gigas,* and *Pseudomonas putida* lines come off from one another very close in time. Their common ancestral line diverged from *Chromatium* and *Chlorobium* before these groups diverged from each other. *B. stearothermophilus* is a facultative aerobe and *P. putida* is a strict aerobe, whereas *D. gigas* is an anaerobic sulfate-reducing bacterium.

5S Ribosomal RNA

The 5S ribosomal RNA (rRNA) molecules are relatively short, about 120 nucleotides long. Figure 3 pictures an evolutionary tree based on 5S rRNA sequences elucidated by a number of workers, including Dyer and Bowman (1979), Nazar et al. (1978), and C.R. Woese (see Schwartz and

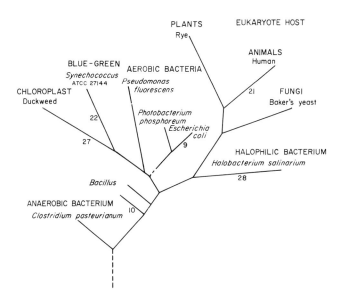

Fig. 3. Phylogenetic tree based on 5S rRNA sequences. Branch lengths are given in
PAMs. The branch leading to *Escherichia coli* and *Photobacterium phosphoreum* could
have diverged as shown or from the *Pseudomonas* branch near its base. The shorter
topology is shown. The lower *Bacillus* divergence leads to *B. stearothermophilus* and
the upper to *B. licheniformis* (5S rRNA). *Synechococcus* ATCC 27144 is also called
Anacystis nidulans. *H. salinarium* is called *H. cutirubrum* by the researchers who sequenced this molecule. (Figure taken with permission from Schwartz and Dayhoff 1980)

Dayhoff 1979c). We have placed the base of the tree on the clostridial
line in conformance with the tree derived from the bacterial ferredoxins. Two branches of the tree are particularly noteworthy. The branch
that leads to animals, plants, and fungi is based on sequences from
cytoplasmic ribosomes and is labeled the eukaryote host. The second
leads to the sequences from *Synechococcus* and from the chloroplast ribosome from duckweed; this branch represents the evolution of the blue-
greens, including the ancestor to the chloroplast of duckweed and the
other higher plants. Clearly, the higher plants are derived from bacteria with two very different evolutionary histories — those that were
ancestral to the eukaryote host and those that led to the chloroplasts.
The possible explanation is that the chloroplast was acquired by a
symbiotic association with the host. We will discuss this more thoroughly later. The detailed picture of the evolution of the blue-greens and
the eukaryote chloroplasts that we will see in the plant-type ferredoxin tree can be pictured as having occurred in the neighborhood of
the *Synechococcus*-duckweed chloroplast divergence on this tree.

Plant-Type Ferredoxins

Although the 2Fe-2S plant-type ferredoxins and the 4Fe-4S bacterial
ferredoxins have different prosthetic groups, they are distantly related in sequence and are both electron transport proteins. The 2Fe-2S
ferredoxin sequences isolated from the blue-greens and from the chloroplasts of eukaryotes are quite similar. A more distant, yet clearly
related, sequence is known from *Halobacterium halobium*. Adrenodoxin from

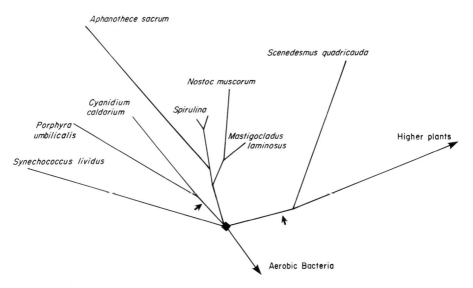

Fig. 4. Phylogenetic tree based on 2Fe-2S ferredoxin sequences. This tree was deduced by scaling the branches for sequences with the two amino-terminal deletions so that they would be comparable to those for sequences without deletions, and then superimposing the two halves of the tree. Arrows indicate organisms that may have independently given rise to eukaryote chloroplasts. (Figure taken with permission from Schwartz and Dayhoff 1980)

mitochondria and putidaredoxin from *Pseudomonas* are very distantly related to the *H. halobium* sequence. Ferredoxin sequences have been determined by a number of workers (Hase et al. 1977, 1978b,c, and pers. comm.; Takruri et al. 1978; Wakabayashi et al. 1978; Takruri and Boulter 1979; D. Borden and E. Margoliash, pers. comm.). See Schwartz and Dayhoff (1979b) for other references. A gene duplication has occurred very near the beginning of the proliferation of the blue-green types. The two genes are clearly distinguished from one another by the occurrence of two deletions in one type. Both types occur in *Nostoc* and *Aphanothece* and probably also in *Spirulina* (Hutson et al. 1978; Cammack et al. 1977; Hase et al. 1976). This duplication provides a temporal ordering for all of the divergences. The tree derived from the two kinds of sequences is shown in Figure 4. An ancestral organism at the radiation of chloroplasts and blue-greens corresponds to the ancestor at the *Synechococcus*-duckweed chloroplast divergence of Figure 3.

The topology of the tree in the immediate vicinity of the ancient gene duplication is not clearly resolved; however, it is clear that *Synechococcus*, the higher plant chloroplasts, and *Spirulina* all diverged from one another at about the same time as the gene duplication occurred. Because the chloroplasts of *Scenedesmus* and the higher plants show no evidence of having the ancestral sequence without the two deletions, we presume that their ancestor lost this gene soon after the duplication. It is probable that the *Porphyra* chloroplast branch represents a separate divergence from the blue-greens slightly later than the gene duplication.

C-Type Cytochromes

The c-type cytochromes are a family of related heme-binding electron transport proteins. Sequences have been done by Ambler et al. (1979),

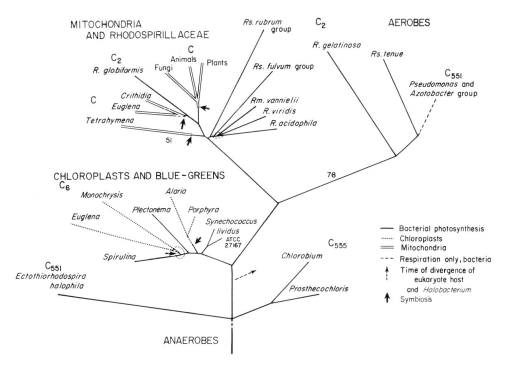

<u>Fig. 5.</u> Phylogenetic tree of major groups from which c-type cytochrome sequences are known. The connections on the tree drawn in a single solid line show the divergences among the free-living prokaryotes. The parallel and dotted lines indicate the mitochondrion and the chloroplast, respectively. The evolutionary connections of the eukaryote organelles need not reflect the history of the cytoplasmic constituents. Arrows indicate branches on which symbioses have occurred. The detailed topologies of subtrees were constructed separately. In order to determine the connections between the subtrees, matrix elements for sequences within each subtree were averaged. The base of the c-c_2 subtree was placed as shown because the c_2 sequences share a unique deletion close to one of their heme-binding cysteines. Where groups are shown, an average of the matrix elements for all species in the group (Schwartz and Dayhoff 1979a) was used

Aitken (1979), and D. Borden and E. Margoliash (pers. comm.). See Schwartz and Dayhoff (1979a) for other references. The phylogenetic tree derived from these selected c-type cytochrome sequences is shown in Figure 5. We have placed the base of the tree on the *Chlorobium* line, in conformance with the ferredoxin tree. There is not yet sufficient information to fix the base of the tree exactly, but it is close to the position shown. The recently determined *Ectothiorhodospira halophila* sequence (Ambler 1979a) falls in the anaerobic portion of the tree as expected for this anaerobic, photosynthetic organism. The base of the chloroplast subtree was placed in conformance with that of the plant ferredoxin tree of Figure 4.

All 12 species of purple, nonsulfur, photosynthetic bacteria, the *Rhodospirillaceae*, that are described in Bergey's Manual of Determinative Bacteriology (Buchanan and Gibbons 1974) are represented, as well as the very interesting sequence from *Rhodopseudomonas (R.) globiformis* recently determined by Richard Ambler (1979b). Apparently, morphology is not

78

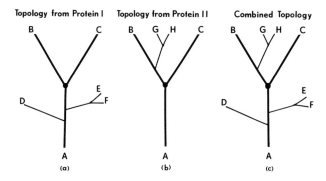

Fig. 6a-c. Combining topologies. The hypothetical topology from protein I (a) is scaled and combined with the hypothetical topology from protein II (b). The result is the combined topology (c) that includes all of the information

an accurate guide to the evolution of these bacteria. Sequences from all three genera of *Rhodospirillaceae* are found intermixed on the tree. The branches to the mitochondrial sequences of ciliates, flagellates, and the plants, animals, and fungi share recent common ancestry with the *R. globiformis* line. From the computations, the *R. globiformis* sequence clearly diverges from the *Euglena* and *Crithidia* branches close to the point of their divergence from one another. The mathematically best solution would have *R. globiformis* coming from the *Crithidia* branch. However, this is probably not the exact order of events. The cytochrome c sequences from *Euglena* and *Crithidia* share a rare substitution of an alanine for one of the heme-binding cysteines, an event that has not been observed in any other cytochrome c sequence examined. We presume that this change occurred once in an ancestor to *Crithidia* and *Euglena*, just subsequent to the divergence of this line from *R. globiformis*.

Evidence from 16S ribosomal RNA catalogues (Woese et al. 1980) confirms that the sequences from *R. gelatinosa* and *Rhodospirillum (Rs.) tenue* form one group, whereas those from *R. sphaeroides*, *R. capsulata*, *Rs. rubrum*, *R. viridis*, *Rhodomicrobium (Rm.) vannielii*, and *R. palustris* form a second group, just as is found for the cytochromes. This supports our hypothesis that the integrity of the genome that codes for structures of basic importance has been conserved.

Cytochrome c_6 is found in the blue-greens and chloroplasts. This electron transport protein has been shown to substitute for plastocyanin in the electron transport chain between photosystems I and II, depending on the availability of copper (Wood 1978). In overall configuration this subtree reiterates that based on 2Fe-2S ferredoxins. Although the position of the base is not known exactly, at least two chloroplast symbioses must be represented to explain the intermixing of prokaryote and eukaryote branches here.

Composite Tree

In order to derive the single phylogenetic tree of species from currently available sequences, trees derived from different proteins or nucleic acids must be combined. If three species appear on two trees, a process of superposition can take place (see Fig. 6). Suppose that proteins I and II are known from three species A, B, and C. The node,

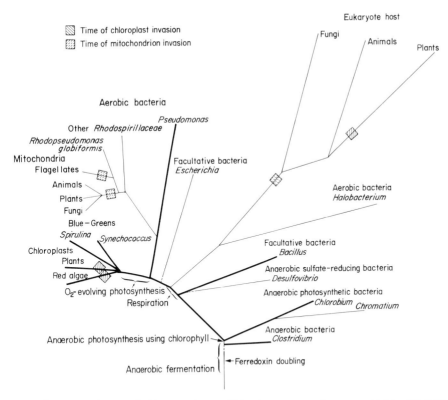

Fig. 7. Composite evolutionary tree. This tree presents an overview of early evolution based on bacterial ferredoxin, 2Fe-2S ferredoxin, c-type cytochrome, and 5S rRNA sequences. The heavy lines represent branches found on two or more of the individual trees. The lighter lines represent branches scaled from a single tree and added to the combined tree. The ancestral organisms participating in the mitochondrial and chloroplast symbioses are indicated by shaded regions

N, between these three branches corresponds to the same ancestral organism and time on each tree, and the same succession of ancestral organisms appears on each branch. Typically, protein I will have been investigated from a number of additional species on one of the branches, NA, as shown in Figure 6a, and protein II will have been investigated from a number of different organisms on branch NB, as shown in Figure 6b. A combined tree of the three branches A, B, and C is formed by averaging the normalized branch lengths. The additional branches from only one tree are scaled according to the branch to which they attach. The two trees are combined to show the additional detail in both branches, as shown in Figure 6c.

In Figure 7, we have made a composite tree by combining the information in trees derived above from the four kinds of molecules. We have first combined the points on the 5S ribosomal RNA tree that are common to the bacterial ferredoxin tree, including *Pseudomonas*, *Bacillus*, and *Clostridium*, as well as the nodal point that depicts the same ancestral organism between them. The 2Fe-2S ferredoxin tree and the c-type cytochrome tree were also mapped onto the composite in a similar way, always maintaining the topology of the composite tree consistent with that of the individual trees. Cytochrome c_{551} sequences clearly describe a close evolutionary relationship among the *Pseudomonas* species, *P. fluorescens*, *P. denitrificans*, *P. mendocina*, *P. stutzeri*, and *P. aeruginosa*. In superimposing branches, we have assumed that *P. putida* is closely related to *P. fluorescens*. We have assumed that *Synechococcus* ATCC 27167 and ATCC 27144 are closely related. Both are classified in subgroup 2 by Rippka et al. (1979) on the basis of DNA base ratio and genome size. We have also assumed that the chloroplasts of duckweed, a higher plant, are similar to the chloroplasts of other higher plants (Dyer and Bowman 1979), including those for which 2Fe-2S ferredoxins are known. Biological evidence as well as 5S rRNA catalogues support this assumption.

The amount of change in a given time can vary considerably in different portions of the trees and for the different sequences. The branch lengths can be only rough estimates of the total change in the organism. The overall rates for the three genes are roughly comparable, so we have averaged the amount of change in PAMs for each of the branches that are found on two or more trees. These branches are shown by heavy lines. The attachment point and length of a branch found on only one tree have been scaled in proportion to the lengths of the neighboring branches on that tree.

The composite tree describes the early evolution starting with the anaerobic, heterotrophic bacteria. Photosynthesis using chlorophyll developed early. Three families of photosynthetic bacteria are represented: Chromatiaceae, Chlorobiaceae, and Rhodospirillaceae. The next major event shown on the trunk of the tree is the development of respiration. The time of divergence of *Bacillus* and *Desulfovibrio* probably marks the presence of major biochemical components of this adaptation. The final elements were evolved separately because these groups differ in their terminal electron acceptor in respiration.

Bacillus and *Escherichia* and the branch that we identify as the eukaryote host all diverged at about the same time. The recently elucidated 5S rRNA sequence from *Halobacterium salinarium* (Dyer and Bowman 1979) diverges from the eukaryote host branch near its base. That *Halobacterium* should be our close relative is further suggested by its use of two biochemical systems that are found in eukaryote cytoplasmic systems but are very rare in bacteria. Methionine transfer RNA rather than N-formylmethionine transfer RNA is used for the initiation of protein synthesis and a light-sensitive system using retinal is found (Bayley 1979).

The blue-greens and chloroplasts of the eukaryotes are capable of oxygen-releasing photosynthesis, and the available sequence data point to this capacity having evolved only once. The chloroplasts are grouped together with free-living prokaryote blue-greens. Because the branches to the eukaryote host and the mitochondrion branched off before the acquisition of oxygen-evolving photosynthesis, humans would never have had an oxygen-producing algal ancestor.

The mitochondria from all species shown are most closely related to the third family of photosynthetic bacteria, the Rhodospirillaceae. Their most recent common ancestors with this group were also ancestors of *R. globiformis*. Therefore, we predict that *R. globiformis* is, on the whole, more similar to the eukaryote mitochondria, particularly those of the flagellates and higher organisms, than are any of the other Rhodospirillaceae shown.

The main stream of evolution was very probably carried by species that performed bacterial photosynthesis, contrary to most classifications

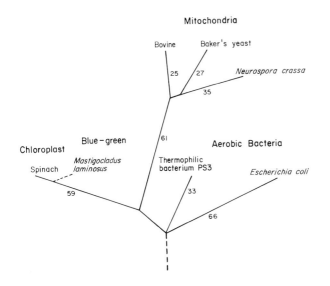

Fig. 8. Evolutionary tree
from ATP synthetase, proteolipid subunit. The root of
the tree has been placed in
accord with that in Fig. 7.
The mitochondrial, chloroplast, and bacterial sequences
form three distinct groups, as
expected. The approximate position of the blue-greens was
deduced from the *Mastigocladus
laminosus* sequence, of which
75% is known

of bacteria. Bacterial photosynthesis was independently lost in a number of lines, including *Desulfovibrio-Bacillus*, *Escherichia coli*, the eukaryote host, and *Pseudomonas*.

ATP Synthetase, Proteolipid Subunit

Of the several thousand genes that may have been present in the free-
living form that has become the mitochondrion, only a few remain in
the DNA of the mitochondrion today. These are transcribed and the messenger translated by the mitochondrial apparatus. An excellent review
of these proteins is given by Tzagoloff et al. (1979). Many other genes
are thought to have originated in the mitochondrion but been transferred to the nucleus. Their messengers are translated by the cytoplasmic apparatus and the mature proteins are transported to the inside of the mitochondrion or to its inner membrane, where they form
functional complexes with the gene products produced in the mitochondrion. The chloroplasts have retained much more of their genetic material, but even so, many genes have been transposed to the nucleus.

Among the sequences retained in the mitochondria and chloroplasts is
ATP synthetase proteolipid subunit (Sebald et al. 1979). The evolutionary tree constructed from these sequences (Fig. 8) is consistent
with that expected from our general scheme of Figure 7. The point of
earliest time and the connection of the eukaryote host should be near
the nodal ancestor of *E. coli* and thermophilic bacterium PS3. The partial sequence from *Mastigocladus laminosus*, a blue-green, has been investigated by W. Sebald and co-workers (personal communication) and indicates a close similarity to the chloroplast (over 85% identity in the
75% of the molecule that has been sequenced). This molecule confirms
the close relationship of these two groups. The proteins that function
in the mitochondria appear together on a separate branch, showing the
same recent evolutionary history as cytochrome c, even though the sequence from baker's yeast is coded in the mitochondrial genome and
translated in the mitochondrion, whereas the sequence in *Neurospora crassa*
is coded in the nucleus and translated in the cytoplasm. The close
similarity of the two fungal sequences supports the view that the *N.*

crassa gene was originally from the mitochondrion. In this case, the place where the protein functions rather than the location of its gene is indicative of evolutionary history. This is also true for both the 2Fe-2S ferredoxin gene and the cytochrome c gene that are coded in the nucleus. Clearly, the amount of change in the mitochondrial line is greater than that in the chloroplast or the bacterial lines.

Symbiotic Origins of Mitochondria and Chloroplasts

There are two schools of thought concerning the origin of eukaryote organelles: one is that they arose by the compartmentalization of the DNA within the cytoplasm of an evolving protoeukaryote (Raff and Mahler 1972; Uzzell and Spolsky 1974); the other is that they arose from free-living forms that invaded a host and established symbiotic relationships with it (Margulis 1970, 1974). In the nonsymbiotic theory, all genes arose within the organism; homologues found in both the nucleus and the organelles arose by gene duplication. There is no reason to think that organelles formed in this way would be viable outside of the cell, and their competitive extracellular existence in evolution would be most unlikely. Therefore, this theory would predict that there would be no prokaryote branches emanating from the chloroplast or mitochondrial lines after these had diverged from the host. The symbiotic theory, on the other hand, proposes that the chloroplast was originally a free-living blue-green; other symbionts include the mitochondrion, which was originally a free-living aerobic photosynthetic bacterium. Their current status as organelles would have gradually evolved from these symbioses. In this theory, mitochondrial, chloroplast, and host genes are expected to show evidence of recent common ancestry with separate types of contemporary free-living prokaryote forms; the host and organelles would occur on different branches that could also contain free-living forms. The tree of Figure 7 clearly supports the symbiotic theory.

Mitochondrial Symbioses

The details of mitochondrial evolution are shown in Figure 5. The group from which mitochondria originated, the Rhodospirillaceae, is shown at the top of the tree. At the divergence of *R. globiformis* from the flagellates, the ancestral species was a free-living form. Very soon thereafter, on the mitochondrial line, a symbiosis occurred. The occurrence of a separate symbiosis among flagellates is consistent with the lack of mitochondria in the Trichomonadida, an anaerobic flagellate group (Müller 1980). This group either diverged before the symbiosis or else the symbiosis was so newly established that the mitochondrion could be easily discarded in an anaerobic niche.

If the evolutionary history is correctly shown in Figure 5, then at least two other symbioses were established, to account for the *Tetrahymena* mitochondrion and that of the multicellular forms. These could both have been established by direct ancestors of *R. globiformis*. More likely they were produced by related forms some time after their divergences from the *R. globiformis* line, as suggested in the figure. In each line photosynthetic ability was lost, either before or soon after the symbiosis.

If there were only a single origin for the mitochondrion the symbiosis would have occurred at the bottom of the eukaryote subtree. *R. globiformis*, a free-living photosynthetic bacterium, would then be descended from a eukaryote cell, a course of events that seems extremely unlikely.

Chloroplast Symbiosis

Some details of chloroplast evolution are pictured on the c-type cytochrome tree (Fig. 5). Regardless of the exact position of the base of
this subtree, the intermixing of eukaryote and prokaryote branches
here requires a minimum of two symbioses: one leading to the chloroplasts of *Alaria* and *Porphyra* and the other leading to those of *Euglena*
and *Monochrysis*. The divergence of *Monochrysis* and *Euglena* being so close
to what we believe (from the 2Fe-2S ferredoxin tree) was a radiation
of blue-greens suggests that these chloroplasts may have arisen from
different blue-greens.

The earliest species divergence on the 2Fe-2S ferredoxin tree of Figure
4 was the one between prokaryotes that were ancestral to *Synechococcus*,
to *Aphanothece*, and to the chloroplasts of higher plants and green algae
such as *Scenedesmus*. Although we cannot be sure that the chloroplasts
of the green algae and the higher plants arose from a single symbiosis,
there was at least one symbiosis on this line. It seems likely that at
least one additional symbiosis led to the chloroplasts of the eukaryote
algae *Porphyra* and *Cyanidium*. The ancestor of these chloroplasts was also
ancestral to *Aphanothece* and other blue-greens and might, therefore,
be expected to resemble contemporary blue-greens. Considering the topologies of both trees, three different symbioses may be represented.

Conclusion

The sequence evidence overwhelmingly favors the origin of the eukaryote mitochondria by endosymbiosis. On the basis of the sequence evidence, there have probably been at least three different symbiotic invasions to produce the mitochondria of the ciliates, the flagellates,
and the higher multicellular organisms (plants, animals, and fungi)
examined here.

The sequence evidence also overwhelmingly favors the origin of the
eukaryote chloroplasts by endosymbiosis. There have been at least two,
and probably three, different invasions in the species examined here
to yield the chloroplast of higher plants, that of the red algae like
Porphyra, and that of the golden-brown algae like *Monochrysis*.

The divergence of the host cell from the prokaryote *Halobacterium salinarium* is more recent than its divergence from *E. coli*, *Bacillus*, or any of
the other prokaryotes found on the tree. Other bacterial and transitional forms will no doubt be found on this line.

Bacterial evolution appears to have started with the clostridia and
similar anaerobic nonphotosynthetic forms, in one line of which photosynthetic ability was established. It probably proceeded through a
mainstream of organisms that could perform photosynthesis under anaerobic conditions. The photosynthetic organisms of the mainstream
have given rise through losses in photosynthetic ability to many lines
of nonphotosynthetic bacteria. One derivative line gave rise to the
blue-greens and chloroplasts that can perform oxygen-evolving photosynthesis. Another line gave rise to the aerobe *Rhodopseudomonas globiformis* and the mitochondria. A third line gave rise to *Halobacterium salinarium* and the eukaryote host.

At the outset, we assumed that there was no exchange between prokaryote
species of the genes for the basic metabolic structures. So far there
has proved to be a single identifiable exception in the eukaryote
lineage, which developed the capacity to engulf and form lasting re-

84

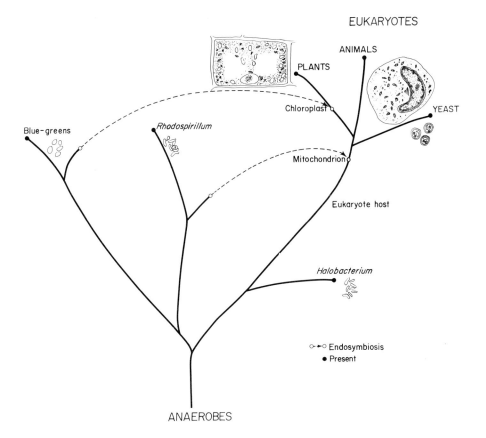

Fig. 9. Schematic representation of the origins of eukaryotes. At least three separate bacterial lines have contributed to this group. The host developed the capacity
to form permanent symbioses with the prokaryotes some time after its divergence from
Halobacterium. An endosymbiosis of the ancestor of higher forms with a relative of
Rhodopseudomonas globiformis produced the mitochondrion. In higher plants, an endosymbiosis with a relative of the blue-greens produced the chloroplast. The mitochondria and chloroplasts are still comparable in size to their prokaryote relative,
whereas the eukaryote host has increased greatly in size and complexity

lationships, even extending through the reproductive cycle, with certain prokaryotes that are now identifiable as organelles within it.
During the formation of the symbiotic association a good deal of the
genetic information of the organelle was incorporated into the nucleus.
In Figure 9, we show schematically the divergences of the lineages
that gave rise to the Rhodospirillaceae and the mitochondria, to the
blue-greens and the chloroplasts, and to the eukaryote host, from
which the higher multicellular organisms arose. The cells and organelles are drawn approximately to scale and it is clear that the Rhodospirillaceae are approximately the same size as the mitochondria and
the blue-greens are approximately the same size as the chloroplasts.
The eukaryote cells, particularly those of plants and animals, are
much larger than those of the prokaryote lines.

Thus, by the use of structural chemistry and mathematical methods, we have been able to probe the main thrust of biological evolution from close to its beginnings.

Acknowledgments. We would like to thank Drs. R. Amber, D. Hall, E. Margoliash, and W. Sebald for supplying us with sequences prior to their publication, and K. Lawson for drafting the figures.

This work was supported by NASA contract NASW-3317 and NIH grants GM-08710 and RR-05681.

References

Aitken A (1979) Eur J Biochem 101:297-308

Ambler RP (1979a) Sequence of cytochrome c_{551} from *Ectothiorhodospira halophila*, strain BN 9626. Submitted to the Atlas of protein sequence and structure reference data collection, August 1979

Ambler RP (1979b) In: Nichols JM (ed) Abstr 3rd Int Symp Photosynthetic Prokaryotes (Oxford), Abstr E17. Univ Liverpool

Ambler RP, Daniel M, Hermoso J, Meyer TE, Bartsch RG, Kamen MD (1979) Nature (London) 278:659-660

Bayley ST (1979) Trends Biochem Sci 4:223-225

Broda E (1975) The evolution of the bioenergetic process. Pergamon Press, Oxford

Bruschi M (1979) Biochem Biophys Res Commun 91:623-628

Bryson V, Vogel HJ (eds) (1965) Evolving genes and proteins. Academic Press, London New York

Buchanan RE, Gibbons NE (eds) (1974) Bergey's manual of determinative bacteriology, 8th edn. Williams and Wilkins, Baltimore

Cammack R, Rao KK, Bargeron CP, Hutson KG, Andrew PW, Rogers LJ (1977) Biochem J 168:205-209

Dayhoff MO (1976) Fed Proc 35:2132-2138

Dayhoff MO (1979) In: Dayhoff MO (ed) Atlas of protein sequence and structure, vol V, suppl 3. National Biomedical Research Foundation, Washington DC, pp 1-10

Dayhoff MO, Schwartz RM (1978) In: Noda H (ed) Origin of life. Proc 2nd ISSOL Meet 5th Int Conf Origin Life, April 1977. Center for Academic Publications Japan/Japan Scientific Societies Press, Tokyo, pp 547-560

Dayhoff MO, Schwartz RM (1979) In: Dayhoff MO (ed) Atlas of protein sequence and structure, vol V, suppl 3. National Biomedical Research Foundation, Washington DC, pp 345-352

Dayhoff MO, Schwartz RM (1980) In: Schwemmler W, Schenk H (eds) Endosymbiosis and cell research. Walter de Gruyter, Berlin, in press

Dayhoff MO, Barker WC, Hunt LT (1976) In: Dayhoff MO (ed) Atlas of protein sequence and structure, vol V, suppl 2. National Biomedical Research Foundation, Washington DC, pp 9-19

Dyer TA, Bowman CM (1979) Biochem J 183:595-604

Hase T, Wada K, Ohmiya M, Matsubara H (1976) J Biochem 80:993-999

Hase T, Wada K, Matsubara H (1977) J Biochem 82:267-276

Hase T, Wakabayashi S, Matsubara H, Ohmori D, Suzuki K (1978a) FEBS Lett 91:315-319

Hase T, Wakabayashi S, Wada K, Matsubara H (1978b) J Biochem 83:761-770

Hase T, Wakabayashi S, Waka K, Matsubara H, Juttner F, Rao KK, Fry I, Hall DO (1978c) FEBS Lett 96:41-44

Hutson KG, Rogers LJ, Haslett BG, Boulter D, Cammack R (1978) Biochem J 172:465-477

Margulis L (1970) Origin of eukaryote cells. Yale Univ Press, New Haven

Margulis L (1974) In: King RC (ed) Handbook of genetics, vol I. Plenum Press, New York, pp 1-41

Müller M (1980) Ann NY Acad Sci, in press

Nazar RN, Matheson AT, Bellemare G (1978) J Biol Chem 253:5464-5469

Raff RA, Mahler HR (1972) Science 177:575-582

Rao KK, Hall DO (1977) In: Leigh GJ (ed) The evolution of metalloenzymes, metalloproteins and related materials. Symposium Press, London, pp 39-65

86

Rippka R, Deruelles J, Waterbury JB, Herdman M, Stanier RY (1979) J Gen Microbiol 111:1-61
Schwartz RM, Dayhoff MO (1978a) In: Ponnamperuma C (ed) Comparative planetology. Academic Press, London New York, pp 225-242 (Proc 3rd College Park Colloq Chem Evol, Sept 1976)
Schwartz RM, Dayhoff MO (1978b) Science 199:395-403
Schwartz RM, Dayhoff MO (1979a) In: Dayhoff MO (ed) Atlas of protein sequence and structure, vol V, suppl 3. National Biomedical Research Foundation, Washington DC, pp 28-43 (The c-type cytochrome sequence literature is reviewed here and our alignment and phylogenetic trees are shown)
Schwartz RM, Dayhoff MO (1979b) In: Dayhoff MO (ed) Atlas of protein sequence and structure, vol V, suppl 3. National Biomedical Research Foundation, Washington DC, pp 45-55 (Ferredoxin sequences and our alignment of them are collected here)
Schwartz RM, Dayhoff MO (1979c) In: Dayhoff MO (ed) Atlas of protein sequence and structure, vol V, suppl 3. National Biomedical Research Foundation, Washington DC, pp 327-336 (5S rRNA sequences and our alignment of them are collected here)
Schwartz RM, Dayhoff MO (1980) In: Proc 6th Int Conf Origins Life. Reidel, Dordrecht The Netherlands, in press
Sebald W, Hoppe J, Wachter E (1979) In: Quagliariello E et al. (eds) Function and molecular aspects of biomembrane transport. Elsevier/North-Holland, Amsterdam New York, pp 63-74
Takruri I, Boulter D (1979) Biochem J 179:373-378
Takruri I, Haslett BG, Boulter D, Andrew PW, Rogers LJ (1978) Biochem J 173:459-466
Tzagoloff A, Macino G, Sebald W (1979) Annu Rev Biochem 48:419-441
Uzzell T, Spolsky C (1974) Am Sci 62:334-343
Wakabayashi S, Hase T, Wada K, Matsubara H, Suzuki K, Takaichi S (1978) J Biochem 83:1305-1319
Woese CR, Gibson J, Fox GE (1980) Nature (London) 283:212-214
Wood PM (1978) Eur J Biochem 87:9-19

Studies of the Maize Chloroplast Chromosome

L. Bogorad, S. O. Jolly, G. Link, L. McIntosh, C. Poulsen, Z. Schwarz, and A. Steinmetz[1]

Introduction

Eukaryotic cells are characterized by their multiple compartmentalized
genomes: nuclear and mitochondrial genomes in animals; nuclear, mitochondrial, and the plastid genomes in plants. Without knowing more than
this, the problem of integration of the expression of these genomes,
i.e., intergenomic integration, already seems important, but the extent of interdependence of these multiple genomes is greater than co-
existence alone suggests.

Genes for multimeric components of organelles are dispersed among at
least two genomes. An example is distribution of genes for ribosomal
proteins of the single chloroplast in the flagellated alga *Chlamydomonas
reinhardii*. The structural gene for one of the chloroplast ribosomal
proteins, the 26 kD (kilodalton) LC 6, has been mapped on linkage
group XI of the nuclear genome of the alga. On the other hand, ribosomal protein LC 4 has been identified as a chloroplast gene product
in the same organism (Mets and Bogorad 1971, 1972; Davidson et al.
1974; Bogorad et al. 1977). Other nuclear loci affecting chloroplast
ribosomes have also been identified (Davidson and Bogorad 1977; Davidson et al. 1978; Hanson and Bogorad 1978). Furthermore, as described
later in this paper, genes for chloroplast rRNAs have been mapped on
cp DNA of maize and other plants. This single example of dispersal of
genes for components of a multimeric organelle component (Bogorad 1975)
emphasizes that to understand eukaryotic cell biology we must learn
the mechanisms of intergenomic integration. It is reemphasized in considering genes for polypeptides of photosynthetic (thylakoid) membranes
of chloroplasts, etc. But to understand the integrative mechanisms we
first need to know about the component genomes — the mechanisms and
regulation of their replication and expression.

Most of the remainder of this paper deals with aspects of the organization of the maize chloroplast genome. The organization of the genome
is of great intrinsic interest and the study of plastid molecular biology provides the opportunity to study devices for regulating gene expression. Higher plants grown in darkness fail to form chlorophyll and
development of their plastids is arrested at the characteristic etioplast stage. Instead of the stacks of thylakoids in chloroplasts, etioplasts contain "paracrystalline" membrane structures — the prolamellar
bodies. Certain thylakoid membrane proteins as well as photosynthetic
pigments needed for photosynthesis are lacking. Upon illumination etioplasts develop into photosynthetically competent green chloroplasts.
Starch storage, for example in potatoes, is in specialized plastids
called amyloplasts. The conversion of green chloroplasts in the skin
of the tomato into carotenoid-bearing chromoplasts is the signal to
the prospective consumer that the fruit is ripe. Still another example

---

[1]The Biological Laboratories, Harvard University, Cambridge, MA 02138, USA

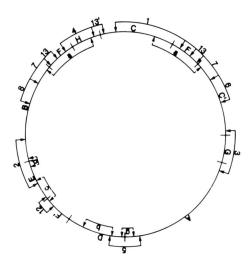

Fig. 1. A map of restriction endonuclease recognition sites on maize chloroplast DNA. Fragments generated by the enzyme Sal I are designated by capital letters. The limits of each fragments are marked by straight lines crossing the circle. All of the recognition sites for this enzyme have been mapped and are shown in this figure .Some of the fragments generated by the enzyme Bam HI are shown. They are designated by Arabic numerals and their limits are shown by arrows pointing inward towards the circle. A few fragments generated by the enzyme Eco RI are shown. They are designated by lower case letters and their limits are shown by arrows pointing outward from inside of the circle. The two fragments marked a are included within the inverted repeated sequences and contain genes for chlorplast RNAs

of this kind is the differentiation of plastids in mesophyll and bundle sheath cells in C-4 plants, as will be discussed below. Each of these is an opportunity for studying development, differentiation, and intergenomic integration.

To summarize: An immediate goal of research into chloroplast genomes is to understand the structure of genes and chromosomes. The longer range goal is to understand the mechanisms for the control of gene expression and in particular mechanisms of intergenomic integration.

The Maize Chloroplast Chromosome
================================

Circular chloroplast DNA molecules were first observed in lysates of *Euglena gracilis* chloroplasts (Manning et al. 1971). Circular supercoiled chloroplast DNA molecules have been observed subsequently in preparations from a number of algae and higher plants including maize.

The maize chloroplast chromosome is 91.1×10^6 daltons in mass, based on mapping of restriction enzyme recognition sites and fragments (Bedbrook and Bogorad 1976) as well as on electron microscope measurements (Kolodner and Tewari 1975; Bedbrook and Kolodner 1979). A striking feature of this chromosome is the presence of two large repeated sequences of 22.5 kbp (kilo base pairs) each. They are set into the circle in inverted orientation to one another. The inverted repeats are 12.6 kbp away from one another at their closest points and 78.5 kbp at their greatest distance apart. Their positions are shown in Figure 2. Inverted repeated sequences have been seen in most cp DNA circles examined in detail except for *Pisum sativum*, which is reported to have a pair of large tandem repeats (Kolodner and Tewari 1975), and *Euglena gracilis* cp DNA, which contains small tandem repeats (Bedbrook and Kolodner 1979).

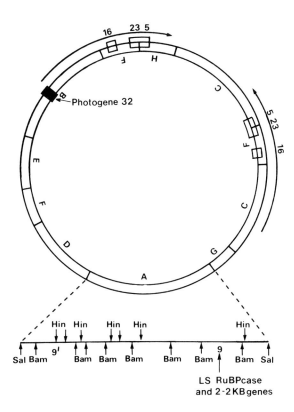

Fig. 2. A map of the maize chloroplast chromosome showing restriction fragments generated by the
enzyme Sal I, the extent of the
two large inverted repeats, and
the locations of genes for 16S,
23S, and 5S rRNAs. Bam fragment 9,
contained within Sal fragment A,
contains the gene for LS RuBPcase.
Photogene 32 lies in Sal fragment
B just beyond the end of the inverted repeats (see Fig. 3)

The 22,500 Base Pair Inverted Repeats of Maize Chloroplast DNA

After all the recognition sites for the restriction enzyme Sal I and
some sites for Eco RI and Bam HI had been mapped on the maize chloroplast chromosome (Fig. 1), a quest was begun for the sites of genes for
chloroplast 16S, 5S, and 23S rRNAs on the chromosome. It had been shown
earlier (L.A. Haff and L. Bogorad, unpublished) that maize cp rRNAs
hybridize to maize cp DNA. Maize chloroplast ribosomes were isolated,
16S and 23S RNAs prepared from them were hydrolyzed slightly, and the
fragments were labeled with ^{32}P at their 5' ends. The radioactive rRNAs
were hybridized against fragments of maize cp DNA generated by digestion with various restriction enzymes and separated according to size
electrophoretically in agarose. Both of the cp rRNAs hybridized to the
largest fragment produced by digestion of the cp DNA with Eco RI. This
fragment, designated a (Bedbrook and Bogorad 1976) was cloned in *E. coli*
using the plasmid vehicle pMB9.

Detailed mapping of 5S as well as 16S and 23S rDNAs on fragment a was
carried out by further restriction mapping of the cloned fragment and
hybridization of individual rRNA species to the fragments. This permitted assignment of 5S, 16S, and 23S RNA genes to positions on a. Two
electron microscopic techniques were employed for detailed mapping.
Heteroduplexes between cloned fragment a and 16S and 23S rRNAs were
formed and measured. These data revealed a 2.1 kpb spacer between the
two genes as well as the positions of the rDNAs on a. The orientations
of the two a fragments on the maize cp chromosome were determined by
locating and measuring the R-loops formed by the presence of rRNAs that

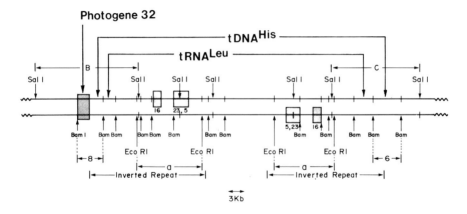

Fig. 3. Details of the arrangements of the rRNA genes and Bam fragment 8. The shaded portion of the latter contains Photogene 32. The unshaded region includes Eco RI fragment ℓ which bears the gene for tRNA^His and part of the 1.6 kb gene described in the text. Bam fragment 7, containing tRNA^Leu, is present in two copies. In this figure they are to the right of fragment 8 and adjacent to and to the left of fragment 6 (see Fig. 1). Bam fragment 7 contains the gene for leucine tRNA

interfered with the two inverted repeats on a single DNA strand from annealing to one another. This orientation of 16S and 23S rDNAs on the chromosome separated by a "spacer" sequence (Fig. 2) is common in cp DNAs. Genes for tRNAs have been identified in the spacer regions (Driesel et al. 1979; Bohnert et al. 1979; Mubumbila et al. 1980; Keller et al. 1980).

tRNAs genes have also been found in maize cp DNA near that end of the inverted repeat closest to the larger separating unique region. Bam fragment 8 overlaps an end of one of the inverted repeats and part of the large intervening unique region. The portion just within the inverted repeat contains a gene for histidine tRNA, as must the corresponding portion of Bam fragment 6 (Figs. 1, 3). The adjacent fragment, Bam 7, contains a gene for leucine tRNA (Schwarz, Steinmetz, and Bogorad, unpublished).

The histidine tRNA in Bam fragment 8 is contained within an Eco RI fragment designated ℓ. Within ℓ, the gene for histidine tRNA overlaps, for a few nucleotides, a gene coding for a 1.6 kb RNA of undetermined function. The latter has a rather long leader and then an open reading frame for translation. From the overlap of these two genes it seems that transcription of one would interfere with transcription of the other. This physical arrangement itself could be a mechanism for regulating gene transcription.

The genes for histidine (GUG) and leucine (CAA) tRNAs and regions near them have been sequenced. The 3' terminal CCA sequence of mature tRNAs is not included in the coding sequence and must be added posttranscriptionally. By contrast, the CCA terminal sequence is included in *E. coli* tRNA genes, but not those for cytoplasmic tRNAs of eukaryotes. Other features of the tRNA genes set them apart both from eukaryotic and *E. coli* tRNAs. The chloroplast tRNA genes so far sequenced have unique combinations of features seen in both of the latter.

The Gene for the Large Subunit of Ribulose Bisphosphate Carboxylase

The gene for the large subunit of the carbon dioxide fixing enzyme ribulose bisphosphate carboxylase (RuBPcase) has been located on maize chloroplast fragment Bam 9 (Fig. 2) and has been cloned in *E. coli* with the vehicle RSF 1030. The cloned DNA directs the synthesis in vitro, using a linked transcription-translation system and ^{35}S-methionine, of a labeled polypeptide which has been identified as the LS (large subunit) of RuBpcase (ribulose bisphosphate carboxylase) (Coen et al. 1977; Bedbrook et al. 1979; Link and Bogorad 1980; McIntosh et al. 1980). The criteria for the identification includes the following: (a) the in vitro product corresponds in size (on sodium dodecyl sulfate polyacrylamide gel electrophoresis) with authentic material isolated from maize seedlings supplied with ^{35}S-methionine; (b) antibody prepared against RuBPcase precipitates the in vitro product whose synthesis is directed by cloned maize cp DNA fragment Bam 9; (c) the immunoprecipitated 52,000 dalton ^{35}S-methionine labeled in vitro and in vivo synthesized polypeptides yield indistinguishable limited proteolysis products (using papain and chymotrypsin); (d) a number of the tryptic fragments procuced by digestion of in vivo and in vitro products are the same. (Some of the tryptic products differ, but this is probably attributable to secondary modification of the peptide.)

With the availability of the cloned 4.2 kbp Bam fragment 9, as well as a 2.7 kbp subfragment which was shown to contain the entire LS gene, it became possible to map the structural gene for LS in detail. Cloned cp DNA sequence Bam 9, or its 2.7 kbp subfragment (designated B 2.7), were hybridized with an unresolved mixture of maize cp RNAs. Chloroplast RNAs complementary to Bam 9 (or B 2.7) would selectively hybridize to the portions of the DNA sequence from which they were transcribed. After hybridization, the mixture was treated with nuclease S1 (which digests all single stranded RNA and DNA). The DNA from the surviving hybrids was then sized by gel electrophoresis (Berk and Sharp 1977). The structural gene for LS was thus judged to be 1600 nucleotide in length and colinear with the RNA, i.e., the gene contains no introns. By other modifications of the S1-nuclease technique the initiation and termination points for transcription from the cp DNA could be determined with fair reliability (Link and Bogorad 1980). In the course of this work a part of a structural gene for a 2.2 kb transcript was discovered lying on the strand opposite the LS gene and about 300 nucleotides away from the position to which the 5' end of LS mRNA hybridizes. These two genes are thus quite close together and transcribed divergently. The function of the 2.2 kb transcript has not been determined.

The fine mapping information on the LS gene has been used as a guide for obtaining the nucleotide sequence of the chromosome complementary to LS mRNA, i.e., the structural gene as well as flanking regions. This is the first entire, i.e., genomic, DNA sequence to be determined for a plant, nuclear or plastidic, encoding a protein.

The DNA sequencing data have shown (L. McIntosh, C. Poulsen, and L. Bogorad, unpublished) with regard to the LS mRNA and the gene from which it is transcribed, that:
a) Translation begins at AUG.
b) A sequence at positions -5 to -9 (i.e., upstream of the AUG translation start) is complementary to the 3' end of maize chloroplast 16S RNA in analogy to the situation described by Shine and Dalgarno (1974) and Steitz (1979) in *E. coli*.
c) It is not known whether chloroplast messages are capped, but no se-

quence analogous to a eukaryotic capping sequence can be found.
d) Transcription is initiated at -60 ± 2 nucleotides from the translation start.
e) All but three of the possible amino acid codons are used in the LS
message.
f) Unlike the situation in fungal and mammalian mitochondria, the stop
codon UGA is not used for tryptophan in this message.
g) No DNA sequences homologous either to a classic bacterial promoter
(Pribnow box) nor its presumed eukaryotic counterpart (the "Hogness"
box) is found upstream from the transcribed portion of the LS gene.
h) The LS message is colinear with the gene in confirmation of our
earlier S1 nuclease data.

The maize LS polypeptide is 475 amino acids in length. Cyanogen bromide
fragments representing approximately half of the barley LS polypeptide
have been sequenced (Poulsen 1979). There are only eight amio acids
different in the deduced amino acid sequence of the maize LS and the
portions of the barley LS polypeptide for which the amino sequences
are known. This is a very high degree of conservation, but whether it
is characteristic of all chloroplast genes has yet to be determined.
Hartman et al. (1978) have identified three amino acid sequences they
judge to be at or near the catalytic site(s) of spinach RuBPcase LS.
These amino acid sequences have now been fitted to the deduced amino
acid sequence of maize LS at positions 166-178, 321-340, and 450-462.
The sequences are well-conserved; neutral replacements occur at two
positions and are widely scattered across the polypeptide. Extensive
folding occurs if these sequences are brought together for function.

Transcriptional Control of the Expression of the Large Subunit-RuBPcase

Gene in Maize Plastids

Enzymology of three general patterns of photosynthetic carbon fixation
has been worked out: C-3, C-4, and CAM. The first to be discovered was
the C-3 or Calvin-Benson path. In this system carbon dioxide is combined with ribulose-1, 5-bisphosphate enzymatically by RuBPcase to produce two molecules of phosphoglyceric acid. This three carbon organic
acid is then reduced to the three carbon sugar phosphoglyceraldehyde
by photosynthetically generated ATP and NADPH. "C-3 plants" carry on
this kind of carbon fixation in all of their chloroplasts. Some other
plants, including maize, fix carbon by the C-4 pathway. In C-4 plants,
a cooperative chloroplast-cytoplasmic system in mesophyll cells fixes
carbon first into a 4 carbon acid. The reaction is: phosphoenolypyruvate
+ CO_2 → oxaloacetate. Malate or aspartate, produced from oxaloacetate,
are transported into cells surrounding the vascular bundles (i.e.,
bundle sheath cells) where they are decarboxylated. The CO_2 released
is refixed by RuBPcase. Mesophyll cells lack RuBPcase, but contain enzymes necessary for C-4 fixation reactions. The availability of the
cloned LS gene made it possible to determine whether control of the
expression of the gene is at the level of transcription or translation.

Maize leaf sections were treated with plant cell wall digesting enzymes.
Mesophyll protoplasts were liberated and collected separately from the
bundle sheath cells that adhered to the vascular bundles. RNA was extracted separately from mesophyll protoplasts and from bundle sheath
strands. The RNA was then made radioactive by addition of ^{32}P. Bam 9
DNA was digested with restriction endonucleases that would produce a
fragment containing only an LS sequence and the fragments were separated
by gel electrophoresis. The RNA samples were next hybridized against
the separated fragments of Bam 9 by the Southern (1975) method. RNA
complementary to the LS gene was abundant in preparations from bundle

sheath cells, but was not detectable in those from mesophyll cells, despite the use of a thousand times more of the latter RNA than the former. Transcripts of the LS gene may be entirely lacking from mesophyll plastids although they contain the gene (Link et al. 1978).

Identification of the gene for the 2.2 kb RNA on Bam 9, about 300 base pairs from the 3' end of the LS gene and transcribed in the opposite direction (Link and Bogorad 1980), made it possible to examine its transcription in the two cell types using the approach described in the paragraph above. In contrast to the observations regarding the LS gene, transcripts of the 2.2 kb gene were readily identified in extracts of both mesophyll and bundle sheath cells (Link and Bogorad, unpublished). The 300 nucleotide untranscribed region that intervenes between the 3' ends of the LS and 2.2 kb genes seems a likely site for DNA signals affecting the differential transcription of these two segments of DNA. Its analysis is in progress.

Photogene 32

As mentioned earlier, seedlings grown in darkness on their reserves are yellow, lack chlorophyll, and their plastids develop to the characteristic etioplast stage. The prolammellar body characteristic of etioplasts has some, but not all, the polypeptides of mature photosynthetically competent chloroplast thylakoids. Upon illumination of dark-grown seedlings, chlorophyll and additional thylakoid membrane proteins are produced and added to those already present. Within a few hours, or less, the membranes become fully competent photosynthetically. One of the polypeptides added during light-induced development of maize has a mass of 32 kD (Grebanier et al. 1979).

Isolated maize chloroplasts produce a 34.5 kD polypeptide which is introduced into the membrane and then processed to the 32 kD size (Grebanier et al. 1978).

The strategy of searching for photogenes (i.e., light-inducible genes) in chloroplast chromosomes was to first prepare RNA from etioplasts, chloroplasts, and from plastids at some stages of light-induced maturation. These RNAs could be translated in vitro and differences in mRNA populations would be (or could be?) reflected in the polypeptide products. On the other hand, hybridization of the RNAs, after being labeled with ^{32}P, against fragments of maize cp DNA separated electrophoretically, could reveal changes in expression of parts of the genome. This is comparable to looking for differences in expression of just two genes on Bam 9 when comparing mesophyll and bundle sheath cell RNAs.

Etioplasts were prepared from dark-grown maize seedlings and their RNA was extracted and made radioactive in vitro. The same sort of preparation was made using developing plastids from leaves of dark-grown plants illuminated for periods ranging from 6 to 16 h. RNA was also prepared from light-grown maize seedlings. Translation of these RNAs in a rabbit reticulocyte lysate using ^{35}S-methionine as a radioactive label demonstrated the inability of the etioplast RNA preparation to direct the synthesis of a 34.5 kD polypeptide, but chloroplast RNA was very active. RNA from plastids of dark-grown plants illuminated for 6 or 16 h was also very active. In a complementary analysis, the RNA obtained from plastids at different stages of development was labeled with ^{32}P (as in the mesophyll-bundle sheath cell experiments) and hybridized against restriction fragments of maize chloroplast DNA separated according to size electrophoretically. Etioplast RNA contains little material that hybridizes against Bam fragment 8, whereas RNA from mature chloroplasts hybridizes very strongly. Plastids at various stages of light-induced greening contain RNA complementary to Bam 8 roughly in proportion to their state of maturation (Bedbrook et al. 1979). This circumstantial evidence suggested that a sequence on maize cp DNA fragment Bam 8 was transcribed into a RNA that can direct the synthesis of a 34.5 kD polypeptide. It was further shown that the in vitro produced polypeptide corresponded in its proteolytic products to the polypeptide synthesized by isolated plastids.

A fragment of Bam 8 (Figs. 2, 3) has been cloned in *E. coli* using the vehicle pBR 322 (L. McIntosh and L. Bogorad, unpublished). The cloned fragment was inserted between the Eco RI and Bam HI sites on the vehicle. The cloned material directs the synthesis of a 34.5 kD polypeptide in a linked transcription-translation system. Proteolytic fragments of the in vitro product identify it as the correct polypeptide (L. McIntosh and L. Bogorad, unpublished). A 1300 nucleotide-long portion of the cloned fragment has been shown to be colinear with a maize chloroplast mRNA (G. Link, L. McIntosh, and L. Bogorad, unpublished).

It is not known whether any photogenes lie adjacent to this photogene, Photogene 32, or whether these are scattered throughout the genome. Saturation hybridization data, using maize chloroplast DNA and etioplast or chloroplast RNA, suggests that there should be several more photogenes (L. Haff and L. Bogorad, unpublished). These are being sought.

Summary of Physical Mapping of Genes on the Maize Chloroplast Genome

So far genes for rRNAs, for several tRNAs (only those for histidine and leucine have been mentioned), for RuBPcase LS, and for Photogene 32 have been mapped with precision on the maize chloroplast chromosome. In several cases, parts or all of the structural genes and adjacent DNA regions have been sequenced. In addition, genes have been identified for a 2.2 kb transcript on Bam 9 (lying next to the LS gene) and for a 1.6 kb transcript on Eco RI fragment ℓ (overlapping the histidine tRNA gene). The gene for the 1.6 kb RNA contains sequences that suggest it may be translated into a polypeptide (which has not yet been identified).

The genes so far identified fit into at least four different expression classes: The LS gene — expressed in bundle sheath but not mesophyll cells; the 2.2 kb gene — transcribed in both mesophyll and bundle sheath cells; Photogene 32 — transcribed in chloroplasts and during light-induced development but not in etioplasts; and rRNA genes whose transcription appears to be qualitatively influenced during light-induced development of etioplasts (Bogorad 1967). One immediate objective of our research is the identification of additional genes of each of these classes, in order to compare sequences flanking the structural genes and to seek sequences characteristic of each class — if they exist.

In Vitro Transcription of Plastid Genes

Another approach to understanding gene expression regulation in chloroplasts is through studies of the interaction of the enzyme of transcription, DNA-dependent RNA polymerase, with cp DNA. Maize chloroplasts contain a unique RNA polymerase (Bottomley et al. 1971; Smith and Bogorad 1974; Kidd and Bogorad 1979). A 26 kD maize plastid polypeptide (the S factor), when added to maize cp RNA polymerase purified through a DEAE column, stimulates the transcription of circular DNA 5- to 15-fold. Of greater interest is the influence of this polypeptide on selective transcription of some chloroplast DNA sequences (Jolly and Bogorad 1980).

In these experiments chloroplast DNA that had been incorporated into a bacterial plasmid and cloned in *E. coli* is used as a template. In one of the cases studied, the transcription of maize cp DNA fragment Eco RI ℓ cloned using the vehicle pMB9, the vehicle and chloroplast sequences are transcribed about equally in the absence of the S factor. However, if the S factor is included and the DNA is in a supercoiled form, the chloroplast DNA sequences are preferentially transcribed by a factor of about 8 (Jolly and Bogorad 1980). We have also found that the 5' end of histidine tRNA isolated from plastids is indistinguishable from the 5' end of the in vitro transcript using cloned cp DNA fragment ℓ as a template and the maize chloroplast RNA polymerase (Z. Schwarz, A. Steinmetz, S. Jolly, and L. Bogorad, unpublished).

The use of cloned cp DNA sequences of genes in different expression classes together with the maize chloroplast RNA polymerase and S, plus any other possible polypeptide factors that we may find, should provide a combination of tools for studying selective transcription in maize chloroplasts and, eventually, permit us to return to the problem of intergenomic integration.

Acknowledgments. The research described in this paper was supported in part by grants from the National Institute of General Medical Sciences, the National Science Foundation, and the United States Department of Agriculture Competitive Research Grants Office. It was also supported in part by the Maria Moors Cabot Foundation of Harvard University. Lee McIntosh was a fellow of the Maria Moors Cabot Foundation, Carsten Poulsen is supported by the Carlsberg Foundation, and Andre Steinmetz is Attache de Recherche at C.N.R.S., France and the recipient of a NATO grant-in aide.

References

Bedbrook JR, Bogorad L (1976) Endonuclease recognition sites mapped on *Zea mays* chloroplast DNA. Proc Natl Acad Sci USA 73:4309-4313

Bedbrook JR, Kolodner R (1979) The structure of chloroplast DNA. Annu Rev Plant Physiol 30:593-620

Bedbrook JR, Coen DM, Beaton AR, Bogorad L, Rich A (1979) Location of the single gene for the large subunit of ribulosebisphosphate carboxylase on the maize chloroplast chromosome. J Biol Chem 254:905-910

Berk AJ, Sharp PA (1977) Sizing and mapping early adenovirus mRNAs by gel electrophoresis of S1 endonuclease-digested hybrids. Cell 12:721-732

Bogorad L (1967) Control mechanisms in plastid development. Dev Biol Suppl 1:1-31

Bogorad L (1975) Evolution of organelles and eukaryotic genomes. Science 188:891-898

Bogorad L, Davidson JN, Hanson MR (1977) The genetics of the chloroplast ribosome in *Chlamydomonas reinhardii*. In: Bogorad L, Weil JH (eds) Nucleic acids and protein Synthesis in plants. Plenum Publ Corp, New York, p 135

Bottomley W, Smith HJ, Bogorad L (1971) RNA polymerases of maize: Partial purification and properties of the chloroplast enzyme. Proc Natl Acad Sci USA 68:2412-2416

Coen DM, Bedbrook JR, Bogorad L, Rich A (1977) Maize chloroplast DNA fragment encoding the large subunit of ribulosebisphosphate carboxylase. Proc Natl Acad Sci USA 74:5487-5491

Davidson JN, Bogorad L (1977) Suppression of erythromycin resistance in *ery*-M1 mutants of *Chlamydomonas reinhardii*. Mol Gen Genet 157:39-46

Davidson JN, Hanson MR, Bogorad L (1973) An altered chloroplast ribosomal protein in *ery*-M1 mutants of *Chlamydomonas reinhardii*. Mol Gen Genet 132:119-129

Davidson JN, Hanson MR, Bogorad L (1978) Erythromycin resistance and the chloroplast ribosome in *Chlamydomonas reinhardii*. Genetics 89:281-297

Grebanier AE, Coen DM, Rich A, Bogorad L (1978) Membrane proteins synthesized but not processed by isolated maize chloroplasts. J Cell Biol 78:734-746

Grebanier AE, Steinback KE, Bogorad L (1979) Comparison of the molecular weights of proteins synthesized by isolated chloroplasts with those which appear during greening in *Zea mays*. Plant Physiol 63:436-439

Hanson MR, Bogorad L (1977) Complementation analysis at the *ery*-Ml locus in *Chlamydomonas reinhardii*. Mol Gen Genet 153:271-277

Hartman FC, Norton LL, Stringer CD, Schloss JV (1978) Attempts to apply affinity labeling techniques to ribulose bisphosphate carboxylase/oxygenase. In: Siegelman HW, Hind G (eds) Photosynthetic carbon assimilation. Plenum Press, New York, pp 245-269

Jolly SO, Bogorad L (1980) Preferential transcription of cloned maize chloroplast DNA sequences by maize chloroplast RNA polymerase. Proc Natl Acad Sci USA 77:822-826

Keller M, Burkard G, Bohnert HT, Mubumbila M, Gordon K, Steinmetz A, Heiser D, Crouse EJ, Weil JH (1980) Transfer RNA genes associated with the 16S and 23S rRNA genes of *Euglena* chloroplast DNA. Biochem Biophys Res Commun 95:47-54

Kidd GH, Bogorad L (1979) Peptide maps comparing subunits of maize chloroplast and type II nuclear DNA-dependent RNA polymerases. Proc Natl Acad Sci USA 76:4890-4892

Kolodner R, Tewari KK (1975) The molecular size and conformation of chloroplast DNA from higher plants. Biochim Biophys Acta 402:372-390

Link G, Bogorad L (1980) Sizes, locations, and directions of transcription of two genes on a cloned maize chloroplast DNA sequence. Proc Natl Acad Sci USA 77:1832-1836

Link G, Coen DM, Bogorad L (1978) Differential expression of the gene for the large subunit of ribulose bisphosphate carboxylase in maize leaf cell types. Cell 15:725-731

Manning JE, Wolstenholme DR, Ryan RS, Hunter JA, Richard OC (1971) Circular chloroplast DNA from *Euglena gracilis*. Proc Natl Acad Sci USA 68:1169-1173

McIntosh L, Poulsen C, Bogorad L (1980) Chloroplast gene sequence for the large subunit of ribulose bisphosphate carboxylase of maize. Nature 288: 556 - 560

Mets LJ, Bogorad L (1971) Mendelian and uniparental alteration in erythromycin binding by plastid ribosomes. Science 174:707-709

Mets LJ, Bogorad L (1972) Altered chloroplast ribosomal proteins associated with two erythromycin-resistant mutants in two genetic systems of *Chlamydomonas reinhardii*. Proc Natl Acad Sci USA 69:3779-3783

Poulsen C (1979) The cyanogen bromide fragments of the large subunit of ribulose-bisphosphate carboxylase from barley. Carlsberg Res Commun 44:163-189

Shine J, Dalgarno L (1974) The 3'-terminal sequence of *Escherichia coli* 16S ribosomal RNA: Complementarity to nonsense triplets and ribosome binding sites. Proc Natl Acad Sci USA 71:1342-1346

Smith HJ, Bogorad L (1974) The polypeptide subunit structure of the DNA-dependent RNA polymerase of *Zea mays* chloroplasts. Proc Natl Acad Sci USA 71:4839-4842

Southern EM (1975) Transfer of DNA fragments to Millipore filters after restriction endonuclease cleavage followed by agarose gel electrophoresis. J Mol Biol 98: 503-517

Steitz JA (1979) Genetic signals and nucleotide sequences in messenger RNA. In: Goldberg RS (ed) Biological regulation and development, vol I. Plenum Press, New York, pp 349-399

The Plastid Chromosomes of Several Dicotyledons

R.G.Herrmann[1], P.Seyer[2], R.Schedel[1], K.Gordon[3], C.Bisanz[1], P.Winter[1], J.W.Hildebrandt[1],
M.Wlaschek[1], J.Alt[1], A.J.Driesel[1], and B.B.Sears[1]

Introduction

The autotrophic eukaryotic cell is a highly complex system; its biogenesis demands a subtle interplay of three genetic compartments: nu-
cleus/cytosol, plastids, and mitochondria. The genetic material of
plastids, the plastome (Renner 1934), possesses unique features which
allow us to probe the cooperation between intracellular genetic compartments as well as gain insight into the evolution of this system.
The primary attributes of plastid DNA (ptDNA) are that it is moderate-
ly complex, structurally defined, and does not appear to encode only
ancillary functions. In vascular plants the plastid genetic information is deposited in a single, highly reiterated circular DNA molecule
of about 150 kilobase pairs (kbp), a size that corresponds roughly to
a coding potential of 200 polypeptides of average molecular weight.
Although the sizes of plastid chromosomes of the vascular plants studied
to date are remarkably uniform (for review see Herrmann and Possingham
1980), they can vary from at least 120 kbp in the liverwort *Sphaerocarpos*
(Herrmann et al. 1980) or the Xanthophycean alga *Vaucheria* (Kowallik
and Hennig unpublished) to 200 kbp in *Chlamydomonas reinhardii* (Behn and
Herrmann 1977).

Although the potential informational content of the plastome appears
well defined, its actual genetic content has been only partially determined. Our present knowledge indicates that the plastid chromosome
encodes part of the organelle-specific translation machinery as well as
polypeptides involved in primary energy conservation. Because they are
essential for plastid function, these genes may be common to all natural plastid DNAs. Little is known about regulatory functions or coding
diversity of plastomes. In each of the cellular genetic compartments
of eukaryotes, regulatory mechanisms probably exist which are analogous
to the fundamental mechanisms of regulation found in prokaryotes. However, it is clear that in the course of evolution intercompartmental
controls have been added. Another kind of complexity has resulted from
the developmental diversity of the organelle, especially with the ad-
vent of multicellular organisms and the concomitant division of labor
between different tissues. Chloroplasts, leucoplasts, and chromoplasts,
for example, are architecturally different, physiologically specialized
modifications of the same type of organelle involved in photosynthesis/
photorespiration, biological storage, or propagation. It is unknown
whether this versatility results from partitioning the developmental
potential of the organelle, perhaps under the influence of the genome,
and/or from qualitative changes in the coding properties of plastomes.

[1]Botanisches Institut der Universität, 4000 Düsseldorf, FRG
[2]Present address: Laboratoire de Biochimie Fonctionelle des Plantes, 13288 Marseille
cedex 2, France
[3]Present address: Department of Biochemistry, University of Adelaide, Adelaide,
Australia 5001

The most informative data about plastome functions come from genetic analyses in the genus *Oenothera*, for which nuclear and plastid genetics are well developed (Kutzelnigg and Stubbe 1974; Kirk and Tilney-Bassett 1978). Interspecific fertility and biparental transmission of plastids within this genus allow the transfer of plastids from one species into the environment of an alien nucleus and vice versa, even when crosses are made between subsections. Many interspecific genome/plastome hybrids show severe developmental disturbances, such as bleaching, impaired fertility, or even lethality, demonstrating that an exchange of plastids and nuclei even between closely related species can disturb harmoniously balanced growth. The prospect that genetic compartments can readily be manipulated for the modification of phenotypic expression in higher plants and the availability of a broad spectrum of well-defined plastome mutants (Kutzelnigg and Stubbe 1974) offers a basis for combined molecular, biochemical, and genetic approaches. This material can be used to study the specificity of genome/plastome cooperation as well as plastome characters that manifest themselves outside the plastid. Phylogenetic divergence in *Oenothera* may also serve as a model for studies of plastome/genome coevolution in speciation (Stubbe 1959), since the phenomenon of reversible hybrid bleaching is widespread in the plant kingdom (Kirk and Tilney-Bassett 1978). The plastome pedigree of the subsection *Euoenothera* is the only instance in which the involvement of plastomes in plant evolution has been amply demonstrated (Stubbe 1959). The five genetically distinguishable plastomes (I-V) of this subsection are associated with six basic genotypes, occurring in morphologically distinct homozygous (AA, BB, CC) or complex heterozygous (AB, AC, BC) species. One phylogenetic trend in this subsection has probably been a closer adaptation of the two genetic compartments, because the fastest multiplying plastid types are specifically adapted to only one or two genotypes.

The development of physical DNA mapping and sequencing techniques as well as immunological methods and recombinant DNA technology has allowed us to probe DNA sequences with more precision than was previously possible. We have employed these physical methods with plastid chromosomes of several dicotelydons including representatives of the genus *Oenothera* and present here the current status of some aspects of our work.

The Status of Plastid DNA

The finding that the five *Euoenothera* plastome types (Stubbe 1959) are represented by unique DNAs (Herrmann 1977) has added a molecular genetic dimension to our studies on ptDNA. The restriction patterns of four of these plastome DNAs are shown in Figure 1a. In 18 of 30 possible *Euoenothera* genome/plastome combinations studied to date these patterns remain constant irrespective of the nuclear background. These results confirm that ptDNA constitutes a distinct genetic element of the autotrophic eukaryote and indicate that it is inherited independently from the nucleus. We found no evidence that ptDNA is altered by nuclear genes.

Although no assignment of functional differences to the different ptDNAs has yet been made, the observation that the extent of DNA differences and the compatibility relationships are correlated supports the contention that these differences contribute to the compatibility traits. For example, plastomes I and II are genetically similar, yet are discernible on the basis of their genome compatibilities and are thus thought to be closely related (Stubbe 1959). Their DNAs displayed no differences following initial Sal I digestion (data not shown), but

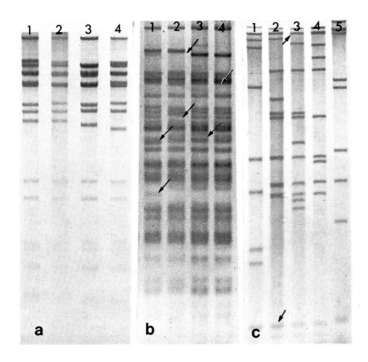

Fig. 1a-c. Agarose gel electrophoresis of restriction endonuclease digests of ptDNAs. (a) Sal I *plus* Pst I restriction patterns of *Euoenothera* DNA from plastome I *(track 1)*, II *(track 2)*, IV *(track 3)* and V *(track 4)* using a 1.0% agarose slab gel (from Herrmann and Possingham 1980). (b) Eco RI restriction patterns of plastome I DNAs from different AA genotypes, from the homozygous species *Oe. hookeri* (Oregon) *(track 1)*, *Oe. jamesii* (Texas) *(track 2)*, *Oe. elata* (Guatemala) *(track 3)* and *Oe. elata* (Mexico) *(track 4)* using a 1% agarose slab gel. Heterogeneities are indicated by *arrows*. (c) Kpn I restriction patterns of ptDNA from spinach *(track 1)*, potato *(track 2)*, tomato *(track 3)*, tobacco *(track 4)*, *Oenothera* plastome IV *(track 5)*. The common two-molar 0.5 Md fragment and the common two-molar band K-2, discussed in the text, are marked by arrows; 0.5% agarose slab gel

at higher resolution could be distinguished by a slight displacement of two bands in the Sal I + Pst I patterns (Fig. 1a). The correlation of ptDNA differences and compatibility traits is not necessarily a strict association. Our current investigations on several hundred wild *Euoenotheras* have revealed that the situation is quite complex. Substantial DNA heterogeneity exists within the basic plastomes, although differences are neither phenotypically apparent nor expressed in the compatibility traits. For example, the genetically homogeneous plastome I, which is associated with several AA genotypes (Stubbe 1959), consists of at least six distinct subtypes of which several are shown in Figure 1b. Thus it appears that the plastome is much more variable than previously thought (Kirk and Tilney-Bassett 1978). Some of the plastome I DNA differences have been physically mapped on the chromosome (see section entitled "Identification and Physical Mapping of Plastome-Coded Genes").

In contrast to the interspecific similarities found in the Euoenotheras, the fragment patterns of evolutionarily more distant organisms are of little or no resemblance at first glance. The Kpn I patterns of ptDNA

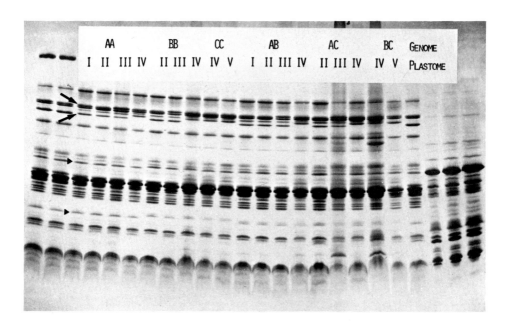

Fig. 2. The thylakoid polypeptides of *Euoenothera* wild species and some of their interspecific genome/plastome hybrids separated by electrophoresis in a denaturing polyacrylamide gradient gel (7.5% - 15%). The genome/plastome combination is given at the top of each track in the usual notation. Wild species: *AA-I (Oe. hookeri)*, *BB-III (Oe. grandiflora)*, *CC-V (Oe. argillicola)*, *AB-II (Oe. biennis)*, *AB III (Oe. erythrosepala)*, *AC-IV (Oe. oakesiana)*, *BC IV (Oe. parviflora)*. The *arrows* indicate (in decreasing molecular weight) the α-subunit of the thylakoid-located ATP synthetase complex and a polypeptide associated with photosystem II (cf. also Fig. 12) which exhibit plastid inheritance. The triangles designate polypeptides of yet unknown identity with nuclear inheritance. (In collaboration with N.H. Chua)

from spinach, tobacco, *Oenothera* plastome IV and potato or tomato have hardly any fragment in common (Fig. 1c). Nevertheless, closer inspection of these DNAs by comparative restriction site mapping and gene localization reveals striking conservation even among these plastid chromosomes (see sections "The Structural Organization of Plastid Chromosomes" and "Identification and Physical Mapping of Plastome-Coded Genes").

The electrophoretic separation of thylakoid polypeptides from *Euoenothera* species and their interspecific hybrids have extended these findings by demonstrating that plastome DNAs are responsible for definite phenotypic traits. The patterns shown in Figure 2 confirm both the distinctness and relatedness of the plastomes and, in addition, clearly reflect the genetically bipartite nature of chloroplast structures. Within plastomes I and II and within plastomes IV and V, the polypeptide patterns are indistinguishable. However, one can readily see from mobility alterations that certain polypeptides are plastome-specific (Fig. 2, arrows), while others are genome-specific (triangles). At least some of the differences reside in structural genes since the distinguishable plastome III α-subunit of the thylakoid-located ATP synthetase can be synthesized from total ptRNA in vitro (data not shown).

Spinacia oleracea

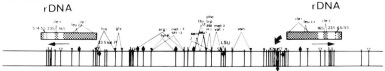

Nicotiana tabacum

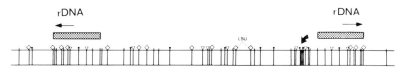

Euoenothera

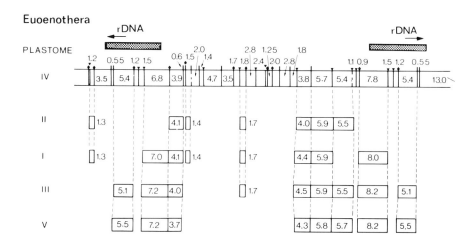

Fig. 3. Comparison of the physical maps of the five basic *Euoenothera* plastome DNAs and the plastid chromosomes of spinach and tobacco. The circular maps were linearized by a cut in the small single-copy region and aligned using the common 0.45 Md primary Sal I fragment *(thick arrow)*. In the *Oenothera* DNAs, this fragment derives from the largest Pst I primary fragment bordered by the 0.9 and 2.0 Md secondary fragments, respectively. The differing fragment sizes of the five plastome DNAs are given in Md (from Gordon et al. 1981a,b). The symbols represent cleavage sites: ▼, Sal I; ◆, Pst I; ▐, Kpn I; ▌, Xho I, ▽, Bgl I, ◇, Pvu II. The fragment order for the spinach ptDNA restriction fragments is from Crouse et al. (1980). The inverted duplication is indicated as a *hatched* region above each map. The locations of the genes for rRNA (spinach, tobacco, *Oenothera*), tRNA species (spinach), the 33.5 kd polypeptide (spinach), and the large subunit (LSU) of Fraction-I-Protein (spinach, tobacco) are indicated. In *Oenothera* ptDNA the latter gene is located at the margin of the conserved part of the large single-copy segment. The polarity of transcription of the rDNA regions is indicated by *arrows*

The Structural Organization of Plastid Chromosomes

In order to learn more about the differences in the restriction patterns, we have constructed physical maps of the five basic *Euoenothera* plastome DNAs (Gordon et al. 1981a,b) and of the plastid chromosomes from spinach (Crouse et al. 1978) and tobacco (Seyer et al., in preparation) using our simple and rapid method for restriction site mapping (Herrmann et al. 1980). These maps are shown in linearized version in Figure 3. They have been aligned with the aid of a small 0.45 megadalton (Md) Sal I primary fragment which seems to be equivalent in all these dicotyledonous ptDNAs with respect to size, map location, and (tested in spinach and tobacco) the existence of an internal Xho I cleavage site.

The most striking point to emerge from this comparison is a remarkable structural similarity of plastid chromosomes in spite of obvious changes in primary sequence. This similarity in organizational detail is matched by a similarity in the overall gene order probed (see sections "Genes for Ribosomal and Transfer RNAs" and "Polypeptide Genes"). There may be a hierarchy of evolutionary conservation of sequence and molecular organization in ptDNA which includes mechanisms or events resulting from the peculiar anatomy of plastid chromosomes as outlined below.

1. The anatomy of the plastid chromosomes from the above-mentioned species is remarkably similar. The circular molecules are organized in two unique-sequence segments differing in size and separating the inversely arranged copies of a large duplication.

2. The overall fragment order in the five *Euoenothera* plastome DNAs is the same, as expected from the gross similarity of the fragmentation patterns (Fig. 1a). The variable sizes of corresponding fragments are mainly due to insertions/deletions rather than inversions. These insertions do not contain recognition sites for Sal I, Pst I and Kpn I, chosen for mapping, since the number of cleavage sites are the same.

Changes occur in all four major segments, but they are nonrandomly distributed along the molecule. Many of the changes are clustered on both ends of the large single-copy segment adjacent to the repeats, although the central part of this segment seems virtually conserved. Large size differences between homologous fragments can result from multiple events.

Notwithstanding the considerable differences in the restriction patterns, conservation of overall organization is also apparent in the more distantly related plastid chromosomes. Such evolutionary conservation was originally indicated by reassociation of mixtures of two ptDNAs followed by thermal dissociation of the reassociates and has now been rigorously proven by comparative restriction site mapping and by hybridization of radioactively labeled RNA species and nick-translated DNA fragments of spinach to Southern blots of the restricted heterologous ptDNAs. A few of many possible examples may illustrate this. In addition to the common 0.45 Md Sal I fragment mentioned above and used to align the maps, the Kpn I restriction patterns of ptDNAs from six species, shown in Figure 1c, resemble each other in a general fashion with regard to cut frequency, size distribution, and location of fragments where mapped. The patterns share a two-molar band K-2 (fragments numbered in decreasing size) which in three instances tested was shown to be composed of two fragments differing in nucleotide sequence. The map locations of these K-2 fragments is similar in all the ptDNAs studied. Furthermore, the K-1 and one of the K-2 fragments span approximately the outer two-thirds of the opposite copies of the inverted duplication, each extending into the large single-copy segment, with K-2 being located close to the previously mentioned 0.45 Md Sal I fragment. Each of these fragments carries a 16S rRNA gene and part of a 23S rRNA gene. In each copy the latter extends into a Kpn I primary fragment of approximately 0.55 Md which is two-molar. Again this fragment is present in all ptDNAs and readily discernible in Figure 1c.

3. We have observed that differences in the inverted duplication are always found in both copies of the repeat, even in the recently diverged subtypes of the *Euoenothera* plastomes. We conclude from this observation that a "rectification" mechanism may exist, resulting in the transposition of a change in one copy to the equivalent site in the other. The reason for the existence of the inverted repeats and such a mechanism remains obscure. Possible modes and implications of such a mechanism will be considered elsewhere (Gordon et al. 1981b). In prokaryotes, inverted repeats are frequently involved in the insertion or transposition of genetic material (Calos and Miller 1980), but similar events have not yet been demonstrated for plastids of higher plants.

Identification and Physical Mapping of Plastome-Coded Genes

Genes for Ribosomal and Transfer RNAs

The unique translation machinery of plastids is partially encoded in the organelle's DNA (for review see Herrmann and Possingham 1980). Comparative mapping of the positions of the rRNA genes in spinach (Crouse et al. 1978; Whitfeld et al. 1978), *Oenothera* (Gordon et al. 1981a,b), and tobacco ptDNA (Seyer et al., in preparation) as well as mapping of tRNA genes of spinach ptDNA (Driesel et al. 1979) by hybridizing the radioiodinated purified RNA species to Southern blots of map fragments has revealed four points of interest.

1. The genes for the four plastid ribosomal RNAs lie in close proximity and are arranged in sets in order 16, 23, 4.5, and 5S rRNA. The set is present twice on the molecule, in inverted orientation, and the two units are widely separated (Fig. 3). The arrangement in sets as well as the intramolecular orientation of the rDNA region are highly conserved traits. As has been demonstrated in spinach, the rDNA regions are transcriptional units, implying that spinach plastids possess the enzymatic equipment for RNA modification (Bohnert et al. 1976). It is likely that polycistronic translation is a common feature of plastid rDNA regions.

2. The genes for the two large rRNAs in each rDNA region are separated by a large spacer. In contrast to the rRNA structural genes, this spacer represents a region of particularly frequent evolutionary change, differing in size even between closely related species and thus contributing to early plastome divergence. The spacer ranges in size from about 1.8 kbp (spinach) to 2.4 kbp *(Oenothera)* and is transcribed as part of the rRNA precursor molecule (Bohnert et al. 1976). In spinach the rDNA spacer carries a gene for tRNA-ile (Driesel et al. 1979; Bohnert et al. 1979).

3. Approximately 40 distinct 4S RNA species have been isolated from spinach plastids (Driesel et al. 1979). These plastids thus contain a set of tRNA species which should be sufficient to read all codons according to Crick's (1966) wobble hypothesis. About 30 4S RNAs have actually been identified as tRNAs specific for 16 amino acids. Most of these are located in the large single-copy segment where they are

104

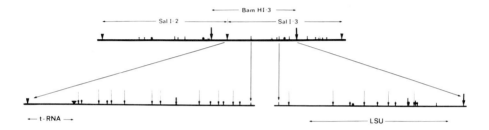

Fig. 4. Fine structure map of the large subunit of Fraction-I-Protein encoding region of spinach ptDNA (*upper part:* detail of Fig. 3). The precise locations of the clustered 4S RNA genes including genes specific for the tRNAs for phe, trp, and ala as well as of the large subunit of Fraction-I-Protein are indicated. The symbols represent cleavage sites for Sal I (▼), Pst I (♦), Kpn I (|), Xho I (ı), Bgl I (∇), Bam HI (↓), Sac I (▼), Pvu I (◊), Hinf I (↓), Pvu II (▼), Taq I (↓), Eco RI (↓), Xba I (↓). The four secondary fragments obtained from digestion of the Bam HI-3 fragment with Kpn I of approximately 0.8, 1.6, 4.2 and 5.1 kbp have been used for the nick-translation experiment described in Fig. 11

largely clustered in two regions (Fig. 3). Several have been mapped to the inverted repeat. Analysis of Hinf I and Taq I fragments from one of the tRNA gene clusters has shown a 500 bp fragment that contains at least five tRNA genes (Fig. 4). These genes must be bordered by regulatory sequences, promotor and terminator and/or processing signals depending on whether they are transcribed individually or as a single precursor and provided they are encoded in the same strand.

4. Isoaccepting tRNAs can be encoded by different loci and may thus represent true isoaccepting species.

Molecular Cloning of Spinach ptDNA

We have inserted DNA fragments obtained from Sal I, Pst I, or Bam HI restriction of spinach ptDNA into the ampicillin- and tetracycline-resistant plasmid pBR 322. The recombinant DNA has been used to transform *E. coli* C 600 and the minicell producing strain P 678-54. A collection of transformants was selected and screened by a rapid three-step procedure using low-gelling-temperature agarose as outlined in the legend of Figure 5. To date, about 40 clones spanning almost the entire spinach circle have been isolated, with some of them preserved more than two years without detectable changes. This collection serves as a sequence library of spinach ptDNA, allowing us to study functions of ptDNA and to investigate interspecific plastome homology.

Polypeptide Genes

After treatment with *Micrococcal* nuclease to reduce endogeneous activity, the rabbit reticulocyte system (Pelham and Jackson 1976) translates ptRNA with high fidelity. More than 60 products with molecular weights ranging from less than 10 to more than 80 kd can be resolved on denaturing polyacrylamide gels (Figs. 6, 7), to check the purity of the ptRNA fractions the more efficiently transcribed total cellular RNA was added. as control (cf. Fig. 8). Omitting RNA completely abolishes the synthetic capabilities of the system with the exception of an artificial band (Pelham and Jackson 1976) at 45 kd.

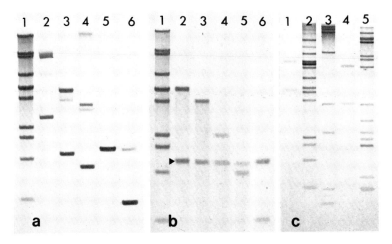

Fig. 5a-c. Three-step characterization of chimeric plasmids containing fragments
from spinach ptDNA. (a) *Step 1:* Insertion-positive plasmids were sized relative to
pBR 322 by electrophoresis in a 0.6% agarose slab gel. *Track 1:* Sal I fragment pattern of spinach ptDNA (for fragment sizes see Herrmann et al. 1980). *Tracks 2 - 4:*
cleared lysates of plasmids containing Sal-6, Sal-7 and Sal-9, respectively. *Track
5:* vector pBR 322 digested with Sal I, *track 6:* pBR 322 (cleared lysate). The lower
of the two major bands in *tracks 2 - 4* and *6* represent supercoiled (ccc)DNA, the upper the relaxed circular form. Little linearized DNA is present. (b) *Step 2:* Insertion-positive plasmids were purified from cleared lysates on tube gels of Seaplaque
agarose (Marine Coll. Inc.) (see Herrmann et al. 1980). The cccDNA bands were excised from the agarose gel, the gel pieces liquefied at 70°C, and the DNA was digested in the presence of the agarose sol at 37°C to split fragment and vector *(arrow)*. The digests were then either fractionated on slab gels alongside restricted
spinach ptDNA *(track 1)* to test for co-migration of the inserted fragments or again
on Seaplaque agarose tube gels to recover the inserted fragment for step 3. *Tracks
2 - 6:* clones containing inserted fragments Sal-5, Sal-6, Sal-7, Sal-9 and Sal-10,
respectively (0.5% agarose slab gel). (c) *Step 3:* The inserted fragment was further
characterized by restriction and comparison with the corresponding authentic ptDNA
fragment. This step is illustrated by analyzing a chimeric plasmid containing the
fragment Pst I-5. *Track 1:* excised Pst-5 fragment secondarily digested with Sal I;
track 2: Sal I *plus* Pst I double digest of total ptDNA; *track 3:* Kpn I digest of
total ptDNA; *track 4:* excised Pst-5 fragment secondarily digested with Kpn I; *track
5:* Pst I *plus* Kpn I double digest of total ptDNA. (0.8% agarose slab gel; for the
mapping relationships of fragments see Herrmann et al. 1980)

In this system we found fidelity of translation for the large subunit
(LSU) of ribulose bisphosphate carboxylase/-oxygenase (Fraction-I-Protein), a key enzyme of the photosynthetic carbon reduction cycle and
photorespiration, the α- and β-subunits (in collaboration with N. Nelson, Haifa), and the DCCD-binding proteolipid (in collaboration with
W. Sebald, Braunschweig) of the thylakoid-associated ATP synthetase
complex (CF_1-CF_0), cytochrome f, as well as a thylakoid-located polypeptide migrating with an apparent M_r of 33.5 kd and exhibiting all
of the properties described for the photogene 32 of *Zea mays* plastids
(Grebanier et al. 1978). Qualitative and quantitative changes in mRNAs
isolated from spinach plastids at different stages of the greening
process suggest that plastid chromosomes of spinach are differentially
expressed during transformation of etioplasts to chloroplasts (Fig. 7).
The in vitro synthesized products have been identified from their co-
migration with isolated authentic product (Fig. 6), double immunodif-

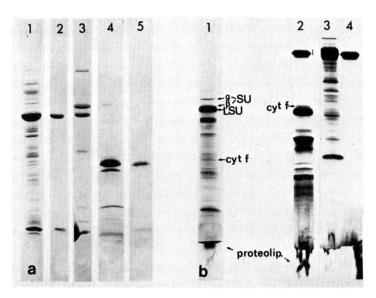

Fig. 6a,b. Characterization of chloroplast polypeptides among products of cell-free translation directed by spinach ptRNA in a rabbit reticulocyte system. SDS/poly-acrylamide gel electrophoresis (7.5% - 15% gel). (a) Polypeptides isolated from chloroplasts for comparison (stained) *track 1:* total soluble polypeptides; *track 2:* large and small subunit polypeptides of ribulose bisphosphate carboxylase/-oxygenase; *track 3:* CF_1 α- to ε-subunits of the thylakoid-associated ATP synthetase complex. The high molecular weight component is probably an aggregation artifact; the lowest band is the tracking dye. *Track 4:* crude cytochrome f fraction; *track 5:* cytochrome f. (b) Fluorography of ^{35}S-methionine labeled products translated from *1:* total ptRNA following 24 hr illumination of etiolated plants (7.5% - 15% gel); *2:* polypeptide pattern obtained from an RNA fraction migrating with an average velocity equivalent to 13S in non-denaturing sucrose gradients; *3:* total ptRNA after 14 hr illumination of etiolated plants; *4:* control, no RNA added. *Tracks 2 - 4:* 10% - 20% gel

fusion (Fig. 8), and two-dimensional immunoelectrophoresis (Fig. 9) against monospecific antibodies and/or by Cleveland's one-dimensional fingerprint technique (Fig. 10) under rigorously controlled conditions. Some of these products are known to be synthesized on plastid ribosomes and thus represent prime candidates for plastome-coded products.

Using nondenaturing sucrose gradient centrifugation, ptRNAs have been separated into size classes. By in vitro translation of these RNA fractions, we have established that the 17S and 15S RNA classes contain the messages for the LSU of Fraction-1-Protein and the 33.5 kd thylakoid polypeptide, respectively. Southern hybridization of these radioactively labeled RNAs has allowed us to determine the transcriptional origin of these two genes on the physical map of spinach ptDNA. The gene for the 33.5 kd protein is situated immediately adjacent to the end of one of the inverted repeats (Driesel et al., submitted, Fig. 3). The LSU gene lies near a cluster of tRNA genes, with most of its coding sequence located on a 1.6 kbp Kpn I primary fragment (Fig. 3). A structural fine map of the LSU gene region is shown in Figure 4. The location of the LSU gene has been verified by hybrid-arrested translation (Paterson et al. 1977) and by programming a coupled transcription/translation system derived from *E. coli* (Zubay et al. 1970; Bottomley and Whitfeld 1979) with cloned Sal I, Pst I, and Bam HI

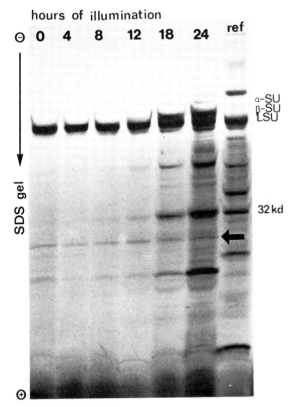

hours of illumination

Fig. 7. Fluorograph of an SDS/polyacrylamide gradient gel (10% - 20%) of ^{35}S-meth-ionine-labeled products synthesized in a rabbit reticulocyte cell-free translation system programmed with total ptRNA from different stages of greening of etioplasts (zero time control) to chloroplasts (24 h). The input RNA was normalized to the translation activity of a component that apparently remained unchanged *(arrow)*. The 32/33.5 kd polypeptide is indicated. The large subunit (LSU) of Fraction-I-Protein, the β-subunit (β-SU) of the ATP synthetase complex, and the artificial band at 45 kd (see text) are poorly resolved under the chosen conditions. Reference RNA is from adenovirus-infected HeLa cells

fragments of spinach ptDNA. The observation that cloned fragments direct the synthesis of LSU indicates that the gene for this polypeptide is colinear and not interrupted by noncoding sequences.

Relying on the phylogenetically conserved nature of the LSU sequence, we have utilized cloned spinach ptDNA fragments to locate this gene and its surrounding sequences on the *Oenothera* and tobacco plastid chromosomes. We have nick-translated the four spinach fragments produced by Bam HI and Kpn I cuts (Fig. 4) as probes along with appropriate controls (pBR 322) and have obtained specific hybridization in all instances (Fig. 11). Again, the striking point which emerges from this study is that although the three plants represent three phylogenetically distinct orders of dicotyledons, their ptDNAs are remarkably similar in structure. A large segment of about 8 kbp which carries the LSU gene has been retained in all three ptDNAs as a single contiguous region, similar with respect to nucleotide sequence, intra-

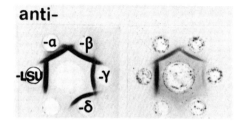

Figs. 8 and 9. Immunochemical identification of ^{35}S-methionine-labeled polypeptides synthesized from ptRNA in a cell-free translation system derived from rabbit reticulocytes. (In collaboration with N.H. Chua, Rockefeller University, and N. Nelson, Haifa)

Fig. 8. Characterization by double immunodiffusion. The central well contained 10 μl ^{35}S-methionine-labeled assay in addition to the purified authentic polypeptides (see Fig. 6) which were present as carriers. The peripheral wells contained monospecific antibodies against the large subunit (LSU) polypeptide of Fraction-I-Protein and the α- to δ-subunits of CF$_1$ of the thylakoid-associated ATP synthetase complex. *Left*, stained sample; *right*, autoradiograph. Note that the cytosolically made δ-subunit is not observed under the radioactively labeled products

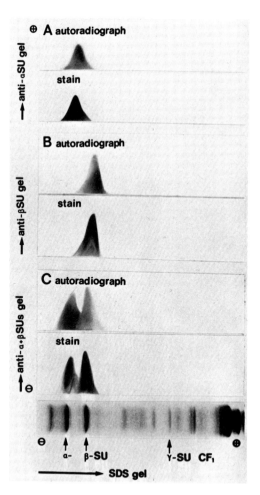

Fig. 9. Characterization by two-dimensional immunoelectrophoresis. A membrane preparation of total thylakoid polypeptides was added as carrier for the initial electrophoretic separation in an SDS/polyacrylamide gradient gel (7.5% - 15%) which composed the first dimension. In the second dimension, the polypeptides were electrophoresed into gels containing specific antibodies. *A:* anti-α, *B:* anti-β, *C:* mixture of anti-α and anti-β of the CF$_1$ subunits of the thylakoid-associated ATP synthetase complex

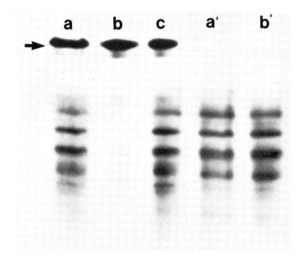

Fig. 10. Comparison of peptide maps of authentic and in vitro synthesized LSU. LSU isolated from spinach plastids and a comigrating radioactively labeled product obtained from ptRNA programmed cell-free translation were digested with *S. aureus* protease and the digestion products were separated on a 15% – 20% polyacrylamide gel under denaturing conditions. The stained gel is shown in tracks *a*, *b*, and *c*. Tracks *a'* and *b'* show the autoradiograph of *a* and *b*, respectively. *Track a:* authentic LSU plus radioactively labeled product; *track b:* radioactively labeled product alone; *track c:* authentic LSU alone

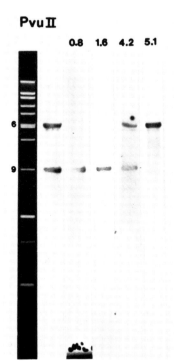

Fig. 11. Hybridization of nick-translated spinach ptDNA fragments to Pvu II fragment patterns of tobacco ptDNA. The Pvu II fragment pattern *(first track)* is followed by autoradiographs showing hybridization of the spinach Bam HI-3 fragment and the four secondary fragments produced by Kpn I digestion of this fragment (cf. Fig. 4). The sizes of these latter fragments are given in kbp above each track. (The 1.6 kbp spinach fragment carries most of the LSU gene.) The localization of the corresponding Pvu II fragment in tobacco on the physical map of tobacco ptDNA is shown in Fig. 3

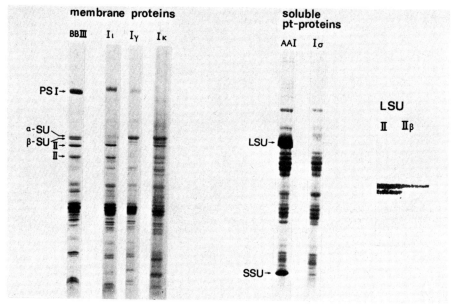

Fig. 12. Plastid polypeptides from *Euoenothera* wildtypes and plastome mutants following SDS/polyacrylamide gel electrophoresis (7.5% - 15%). From the preparations of thylakoid proteins, polypeptides associated with photosystem I, II as well as the α- and β-subunits of the thylakoid-associated ATP synthetase complex are indicated. Plastid membranes were prepared from *(left to right)* BB-III *Oe. grandiflora* (cf. Fig. 2); mutant iota of plastome I, deficient in ATP synthetase activity; mutant gamma of plastome I, deficient in photosystem II activity; mutant kappa of plastome I, deficient in photosystem I activity. All mutant plastomes associated with genotype AA. Soluble proteins were prepared from plastids isolated from *(left to right)*: AA-I *Oenothera hookeri* (cf. Fig. 2); mutant sigma of plastome I which lacks ribulose bisphosphate carboxylase/-oxygenase activity. The electrophoretic mobility of LSU from wildtype plastome II is compared with LSU isolated from mutant beta of plastome II. In the latter, purified LSU exhibits only the larger of two discernible subunit modifications of the wildtype (Schmitt 1978)

chromosomal position, and polarity relative to the 0.45 Md Sal I fragment used to align the maps (see section "The Structural Organization of Plastid Chromosomes").

A complementing approach to gene identification and mapping would utilize interspecific hybrids and plastome mutants and could be particularly valuable for the characterization of nonstructural plastome genes. However, genetic methods for localizing markers on fragment maps have not yet been employed due to the lack of suitable materials and to difficulties encountered in the propagation of mutant material and in subcellular fractionation of labile mutant organelles (Schmitt and Herrmann 1977).

About 60 pigment-deficient mutants have been isolated from the five basic *Euoenothera* plastomes. These non-Mendelian mutants have arisen spontaneously (Kutzelnigg and Stubbe 1974) or have been induced by a nuclear "mutator" gene isolated by Epp (1973). As pure mutant plants, they are generally not viable on soil; mutant leaf sectors require nourishment from wild-type tissue (variegation). However, we have recently developed a tissue culture system through which genetically stable material can now be readily obtained in sufficient quantity for biochemical analysis (Mehra-Palta and Herrmann 1975).

Most of the *Oenothera* plastome mutants have been characterized physiologically (Hallier et al. 1978), biochemically, and electron microscopically (Kutzelnigg et al. 1975), and several of them are of particular interest because their genetic defects are probably specific rather than pleiotropic. Mutants deficient in photosystem I or II, cytochrome f, or ATP synthetase activitities have been shown to lack the corresponding polypeptides. A few examples are illustrated by Figure 12.

Two classes of plastome mutants defective in Fraction-I-Protein have been found (Schmitt 1978): those containing no detectable Fraction-1-Protein and those with an inactive or altered protein. The mutant "sigma" of plastome I represents the former, the mutant "beta" of plastome II is a member of the latter category (Fig. 12). Plastids of the mutant I sigma contain functional machineries for translation and photosynthetic electron transport, the enzymes ribose-5-phosphate isomerase, phosphoribulose kinase, and NADP-dependent glyceraldehyde phosphate dehydrogenase, but neither large nor small subunits of ribulose bisphosphate carboxylase/-oxygenase are detectable by immunological methods. Since we have found no differences between the LSU encoding regions of the plastid chromosome of mutant I sigma and wildtype plastome I (Gordon et al. 1980), the mutation might involve control of the nuclearly-coded SSU. In the other group of plastome mutants for Fraction-I-Protein, mutant II beta contains only the slower migrating of two LSU modifications resolvable under stringent electrophoretic conditions. This suggests that the LSU may exist as a precursor polypeptide which is able to associate with the small subunit to form a holoenzyme. If correct, this type of precursor needs to be defined.

Even though our analyses of these two mutants are preliminary, they indicate that ptDNA may not only encode the structural gene for the larger, catalytic subunit of ribulose bisphosphate carboxylase/-oxygenase, but may also be involved in other processes in the biogenesis of the enzyme. By pin-pointing the location of these plastome markers, we hope to test this hypothesis critically.

Acknowledgments. The expert technical assistance of Ms. Barbara Schiller and Ms. Monika Streubel is gratefully acknowledged. This work was supported by the Deutsche Forschungsgemeinschaft (grant He 693) and by Forschungsmittel des Landes Nordrhein/ Westfalen.

References

Behn W, Herrmann RG (1977) Circular molecules in the β-satellite DNA of *Chlamydomonas reinhardii*. Mol Gen Genet 157:25-30

Bohnert HJ, Driesel AJ, Herrmann RG (1976) Characterization of the RNA compounds synthesized by isolated chloroplasts. In: Bücher Th, Neupert W, Sebald W, Werner S (eds) Genetics and biogenesis of chloroplasts and mitochondria. Elsevier/North Holland, Amsterdam, pp 629-636

Bohnert HJ, Driesel AJ, Crouse EJ, Gordon K, Herrmann RG, Steinmetz A, Mubumbila A, Keller M, Burkard G, Weil JH (1979) Presence of a transfer RNA gene in the spacer sequence between the 16S and 23S rRNA genes of spinach chloroplast DNA. FEBS Lett 103:52-56

Bottomley W, Whitfeld PR (1979) Cell-free transcription and translation of total spinach chloroplast DNA. Eur J Biochem 93:31-39

Calos MP, Miller JH (1980) Transposable elements. Cell 20:579-595

Crick FHC (1966) Codon-anticodon pairing: The wobble hypothesis. J Mol Biol 19: 548-555

Crouse EJ, Schmitt JM, Bohnert HJ, Gordon K, Driesel AJ, Herrmann RG (1978) Intra-
    molecular heterogeneity of *Spinacia* and *Euglena* chloroplast DNAs. In: Akoyunoglou
    G, Argyroudi-Akoyunoglou JH (eds) Chloroplast development. Elsevier/North Holland,
    Amsterdam, pp 565-572
Driesel AJ, Crouse EJ, Gordon K, Bohnert HJ, Herrmann RG, Steinmetz A, Mubumbila M,
    Keller M, Burkard G, Weil JH (1979) Fractionation and identification of the indi-
    vidual spinach chloroplast transfer RNAs and mapping of their genes on the restric-
    tion endonuclease cleavage site map of chloroplast DNA. Gene 6:285-306
Driesel AJ, Speirs J, Bohnert HJ (1980) Spinach chloroplast mRNA for a 32 kd poly-
    peptide - size and localization on the physical map of the chloroplast DNA. Bio-
    chim Biophys Acta in press
Epp M (1973) Nuclear gene-induced plastome mutations in *Oenothera hookeri*. I. Ge-
    netic analysis. Genetics 75:465-483
Gordon KHJ, Crouse EJ, Bohnert HJ, Herrmann RG (1981a) Restriction endonuclease
    cleavage site map of chloroplast DNA from *Oenothera parviflora* (plastome IV).
    Theor Appl Genet in press
Gordon KHJ, Crouse EJ, Bohnert HJ, Herrmann RG (1981b) Physical mapping of differences
    in DNA of the five wild-type plastomes in *Oenothera* subsection *Euoenothera*. Theor
    Appl Genet submitted
Gordon KH, Hildebrandt JW, Bohnert HJ, Herrmann RG, Schmitt JM (1980) Analysis of
    the plastid DNA in an *Oenothera* plastome mutant deficient in ribulose bisphosphate
    carboxylase. Theor Appl Genet 57:203-207
Grebanier AE, Coen DM, Rich A, Bogorad L (1978) Membrane proteins synthesized but
    not processed by isolated maize chloroplasts. J Cell Biol 78:734-746
Hallier UW, Schmitt JM, Heber U, Chaianova SS, Volodarsky A (1978) Ribulose-1,5-
    bisphosphate carboxylase-deficient plastome mutants of *Oenothera*. Biochim Biophys
    Acta 504:67-83
Herrmann RG (1977) Studies on *Oenothera* plastid DNAs. In: Prog Regul Dev Proc Plants.
    Halle/Saale 1977, p 48
Herrmann RG, Possingham JV (1980) Plastid DNA-The Plastome. In: Reinert J (ed) Re-
    sults and problems in cell differentiation, vol X. Springer, Berlin Heidelberg
    New York, pp 45-96
Herrmann RG, Palta HK, Kowallik KV (1980a) Chloroplast DNA from three archegoniates.
    Planta 148:319-327
Herrmann RG, Whitfeld PR, Bottomley W (1980b) Construction of a Sal I/Pst I restric-
    tion map of spinach chloroplast DNA using low-gelling-temperature-agarose electro-
    phoresis. Gene 8:179-191
Kirk JTD, Tilney-Bassett RAE (1978) The plastids — their chemistry, structure,
    growth and inheritance, 2nd edn. Elsevier/North Holland, Amsterdam
Kutzelnigg H, Stubbe W (1974) Investigations on plastome mutants in *Oenothera*. 1.
    General considerations. Sub-Cell Biochem 3:73-89
Kutzelnigg H, Meyer B, Schötz F (1975) Untersuchungen an Plastom-Mutanten
    von *Oenothera*. III. Vergleichende ultrastrukturelle Charakterisierung der Mu-
    tanten. Biol Zentralbl 94:527-538
Mehra-Palta A, Herrmann RG (1975) Controlled morphogenesis in tissue cultures of
    wildtype and plastome mutants of *Oenothera*. In: Prog Int Bot Congr, Leningrad,
    p 301
Paterson BM, Roberts BE, Kuff EL (1977) Structural gene identification and mapping
    by DNA-mRNA hybrid-arrested cell-free translation. Proc Natl Acad Sci USA 74:
    4370-4374
Pelham HRB, Jackson RJ (1976) An efficient mRNA-dependent translation system from
    reticulocyte lysates. Eur J Biochem 67:247-256
Renner O (1934) Die pflanzlichen Plastiden als selbständige Elemente der genetischen
    Konstitution. Ber Verh Saechs Akad Wiss Leipzig Math Phys Kl 86:241-266
Schmitt JM (1978) Zur Funktion des genetischen Apparates der Plastiden: Nuklein-
    säuresynthese und plastomcodierte Proteine. Diss Univ Düsseldorf
Stubbe W (1959) Genetische Analyse des Zusammenwirkens von Genom und Plastom bei
    Oenothera. Z Indukt Abstamm Vererbungsl 90:288-298
Whitfeld PR, Herrmann RG, Bottomley W (1978) Mapping of the ribosomal RNA genes on
    spinach chloroplast DNA. Nucleic Acid Res 5:1741-1751
Zubay G, Chambers DA, Cheong LC (1970) Cell-free studies on the regulation of the *lac*
    operon. In: Beckwith JR, Zipser D (eds) The lactose operon. New York, Cold Spring
    Harbor Lab, pp 375-391

Synthesis, Transport, and Assembly of Chloroplast Proteins

N.-H. Chua[1], A. R. Grossman[1], S. G. Bartlett[1], and G. W. Schmidt[2]

Introduction

Intact chloroplasts of higher plants have three separate compartments:
envelope, stroma, and thylakoids. These three subchloroplastic compartments may be purified, with minimal cross-contamination, by fractionation of a lysed chloroplast preparation on discontinuous sucrose gradients (cf. Douce and Joyard 1979). Electrophoretic analysis by SDS
polyacrylamide gels reveals that each subchloroplastic fraction contains a distinctive set of constituent polypeptides (Pineau and Douce
1974; Joy and Ellis 1975; Morgenthaler and Mendiola-Morgenthaler 1976).
During the last decade, studies carried out in several laboratories
have demonstrated that only three envelope polypeptides, 1 - 3 stromal
polypeptides, and 10 - 15 thylakoid membrane polypeptides are synthesized inside the organelle (cf. Ellis 1977). The great majority of
chloroplast polypeptides are synthesized on cytosolic ribosomes and
must be imported into the organelle in order to reach their final location. Our laboratory is interested in the biosynthetic pathways of
this particular group of polypeptides. Here, we summarize recent results on the sequence of events surrounding the cytosolic synthesis
of chloroplast polypeptides and their subsequent transport into the
organelle. Results of earlier work can be found in a previous review
dealing with the same topic (Chua and Schmidt 1979).

Synthesis, Transport, and Assembly of the Small Subunit of Ribulose-1,5-bisphosphate Carboxylase

Ribulose-1,5-bisphosphate carboxylase (E.C. 4.1.1.39) is the most abundant enzyme in the chloroplast stroma and accounts for up to 10% of the
total soluble proteins in higher plants (cf. Kawashima and Wildman
1970). The holoenzyme is made up of eight copies each of two noniden-
tical subunits designated large and small. Since the small subunit (S)
is a product of cytosolic protein synthesis (cf. Kung 1977) and the
holoenzyme is so abundant, it was selected for studies on the biosynthesis and transport of chloroplast proteins.

As first steps toward elucidating the transport mechanism of S it is
necessary to: (1) determine whether S is synthesized on free or membrane-bound ribosomes, and (2) identify the primary translation product
of the mRNA coding for S. In *Chlamydomonas reinhardii*, a unicellular green
alga, S (mol.wt. 16,000) is synthesized on free polysomes as a precursor (pS) 4000 - 5000 daltons larger than the mature form (Dobberstein
et al. 1977). The synthesis of pS on free polysome suggests that its
import into chloroplasts occurs after polypeptide chain termination.

[1]Department of Cell Biology, The Rockefeller University, New York, New York 10021, USA
[2]Present address: Botany Department, University of Georgia, Athens, Georgia, USA

Dobberstein et al. (1977) postulated that pS is an extrachloroplastic
form of S and that the additional sequence (transit sequence) of pS
plays a role in posttranslational transport. The inability to isolate
intact chloroplasts from *Chlamydomonas* has precluded a direct demonstration of the transport of pS by in vitro reconstitution experiments in
this organism.

Subsequent work with higher plants confirmed the finding that S is
synthesized as a larger precursor (pS) in cell-free translation systems programmed with poly(A) RNA from spinach (Chua and Schmidt 1978a,
b), peas (Highfield and Ellis 1978; Cashmore et al. 1978; Chua and
Schmidt 1978a,b), and duckweed (Tobin 1979). In addition the ability
to prepare intact chloroplasts from higher plants has provided the opportunity to examine the physiological role of higher plant pS. Thus,
polypeptide transport may be investigated in vitro by incubation intact chloroplasts with ^{35}S-labeled translation products synthesized
from poly(A) RNA template in a wheat germ cell-free system. Such in
vitro reconstitution systems were utilized to show that pS, which is
present in postribosomal supernatant of translation mix, can be imported by intact chloroplasts in the absence of protein synthesis
(Highfield and Ellis 1978; Chua and Schmidt 1978a,b). The precursor
is processed either during or shortly after passage through the chloroplast envelope, and more than 80% of the newly imported S assembles
with endogenous large subunit to form the 18S holoenzyme (Chua and
Schmidt 1978a,b). Since the large subunit is synthesized in the chloroplast stroma (cf. Ellis 1977), these results show that pS has traversed both envelope membranes and that assembly of the holoenzyme is
likely to be a stromal event. The processing enzyme (transit peptidase)
that converts pS to S is soluble (Dobberstein et al. 1977) and has
been localized to the stroma (Smith and Ellis 1979). Purification of
the transit peptidase should allow its further characterization such
as precursor specificity and sensitivity to inhibitors, etc.

Synthesis, Transport, and Assembly of the CP II Polypeptides

The demonstration that pS enters chloroplasts after protein synthesis
raises the question whether a similar posttranslational mechanism also
applies to the transport of other chloroplast proteins that are made
on cytosolic ribosomes. The synthesis, transport, and assembly of integral thylakoid polypeptides post interesting problems because, in contrast to S, they are hydrophobic and are localized in a different subchloroplastic compartment. After synthesis in the cytosol and transport
across the envelope membranes, the polypeptides must be inserted into
yet another membrane, the thylakoids.

The most abundant proteins of the thylakoid membrane are the apopro-
teins of chlorophyll-protein complex II (CP II), which functions in
harvesting light energy (cf. Thornber 1975). Using the techniques of
in vitro translation and immunoprecipitation with specific antibodies,
Apel and Kloppstech (1978) first reported that the barley CP II apoprotein (mol.wt. 25,000) is synthesized as a larger precursor (mol.wt.
29,500). In peas, the CP II complex contains at least two polypeptides,
designated 15 (mol.wt. 28,000) and 16 (mol.wt. 27,000) and both are
also synthesized as larger forms, p15 (mol.wt. 33,000) and p16 (mol.wt.
32,000), in cell-free translation systems (Schmidt et al. 1980). Interestingly, centrifugation of cell-free translation mixtures at
140,000 x g for 1 h does not sediment p15 and p16, indicating that
they are soluble precursors of insoluble membrane polypeptides. Upon
incubation of a postribosomal supernatant containing p15 and p16 with
intact chloroplasts in the light, the precursors are taken up and pro-

cessed correctly to yield the mature polypeptides. The newly imported CP II polypeptides are indistinguishable from the thylakoid forms by the following criteria: (1) solubility in a 2:1 (v/v) mixture of chloroform:methanol; (2) resistance to extraction by 0.1 M NaOH, which releases peripheral membrane polypeptides; (3) identical dispositions in the membrane as probed by limited proteolysis of isolated thylakoid membrane vesicles; and (4) ability to bind chlorophyll a and b to form the CP II complex (Schmidt et al. 1980).

The availability of a reconstitution system for uptake and processing of p15 and p16 in vitro provides the opportunity to determine the assembly pathway of CP II in intact chloroplasts. The complex may be assembled by two possible pathways:

1. The newly transported polypeptides 15 and 16 combine first with newly synthesized chlorophyll a and b as a chlorophyll-protein complex before insertion into the thylakoids. Thus, the formation of the chlorophyll-protein complex is considered to be an obligatory intermediate step prior to membrane insertion.

2. Thylakoid insertion of polypeptides 15 and 16 occurs independently of new chlorophyll synthesis. The inserted polypeptides may then recruit preexisting chlorophylls in the thylakoid membranes to form the CP II complex.

In an attempt to discriminate between these two pathways, p15 and p16 were incubated with intact chloroplasts in the dark, a condition which prevents chlorophyll synthesis in higher plants. Uptake and processing of the precursors and assembly of the imported polypeptides 15 and 16 can occur in the dark, although at a reduced rate (Schmidt et al. 1980). These results demonstrate that assembly of the CP II complex occurs in the thylakoid membrane after insertion of the apoprotein.

Further experiments are required to clarify some ambiguities in the assembly pathway for the CP II complex. Precursor processing presumably precedes thylakoid insertion; however, the possibility that the precursors are processed only after their assembly with chlorophylls in the thylakoid membranes cannot be ruled out. Furthermore, the precursor segment of these integral membrane polypeptides may contain information for both transport across the chloroplast envelope and integration into the thylakoid membranes. Thus, p15 and p16 may be processed in two steps. Experiments to date have yielded no convincing data on these points. Attempts to process p15 and p16 with subchloroplastic fractions have been unsuccessful so far (unpublished data). Therefore, the localization of the transit peptidase for these precursors remains to be established.

Transport of Other Soluble and Thylakoid Membrane Polypeptides

The posttranslational uptake of polypeptides by chloroplasts is not restricted to pS, p15, and p16. Under optimal conditions in the light, 100 - 200 stromal polypeptides and 20 - 30 thylakoid membrane polypeptides enter intact chloroplasts in vitro in the absence of protein synthesis (Grossman et al. 1980b, unpublished data). Since few of these polypeptides have been identified, we do not know whether synthesis as a larger precursor is a general rule. The general lack of correlation between profiles of translation products and stroma polypeptides in SDS gels suggests that this is the case. Irrespective of their extrachloroplastic forms, the results show conclusively that the posttranslational mode of polypeptide transport applies to most, if not all, chloroplast polypeptides.

Mechanism of Posttranslational Transport

Some progress has been made toward elucidating the molecular mechanism of posttranslational uptake of chloroplast proteins. Uptake presumably is mediated by specific receptors, and indirect evidence suggests that the putative receptors are protein of the envelope outer membrane since the capacity for polypeptide uptake can be abolished by prior treatment of intact chloroplasts with proteases (Chua and Schmidt 1978a,b). The protease-sensitive polypeptides of the outer envelope membrane have not yet been identified. Binding of precursors to envelope receptors and transfer of the bound precursors across the chloroplast envelope are two separate events that can be dissociated at low temperature (Grossman et al. 1980b). Precursor binding occurs at 4°C, while both processes proceed when the temperature is raised to 25°C. These results suggest that precursor-receptor interaction does not require energy whereas passage of the precursors through the envelope membrane might be an energy-dependent step.

Because of the posttranslational mode of transport involves, the transfer of a completed polypeptide chain which has attained a three-dimensional configuration energy might be required to effect a conformational change in the protein in order to expedite membrane traversal. Indeed, recent results show that polypeptide transport into chloroplasts in vitro is greatly stimulated by light (Grossman et al. 1980a). The light-stimulated uptake is inhibited by uncouplers, but not by the electron transport inhibitor dichlorophenyldimethylurea or the protein synthesis inhibitor chloramphenicol, suggesting that ATP produced by photosynthetic phosphorylation in the light is responsible for the elevated polypeptide transport. This conclusion is supported by the observation that addition of ATP to the uptake mixture in the dark mimics the light stimulation of transport. Taken together, these data provide unequivocal evidence that the uptake of cytoplasmically synthesized polypeptides requires ATP inside the chloroplast. This finding suggests a regulatory mechanism through which the organelle may determine its own fate. By adjusting the endogenous ATP level the chloroplast can regulate the intake of cytoplasmically synthesized polypeptides that are necessary for its biogenesis.

Conclusion

Approximately 80% by weight of chloroplast proteins are synthesized outside the organelle (cf. Ellis 1977). The biosynthesis of this particular group of protein begins with their synthesis in the cytosol and ends with their specific localization in the correct subchloroplastic compartment. A reconstitution system, consisting of intact chloroplasts and cell-free translational products of poly(A) RNA, has been developed to study the major biosynthetic events (synthesis, transport, processing, and assembly) in vitro (Chua and Schmidt 1978a,b; Schmidt et al. 1980; Grossman et al. 1980a,b). We have used this in vitro reconstitution system to demonstrate that cytoplasmically synthesized polypeptides are imported by a posttranslational process dependent on ATP inside the chloroplast. At least in the case of the small subunit of carboxylase and the CP II polypeptides, the imported polypeptides assemble correctly into the appropriate macromolecular structures. Thus, the in vitro reconstitution system faithfully reconstructs the major biosynthetic events that occur in vivo, making this system invaluable for investigating factors that regulate the transport, processing, and assembly of chloroplast polypeptides.

Acknowledgments. This work was supported by NIH research grants GM 21060 and GM 25114, NIH research career development award GM 00223 to N.-H.C., and NIH postdoctoral fellowships GM 06444 and GM 06678 to A.R.G. and S.G.B., respectively.

References

Apel K, Kloppstech K (1978) The plastid membranes of barley *(Hordeum vulgare)*. Light-induced appearance of mRNA coding for the apoprotein of the light-harvesting chlorophyll a/b protein. Eur J Biochem 85:581-588

Cashmore AR, Broadhurst MK, Gray RE (1978) Cell-free synthesis of leaf protein: Identification of an apparent precursor of the small subunit of ribulose-1,5-bisphosphate carboxylase. Proc Natl Acad Sci USA 75:655-659

Chua N-H, Schmidt GW (1978) Post-translational transport into intact chloroplasts of a precursor to the small subunit of ribulose-1,5-bisphosphate carboxylase. Proc Natl Acad Sci USA 75:6110-6114

Chua N-H, Schmidt GW (1979) Transport of proteins into mitochondria and chloroplasts. J Cell Biol 81:461-483

Chua N-H, Schmidt GW (1980) In vitro synthesis, transport, and assembly of ribulose-1,5-bisphosphate carboxylase subunits. In: Siegelman HW, Hind G (eds) Photosynthetic carbon assimilation. Plenum Publishing Corp, New York, p 325

Dobberstein B, Blobel G, Chua N-H (1977) In vitro synthesis and processing of a putative precursor for the small subunit of ribulose-1,5-bisphosphate carboxylase of *Chlamydomonas reinhardii*. Proc Natl Acad Sci USA 74:1082-1085

Douce R, Joyard J (1979) Structure and function of the plastid envelope. Adv Bot Res 7:1-116

Ellis RJ (1977) Protein synthesis by isolated chloroplasts. Biochim Biophys Acta 463:185-215

Grossman AR, Bartlett SG, Chua N-H (1980a) Energy-dependent uptake of cytoplasmically synthesized polypeptides by chloroplasts. Nature (London) 285:625-628

Grossman AR, Bartlett SG, Schmidt GW, Chua N-H (1980b) Post-translational uptake of cytoplasmically-synthesized proteins by intact chloroplasts in vitro. In: Zimmerman M, Mumford RA, Steiner DF (eds) Precursor processing and biosynthesis of proteins. New York Acad Sci, New York, p 266

Highfield PE, Ellis RJ (1978) Synthesis and transport of the small subunit of chloroplast ribulose-bisphosphate carboxylase. Nature (London) 271:420-424

Joy KW, Ellis RJ (1975) Protein synthesis in chloroplasts. IV. Polypeptides of the chloroplast envelope. Biochim Biophys Acta 378:143-151

Kawashima N, Wildman SG (1970) Fraction I protein. Annu Rev Plant Physiol 21:325-358

Kung SD (1977) Expression of chloroplast genomes in higher plants. Annu Rev Plant Physiol 28:401-437

Morgenthaler JJ, Mendiola-Morgenthaler L (1976) Synthesis of soluble, thylakoid, and envelope membrane proteins by spinach chloroplasts purified from gradients. Arch Biochem Biophys 172:51-58

Pineau B, Douce R (1974) Analyse electrophoretique des proteines de L'envelope des chloroplastes d'epinard. FEBS Lett 47:255-259

Schmidt GW, Bartlett SG, Grossman AR, Cashmore AR, Chua N-H (1980) In vitro synthesis, transport, and assembly of the constituent polypeptides of the light-harvesting chlorophyll a/b protein complex. In: Leaver CJ (ed) Genome organization and expression in plants. Plenum Press, New York, p 337

Smith SM, Ellis RJ (1979) Processing of small subunit precursor of ribulose bisphosphate carboxylase and its assembly into whole enzyme are stromal events. Nature (London) 278:662-664

Thornber JP (1975) Chlorophyll-proteins: Light-harvesting and reaction center components of plants. Annu Rev Plant Physiol 26:127-158

Tobin EM (1978) Light regulation of specific mRNA species in *Lemna gibba* L. G.3. Proc Natl Acad Sci USA 75:4749-4753

Role of the Golgi Complex in Intracellular Transport[1]

V. Herzog[2]

Introduction

In 1898, the Italian histologist Camillo Golgi (1898a,b) described
reticular structures in the cytoplasm of nerve cells which he called
"apparato reticolare interno". He used a capricious and little-under-
stood silver staining technique for its visualization. The first direct evidence that this structure is not due to a staining artifact
came from electron microscopic observations (Dalton and Felix 1954).
With the technical progress in cell-biological areas such as cell fractionation and electron microscope-cytochemistry and -autoradiography,
the analysis of structural, kinetic and chemical characteristics of
the Golgi complex became possible. Since its discovery, the Golgi complex has been looked upon as functioning in the secretory process.
Therefore, knowledge on the structure and function of the Golgi complex accumulated with the increasing interest in the analysis of the
secretory process. This was studied primarily in secretory cell types
such as the exocrine pancreas. However, all eukaryotic cells release
secretory products permanently, intermittently, or at some stage of
their development. Because of the great variability in the specific
function among cells, numerous variations on the common theme of the
secretory process (Palade 1975) and, in particular, on the functions
of the Golgi apparatus, were discovered. It became clear that the Golgi
apparatus is a station in the secretory pathway where the secretory
product is modified while in transit. Because secretory products are
segregated from the cytoplasm by a membrane at all stations, it has
been assumed that their intracellular transport implies the concomitant
transfer of membranes or of membrane constituents.

In this paper, it is attempted to review[1] the current state of the
knowledge on the function of the Golgi complex in the intracellular
transport of secretory proteins and to assess its membrane interactions
during this process.

Structure of the Golgi Apparatus

Secretory epithelial cells exhibit a polarized intracellular organization, in which the Golgi complex is characteristically interposed between the basally located endoplasmic reticulum and the apical cytoplasm
which usually contains secretion granules (Figs. 1 and 2). It consists
of a variable population of (usually 4-6) closely apposed flattened

---

[1]This review is restricted to the Golgi complex in mammalian secretory cells. For
recent reviews on the Golgi complex in nonmammalian cells including plant cells,
see Brown and Willison (1977) and Morré (1977)

[2]Institut für Zellbiologie der Universität München, Goethestr. 33, D 8000 München 2
FRG

120

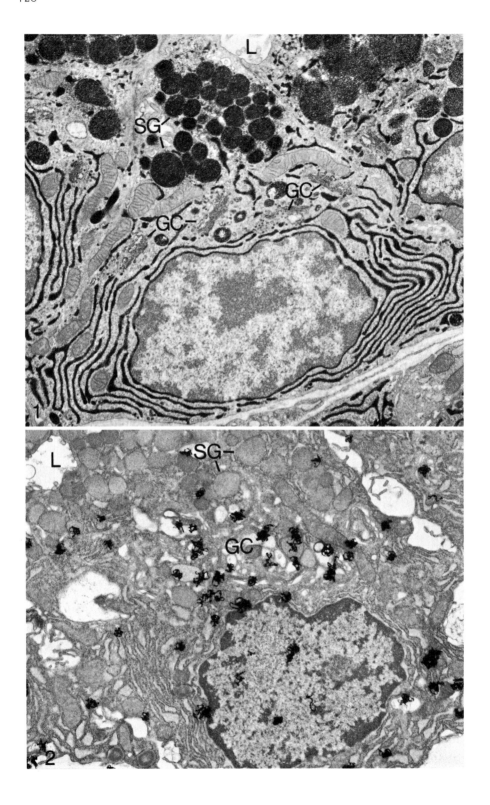

cisternae the membranes of which, in contrast to the rough endoplasmic reticulum, do not carry ribosomes on their cytoplasmic surfaces. This stack of Golgi cisternae is oriented with its convex or cis- (also entry- or forming-) face towards the rough endoplasmic reticulum (RER) and with its concave or trans- (also exit- or mature-) face towards the secretion granules (Figs. 3 and 5). In the cytoplasm between the cis-face and the RER, a population of small, usually smooth-surfaced but in part also coated vesicles (60-80 nm) is found which are variously referred to as shuttling or peripheral (Jamieson and Palade 1967a,b), transitional (Farquhar 1978a) or transporting (Jamieson 1978) vesicles (Figs. 3, 12, and 15). They are believed to derive from transitional elements (Jamieson and Palade 1967a,b) which are special ribosome-free regions of the RER in close proximity to the cis-Golgi-cisterna. From the transitional elements, peripheral vesicles appear to bud off and move to the adjacent cis-Golgi-cisterna which may be formed in part by coalescence of such vesicles (Fig. 3). Alternatively, direct connections between the endoplasmic reticulum and the Golgi cisternae in the form of twisted tubules are discussed (Claude 1970; Rambourg et al. 1974; Morré 1977). On the trans-face, condensing vacuoles (Figs. 1, 3, 5, and 12) arise which are partially formed by the transmost cisterna of the Golgi stack. The membranes of Golgi cisternae and condensing vacuoles appear morphologically similar but differ from the membranes of mature secretion granules as revealed by freeze fracture observations: Heterogeneities and mosaic organizations in the distribution of intramembranous particles were observed in Golgi cisternae and in condensing vacuoles, whereas mature secretion granules are characterized by a low particle density (De Camilli et al. 1976). The cisterna on the trans-side of the Golgi complex is occasionally more narrow than the other stacked cisternae and sometimes observed at a distance from the stack. This special region of the Golgi apparatus has been claimed to be continuous with the smooth endoplasmic reticulum; because of its possible involvement in the formation of primary lysosomes, it is also called GERL, an acronym derived from *G*olgi, *E*ndoplasmic *R*eticulum, *L*ysosomes (Novikoff 1976). In some but not all cell types, the membrane thickness of the Golgi cisternae varies across the stack; cis-cisternae have a thin ER-like membrane of ~50 Å, and membranes of the trans-cisternae are of a width (~75 Å) comparable to the plasma membrane (Morré et al. 1974). This structural detail, together with the tubular continuities between RER and Golgi cisternae mentioned before, has been interpreted as being supportive for the concept of membrane flow and differentiation (Morré et al. 1974, 1979) which assumes that cell membranes and the enclosed content move and differentiate from RER to the cis-cisterna and through the stacked cisternae of the Golgi complex to its trans-face from where they reach the plasmalemma.

Figs. 1 and 2. The secretory pathway in rat lacrimal gland as visualized cytochemically by demonstration of secretory peroxidase (Fig. 1) or autoradiographically by localization of the radioactive wave of newly synthesized secretory protein (Fig. 2). Note that cytochemical visualization reveals all compartments along the secretory route, simultaneously, including the Golgi complex (*GC*) which is characteristically interposed between the rough endoplasmic reticulum and the secretion granules (*SG*) (Fig. 1). By autoradiography, only a certain station of the secretory pathway is demonstrated at a given time point after a 1-min pulse with L-[³H]leucine; for example, after a chase period of 20 min, the wave of pulse-labeled secretory proteins has reached the Golgi complex (*GC*) with most silver grains located over Golgi cisternae (Fig. 2). (Herzog and Miller 1972; with permission of Rockefeller University Press).
Fig. 1, x 11,000; Fig. 2, x 10,000

The structural cis-trans-polarized organization is also expressed by the heterogeneity of its elements as revealed by cell fractionation and by cytochemistry but is still far from being understood. Cytochemically, 5'-nucleotidase is located preferentially in the cisternae of the cis-side (Farquhar et al. 1974), nicotinamide adenine dinucleotide phosphatase activity is concentrated in intermediate Golgi cisternae (Smith 1980), whereas acid phosphatase and thiamine pyrophosphatase are found in most cell types in trans-elements of the Golgi complex (Farquhar et al. 1974). In several cell types, both enzymes are located in different areas with thiamine pyrophosphatase found in trans-Golgi-cisternae and acid phosphatase being restricted to GERL (Novikoff 1976).

Structural and topological characteristics of the Golgi complex may change under experimental conditions. This is exemplified by observations in the parotid gland and the exocrine pancreas. During stimulated exocytosis, the Golgi apparatus of the exocrine pancreas cell may transiently increase in size (Jamieson and Palade 1971b). The position of the Golgi apparatus and its relation to the plasma membrane is also modified dramatically during stimulated exocytosis. In rat parotid acinar cells a massive burst of exocytosis with almost complete degranulation can be induced by stimulation with isoproterenol. As a consequence, the stacked Golgi cisternae come into close proximity to the apical plasmalemma — i.e., as close as ~0.5 μm (Herzog and Farquhar 1977). Under normal conditions, however, the surface area of the compartments participating in the secretory process including the plasma membrane remain remarkably constant (Palade 1975; for stereological estimations of the corresponding surface areas see Table 1).

Table 1. Relative cytoplasmic volumes and membrane surface areas of secretory compartments in resting guinea pig pancreatic exocrine cells. (From Jamieson and Palade 1977; with permission of Rockefeller University Press)

Compartment	Relative cytoplasmic volume	Membrane surface area
	%	μm^2/cell
RER	20	8,000
Golgi complex	8	1,300
Condensing vacuoles	2	150
Secretory granules	20	900
Apical plasmalemma		30
Basolateral plasmalemma		600

Data compiled from Bolender 1974, and Amsterdam and Jamieson 1974.

Percent relative to cytoplasmic volume exclusive of the nucleus.

The Golgi Complex as a Station in the Intracellular Route of Secretory Proteins

The principal steps of the secretory route through the cell and the participation of the Golgi complex as a station in this pathway appear now well established (Palade 1975; Jamieson and Palade 1977). Detailed

knowledge of the pathway of secretory proteins through the Golgi complex is still missing and only scant evidence for the possibility of
sequential movement from cis- to trans-Golgi elements is available.

Results from Cytochemical Studies

The cytochemical or immunocytochemical localization of secretory proteins and of membrane proteins does not allow conclusions on the kinetics of their intracellular route, but the compartments participating
in the secretory process can be identified with high spatial resolution. Among the cytochemical techniques, the 3,3'-diaminobenzidine
(DAB)-procedure (Graham and Karnovsky 1966) and its numerous modifications for electron microscope cytochemical visualization of secretory
hemoproteins such as lacrimal or salivary gland peroxidases have been
particularly successful (for recent review see Herzog 1979). Secretory
peroxidases of several cell types have been shown to be present in
stacked Golgi cisternae, suggesting their involvement in transport and
in processing of this enzyme (Figs. 1 and 3). Transfer of secretory
peroxidase may be mediated by small peripheral vesicles which appear
to bud off from transitional elements and seem to follow two routes,
(1) to the cis-Golgi-cisterna and (2) directly to condensing vacuoles
on the trans-side of the Golgi complex (Fig. 3). In rat parotid and
lacrimal gland acinar cells (Herzog and Miller 1970, 1972) and in eosinophil leukocytes (Miller and Herzog 1969; Bainton and Farquhar 1970),
peroxidase was present in all Golgi cisternae with the reaction product occasionally forming a gradient of increasing density from the cis-
to the trans-side of the Golgi stacks. A similar gradient along the
cis-trans-axis of stacked Golgi cisternae has also been shown for glycoproteins (Leblond and Bennett 1977). This could imply sequential
concentration while the secretory product is moved from the cis- to
the trans-Golgi elements.

Sequential drainage from RER and from Golgi cisternae occurs when synthesis of peroxidase ceases. Thus, peroxidase disappears with increasing maturation of eosinophil leukocytes first from the RER and subsequently from Golgi cisternae (Miller and Herzog 1969; Bainton and
Farquhar 1970).

Immunocytochemical studies clearly demonstrated the participation of
all Golgi cisternae in the intracellular transport of secretory proteins
in exocrine pancreas (Kraehenbuhl et al. 1977; Geuze et al. 1979) and
of immunoglobulins in plasma cells and myeloma cells (Leduc et al. 1968;
Ottosen et al. 1980) independent of the degree of glycosylation (Kraehenbuhl et al. 1977).

Results from Kinetic Studies

Studies on the kinetics of the intracellular transport of secretory
proteins involve autoradiographic and cell fractionation techniques.
Both approaches have been effectively used either separately or in
conjunction for tracking the general intracellular secretory pathway.
Details of the route within the Golgi complex, however, have not been
clarified yet by either technique.

Autoradiography

The site of synthesis and the intracellular pathway of secretory proteins are analyzed by the use of precursors labeled with radioactive
isotopes. Tissue sections coated with photographic emulsion reveal

124

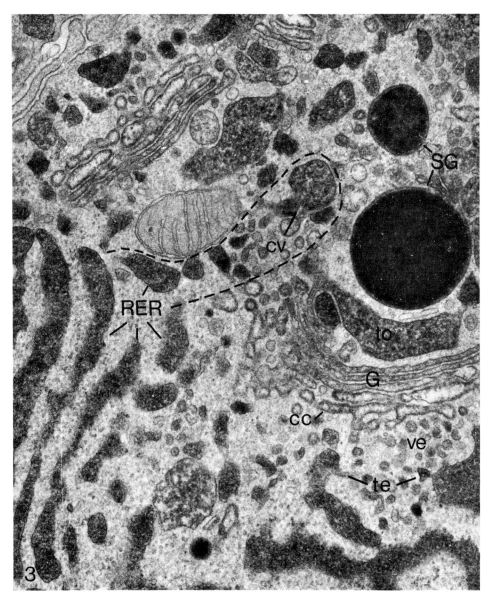

Fig. 3. Golgi complex of an acinar cell from rat parotid gland after cytochemical visualization of secretory peroxidase. Reaction product is visible in cisternae of RER including the transitional elements (*te*) from which peripheral Golgi vesicles (*ve*) carrying peroxidase appear to bud off and to move to the cis-cisterna (*cc*) of the Golgi stack (*G*) or directly to condensing vacuoles (*cv*) on the trans-side of the stack (*dotted line*). Note the gradient of peroxidase density increasing from the cis- to the trans-Golgi cisterna (*tc*). Condensation continues in condensing vacuoles (see also Table 2) and highest peroxidase concentration is seen in mature secretion granules (*SG*). (Herzog and Miller 1970; with permission of Springer Verlag). x 53,000

125

silver grains over radiolabeled cell components with an optimal resolution (half distance as determined with a line source) of ~150 nm
(for review on the technique see Salpeter and Bachmann 1972). Because
of the limited resolution, the kinetics of transport within the Golgi
complex are still contradictory. Following pulse labeling of guinea
pig exocrine pancreas slices with L-[³H]leucine, Jamieson and Palade
(1967a,b) found no evidence for the involvement of stacked Golgi cisternae in the transport of the radiolabeled secretory proteins. In
other cell types, such as parotid gland (Castle et al. 1972), pituitary
prolactin cells (Hopkins and Farquhar 1973), or lacrimal gland cells
(Fig. 2) a wave-like movement of pulse labeled secretory proteins from
RER via Golgi complex to secretion granules was observed (Fig. 4).
Neutra and Leblond (1966) have demonstrated that incorporation of glucose and of galactose into intestinal mucin occurs first in stacked
Golgi cisternae of goblet cells. In thyroid follicle cells, [³H]galac-
tose was shown by Whur et al. (1969) to be incorporated into thyroglobulin within Golgi cisternae (Fig. 7). This is in agreement with the
proposed structure of the large side chain of thyroglobulin (Toyoshima
et al. 1972) which carries galactose in its terminal regions (Fig. 8).
Studies of the intracellular transport of glycoproteins in goblet cells
show that incorporated terminal sugars move sequentially from the cis-
face towards the trans-face through the stacked Golgi cisternae. It
has been estimated that every 2-4 min a Golgi cisterna on the trans-
side is transformed into a mucin granule, thus implying continuous re-
newal of Golgi stacks (Neutra and Leblond 1966).

Cell Fractionation

In the analysis of the secretory process, cell fractionation procedures
have been used either alone or in conjunction with autoradiography
after pulse-labeling of exportable proteins. A major advantage is the
possibility for chemical analysis of secretory proteins and of the
membrane composition in tissue fractions. Some disadvantages are, however, that the kinetics obtained from tissue fractions are less sharply
defined and that usually a considerable amount of secretory proteins
is recovered in the postmicrosomal supernatant presumably due to leakage from compartments damaged by the homogenization procedure. A consequence of this artifact may be readsorption of leaked secretory proteins to the membranes of isolated compartments (Scheele et al. 1978).

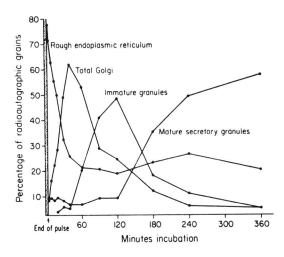

Fig. 4. Wavelike movement of pulse-labeled secretory protein through the intracellular compartments in rabbit parotid acinar cells. Summation of percentages at each time point shows that 75%-80% of the incorporated label moves with the wave through the compartments along the secretory pathway including the Golgi complex. (From Castle et al. 1972; with permission of Rockefeller University Press)

In all fractionation techniques, the homogenization step is critical because the Golgi complex is easily disrupted. The major techniques are as follows: (1) Extensive homogenization of the Golgi complex yields ribosome-free vesicles which sediment with the microsome fraction. In tissues which possess only minute amounts of smooth ER (e.g., exocrine pancreas, parotid or lacrimal gland), crude microsome fractions can be separated into a rough microsomal fraction and into smooth microsomes which consist mainly of ruptured Golgi elements (Tartakoff and Jamieson 1974). (2) Gentle (Polytron-) homogenization and the application of a tissue-specific combination of centrifugal forces and time periods necessary for sedimentation, may provide Golgi fractions in which the stacked nature of Golgi cisternae is still preserved (Morré 1971). (3) Modification of the normal density of Golgi elements by the presence of very low-density lipoproteins (VLDL) in livers of ethanol-treated rats (Ehrenreich et al. 1973).

In cell fractions, the morphological identification of components as Golgi elements is based on negative staining (Morré et al. 1970), on osmium impregnation of isolated vesicles (Fleischer et al. 1969), on the preservation of the stacked nature of Golgi cisternae (Morré et al. 1970) or on the presence of VLDL (Ehrenreich et al. 1973).

Biochemical identification of Golgi fractions depends on specific markers. The concept of markers is based on the premise that one or several constituents are unique and restricted to a morphologically distinguishable cell component (De Duve 1975). There is no definite marker restricted to elements of the Golgi complex. The reason for the absence of specific Golgi markers may be related to interactions of this organelle with the RER, the plasma membrane, and with lysosomes. Consequently, markers thought to be specific for RER may also occur in the Golgi fraction. For example, glucose-6-phosphatase and NADPH-cytochrome c reductase were long thought to be restricted to the endoplasmic reticulum; recent observations suggest, however, that both enzymes are also indigenous components of Golgi membranes (Howell et al. 1978). Although specific Golgi markers are unknown, several proteins such as glycolipid glycosyltransferases (Keenan et al. 1974), glycoprotein glycosyltransferases (Schachter 1974; Bretz et al. 1980) or thiamine pyrophosphatase (Cheetham et al. 1971) appear enriched in Golgi fractions.

Presence of integral membrane proteins in several membrane fractions does not necessarily imply that these constituents are synthesized on membrane-bound ribosomes and transferred from one compartment to the other due to interactions during the secretory process. For example, cytochrome b-5 is an integral membrane protein present in several organelles of rat liver. Apparently, this protein is exclusively synthesized on free ribosomes and inserted into the plasma membrane and into membranes of RER, Golgi complex, and mitochondria (Rachubinski et al. 1980).

Numerous studies have demonstrated the usefulness of cell fractionation procedures for isolation of Golgi elements and contributed to the knowledge about the general role of Golgi cisternae in the intracellular transport of exportable proteins. However, only little additional information on the possible sequential movement of secretory proteins across the Golgi complex is available. The technique of Ehrenreich et al. (1973) resolves the Golgi complex of rat hepatocytes into three successive fractions: Heavy cis- (GF_3), intermediate-(GF_2) and light trans-(GF_1) Golgi elements. Using this technique, Bergeron et al. (1978) observed that a peak of L-[^{3}H] leucine-labeled hepatic secretory proteins reaches first heavy Golgi fractions (GF_3) and 4 or 7 min later the intermediate (GF_2) or light (GF_1) Golgi fractions, respectively. This would suggest that secretory proteins move indeed sequentially from the cis- to the trans-Golgi face before being discharged by the cell.

Modifications of Secretory Proteins during Their Transit through the Golgi Complex

Whereas the detailed traffic of secretory proteins through the Golgi complex is still enigmatic, cytochemical, cell fractionation and autoradiographic studies have contributed to the knowledge on the condensation and on the chemical transformations of secretory proteins which take place in part in the Golgi complex.

Condensation of Secretory Products

In "nonregulated" cell types (Tartakoff and Vassalli 1978) such as plasma cells, secretory products appear to be packaged in dilute solution into small vesicles with no or only little concentration and to be released continually by exocytosis without previous intracellular storage. In other cells with periodically occurring exocytosis, e.g., after appropriate stimulation of a resting cell, secretory products are concentrated and stored in morphologically recognizable secretion granules until needed. Here, secretory proteins which arrive in the Golgi complex in dilute solution may be concentrated by a large factor up to 50-150 fold (Palade 1975; Jamieson and Palade 1977; Farquhar et al. 1978). Concentration may begin already in stacked Golgi cisternae as revealed morphologically (Farquhar 1971) or by cytochemical (for recent review see Herzog 1979) and immunocytochemical (Geuze et al. 1979) techniques. Concentration continues in condensing vacuoles which arise at the trans-face of the Golgi complex. The site of concentration within the Golgi complex may vary considerably according to the cell-type, the species and the functional state of the cell (see Table 2).

Table 2. Sites of concentration of secretory proteins in the Golgi complex

Exocrine Cell Type	Species	Functional State of the Cells	Golgi Cisternae[a]	Condensing Vacuoles	References
Exocrine pancreas	Guinea pig	Unstimulated	−	+	Jamieson and Palade 1967a,b
Exocrine pancreas	Guinea pig	Hyperstimulated	+	−[b]	Jamieson and Palade 1971b
Exocrine pancreas	Rat	Unstimulated	+	+	Geuze et al. 1979
Exocrine pancreas	Rat	Hyperstimulated	+	−[b]	Kern et al. 1979
Parotid gland	Rat	Unstimulated	+	+	Herzog and Miller 1970; Castle et al. 1972

[a]Concentration occurs preferentially within the trans-Golgi-cisternae

[b]in hyperstimulated cells, the usual condensing vacuoles are absent

Sites of concentration (+) within the Golgi complex may vary according to the cell type, the species and the functional state

128

As shown by Jamieson and Palade (1971a,b) concentration does not require energy and is independent of protein synthesis. The authors postulated that concentration results from a progressive reduction in the osmotic activity by ionic interaction among secretory macromolecules within condensing vacuoles. This reduction of osmotic activity may lead to loss of water through the granule membrane into the surrounding cyto- plasm and could be caused by the presence of a sulfated polyanion, pre- sumably a sulfated peptidoglycan (Reggio and Palade 1978). In vitro observations indicate that sulfated polyanions (chondroitin sulfate A.B.C) may form large aggregates with cationic pancreatic proteins such as chymotrypsinogen A (Fig. 6), a process which requires Ca^{2+} and is pH-dependent (Reggio and Dagorn 1978). Concentration of secretory pro- teins may be linked, therefore, in part to the metabolism of sulfate and of Ca^{2+} in secretory cells.

In vivo incorporation of $^{35}SO_4^{2-}$ into unidentified macromolecules has been demonstrated by autoradiography in the Golgi complex of a wide variety of cell types (Young 1973) and in Golgi fractions of rat liver (Katona 1976). In pancreatic acinar cells of mouse (Berg and Young 1971)

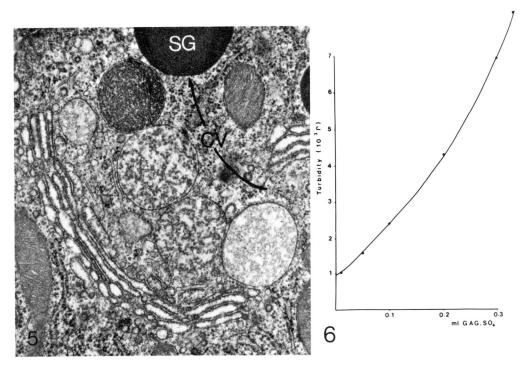

Figs. 5 and 6. Concentration of secretory proteins in situ (Fig. 5) and in vitro (Fig. 6)

Fig. 5. Golgi complex from a rat parotid acinar cell showing increasing density of secretory content in condensing vacuoles (cv, arrow) with the highest concentration in mature secretion granules (SG). x 27,000

Fig. 6. Aggregate formation between chymotrypsinogen and different amounts of sul- fated glucosaminoglycan isolated from guinea pig secretion granules. (Reggio and Dagorn 1978; with permission of Rockefeller University Press)

and guinea pig (Tartakoff et al., 1975), the incorporated $^{35}SO_4^{2-}$ has been recovered in the discharged secretory product. In all cases, the Golgi complex has been shown to be the major site of initial sulfate incorporation.

The possible intracellular function of Ca^{2+} in the packaging of exportable proteins has been investigated in several secretory cells including rat parotid gland (Wallach and Schramm 1971) and guinea pig exocrine pancreas (Clemente and Meldolesi 1975). In the exocrine pancreas, Ca^{2+} was mainly associated with the secretory proteins; in part it appeared also bound to the inner surface of the secretion granule membrane and to the outer surface of the plasma membrane. The initial site where Ca^{2+} becomes associated with secretory proteins may be the Golgi complex including the condensing vacuoles (Clemente and Meldolesi 1975). It is known that several secretory proteins have Ca^{2+}-binding sites; for example, amylase contains one atom of calcium per mol of enzyme (Stein et al. 1964). In addition, the calcium in the granule matrix could function in aggregate formation during the condensation process. Finally, it has been speculated that calcium associated with the inner surface of the granule membrane may play a role in anchoring the granule content to the membrane by the formation of Ca^{2+}-bridges (Clemente and Meldolesi 1975). Sulfated polyanions seem to be, at least in part, also membrane-associated (Reggio and Palade 1978) and could constitute, therefore, part of the membrane-associated binding sites for Ca^{2+}.

Proteolytic Processing of Secretory Proteins

Many peptide hormones are synthesized as larger precursors which undergo proteolytic processing during their intracellular transport from RER to secretion granules (Patzelt et al. 1978). The signal peptide of nascent, uncompleted presecretory proteins which functions in the vectorial transfer from the large ribosomal subunit across the membrane of the RER is cleaved proteolytically in the cisternal space of RER (Blobel and Dobberstein 1975). The resulting prosecretory peptides appear to be further processed by a special group of membrane-bound proteases located in the Golgi complex. The best-documented system are the beta-cells in the pancreatic islets (Steiner et al. 1972). The conversion of proinsulin to insulin requires its transport to the Golgi apparatus where the cleaving enzymes are first brought in contact with proinsulin. It has been speculated that the connecting peptide segment located between the A- and the B-chain of proinsulin functions as a spacer, so that during segregation the entire peptide can span the distance from the site of its synthesis between the large and the small ribosomal subunit and the cisternal face of the RER membrane. Hence, many other small secreted peptides with less than 50 amino acid residues would require a similar "spacer" which after segregation into the cisternae of RER may be cleaved in the Golgi complex (for recent review see Tager et al. 1979).

Terminal Glycosylation

The detection of carbohydrates in elements of the Golgi complex, as revealed by the periodic acid-silver methenamine technique, has led to the concept that this organelle plays a major role in the addition of carbohydrate residues to glycoproteins which are constituents either of the secretory product or of the plasma membrane (Rambourg et al. 1969). Immunocytochemical studies on the localization of various secretory proteins in bovine exocrine pancreas have indicated that the passage through the Golgi complex represents an obligatory route for both glycosylated and nonglycosylated secretory proteins in which all Golgi cisternae participate (Kraehenbuhl et al. 1977). In the thyroid gland, it appears established that mannose and N-acetylglucosamine (GlNAc) are incorporated in the RER. In contrast, galactose and sialic acid, the terminal sugars of thyroglobulin (Toyoshima et al. 1972; see Fig. 8), are added in the Golgi cisternae (Fig. 7, for recent review see Leblond and Bennet 1977). It has been suggested that the sugars are added sequentially in the different successive cisternae beginning at the cis-face and ending at the trans-side of the stacked Golgi cisternae (Matsukawa and Hosoya 1979). However, localization of the corresponding glycosyltransferases in separate successive Golgi cisternae has not been demonstrated in this gland. In hepatocytes of ethanol-treated rats, a structurally characteristic series of Golgi fractions deriving from cis- (GF_3), intermediate (GF_2) and trans- (GF_1) Golgi elements can be isolated (Ehrenreich et al. 1973). Studies on the localization of three glycosyltransferases, i.e., sialyltransferase, galactosyltransferase, and GlNAc-transferase, indicate that their concentrations increase in the cis-trans-direction but that none of the three enzymes is enriched over the others in any of the three successive Golgi fractions in rat liver (Bretz et al. 1980). The sequential character of terminal glycosylation (Schachter 1974) can, therefore, not yet be explained by a serial localization of different glycosyltransferases in successive Golgi cisternae.

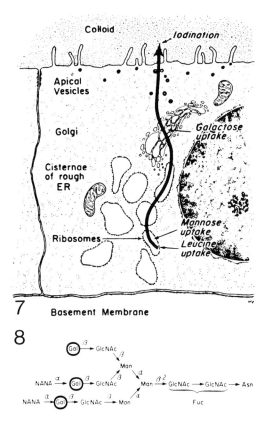

7

8

Figs. 7 and 8. Terminal glycosylation of thyroglobulin in the Golgi complex of thyroid follicle cells. Incorporation of galactose occurs within the stacked Golgi cisternae before packaging of thyroglobulin into secretion granules (Fig. 7). The terminal position of galactose in the large side chain of thyroglobulin is shown in Fig. 8. (Whur et al. 1969 and Toyoshima et al. 1972; with permission of Rockefeller University Press and American Chemical Society, respectively)

Routing of Proteins to Different Intracellular Compartments

A major problem in the understanding of the secretory process is the
question how eukaryotic cells separate the intracellular pathway of
lysosomal enzymes from the route of exportable proteins. A typical example is the thyroid follicle cell which synthesizes lysosomal hydrolases and thyroglobulin, both of which are transported through the
Golgi complex for terminal glycosylation. Subsequently, lysosomal enzymes and thyroglobulin have to be segregated from each other to avoid
precocious proteolysis which normally takes place only after re-entry
of thyroglobulin into the cells by endocytosis and after its transfer
to lysosomes (see Fig. 16). It is unknown at present how and in which
intracellular compartment this segregation occurs.

Cytochemical work of Novikoff and coworkers (1976, 1977) suggests that
the segregation of lysosomal enzymes and lysosome formation occur in
GERL which is considered a special cisterna on the trans-side of the
Golgi complex and thought to be in continuity with and, therefore,
part of the endoplasmic reticulum. Hence, the sorting of lysosomal enzymes from proteins destined for export could be handled by distinct
regions of the Golgi complex (for critical evaluation of the GERL concept see Farquhar 1978a).

Heterotopic packaging of proteins occurs in polymorphonuclear leukocytes, which form (at least) two different granule populations of distinct enzyme composition, (1) azurophil granules which are primary
lysosomes and (2) specific granules which do not contain lysosomal
hydrolases (Bainton and Farquhar 1966, 1968). Here, the problem of
sorting is solved by separating granule formation in time and in space:
The azurophil granules are formed in the trans-cisternae of the Golgi
apparatus during the promyelocyte stage, whereas the specific granules
arise from the cis-cisterna during the myelocyte stage.

An interesting hypothesis has been proposed by Sly and Stahl (1978) for
routing lysosomal enzymes in fibroblasts: According to Hickman and Neufeld (1972) lysosomal enzymes share a common recognition marker which
proved to be in some cases a 6-phosphomannosyl-residue (Kaplan et al.
1977). This recognition marker may play a role in segregating acid
hydrolases — which are normally not secreted — from exportable proteins
and may route enzymes to lysosomes by binding to a specific receptor
on a membrane component destined to be incorporated into lysosomal
membranes (Sly and Stahl 1978).

Finally, Blobel et al. (1979) proposed another mechanism by which specific "sorting" sequences within the polypeptide chain would guide proteins to their appropriate compartment. This specific sequence would
be cleaved upon completion of the sorting process.

Membrane Interactions of the Golgi Complex during the Secretory Process

After their co-translational segregation across the membrane into the
cisternal space of RER, exportable proteins remain membrane-bounded
throughout the secretory pathway until their release by exocytosis.
Accordingly, intracellular transport of secretory proteins involves
considerable membrane interactions between the compartments participating in secretion.

It has been assumed that the segregating membrane migrates together
with secretory proteins from RER to the Golgi complex and from there
to the plasmalemma. The idea of membrane movement, either as bulk mig-

ration of membrane or membrane-domains or in the form of selective
traffic of single membrane proteins, is included in the hypothesis of
"membrane flow and differentiation" (Morré 1977; for recent review see
Morré et al. 1979). Early kinetic experiments which suggested that membrane proteins may be synthesized and subsequently travel down the secretory pathway at a rate comparable to that of secretory proteins
(Evans and Gurd 1971; Franke et al. 1971) were in support of this hypothesis. However, proteins of membranes participating in the intracellular transport and in the discharge of secretory proteins turn over
at rates much slower than the proteins destined for export (Winkler et
al. 1972; Meldolesi 1974; Wallach et al. 1975; for reviews see also
Siekevitz 1972; Schimke 1975; Tweto and Doyle 1977). From these observations, it was concluded that the membranes are not destroyed after
fusion but recycled within the cell (Palade 1975). Because each compartment retains a characteristic membrane composition, traffic of
membranes through the Golgi complex to the cell surface and retrieval
of membranes during endocytosis must avoid random mixing of the interacting membranes and operate selectively with regard to some and re-
strictively with regard to other membrane constituents (Meldolesi et
al. 1978).

Membrane Interactions of the Golgi Complex Concomitant with the Transfer of Exportable Proteins

Sufficient support is available to conclude that the Golgi complex
functions in the vectorial transfer of secretory proteins from RER to
secretion granules. This activity involves membrane interactions of
the Golgi complex with membranes of the RER and of secretion granules.
Indirectly, studies on the intracellular transport of secretory proteins performed first in the exocrine pancreas and subsequently in a
variety of other cell types indicate that membrane interactions between the endoplasmic reticulum and the Golgi complex and the concomitant transfer of secretory proteins require energy supplied by oxidative phosphosylation because in the absence of ATP the secretory proteins remain in the RER (Jamieson and Palade 1971a; Howell and Whitfield 1973; Kruse and Bornstein 1975; Chu et al. 1977; Tartakoff and
Vasalli 1978). The energy-requiring steps during the interactions between RER and Golgi membranes are unknown. However, it has been shown
that clathrin-like proteins associate with the membranes of peripheral
Golgi vesicles during protein transport from the RER to the Golgi complex and dissociate from the carrier membrane when ATP production is
blocked (Palade and Fletcher 1977). When the energy-requiring lock is
open again, secretory proteins flow vectorially to the Golgi complex.
No such energy lock has been demonstrated during granule formation on
the trans-side of the Golgi complex. Once the Golgi complex is reached
by secretory proteins, further condensation and transformation into
mature secretion granules continues even in the absence of energy.

There is only little direct information available on the flow of membranes through the Golgi complex.

1. Thyroid peroxidase which is an integral membrane protein (Rawitch
   et al. 1979) may be transported through the Golgi complex because
   it has been demonstrated cytochemically in the endoplasmic reticulum, in one or two of the stacked cisternae on the trans-side of
   the Golgi complex, in vesicles located in the apical portion of the
   cell and in domains of the luminal plasma membrane (Tice and Wollman 1974). It is assumed that peroxidase reaches the luminal plasmalemma by fusion of thyroperoxidase containing vesicles (Tice and

Wollman 1974). However, the kinetics of the intracellular transport of this membrane protein are unknown.

2. Plasma membrane glycoproteins become first labeled 10 min after a pulse with [³H]-fucose in columnar cells of the large intestine. The wave reached vesicles carrying the radiolabeled glycoprotein after 30 min and the luminal plasma membrane after 4 h (Figs. 9 and 10; Michaels and Leblond 1976; Leblond and Bennett 1977). This indicates that the Golgi complex functions also in the synthesis (glycosylation) of plasmalemmal constituents which may be subsequently transferred via carrier-vesicles to the apical cell surface where they are inserted by a process of membrane-fusion and -fission similar to exocytosis.

3. The spike protein of vesicular stomatitis virus is synthesized and subsequently glycosylated as a transmembrane protein in the RER (Katz et al. 1977). Terminal glycosylation occurs, later, in the smooth membrane fraction possibly derived from Golgi elements; thereafter, the glycosylated spike protein is inserted into the plasma membrane (Rothman and Lodish 1977).

Further indications for involvement of Golgi cisternae in plasma membrane biogenesis have been deduced from observations on the presence of adenylate cyclase (Cheng and Farquhar 1976) and of the insulin re-

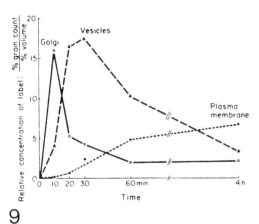

9

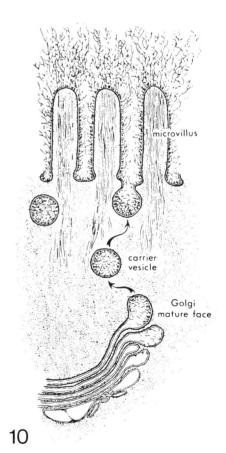

10

Figs. 9 and 10. Intracellular pathway of [³H] fucose-labeled cell coat in epithelial cells of mouse ascending colon. Carrier vesicles arise from distended portions of Golgi cisternae (Fig. 10) where the inital peak of radioactivity is found (Fig. 9). Subsequently, the wave of radioactivity is found over vesicles which migrate to the cell apex. Here, their membrane carrying the cell coat is inserted into the luminal plasmalemma (Fig. 10) which is reached by the wave of radioactivity by ~4 h (Fig. 9). (Leblond and Bennett 1977 and Michaels and Leblond 1976; with permission of Rockefeller University Press and Société Française de Microscopie Électronique, respectively)

ceptor (Bergeron et al. 1973) in Golgi fractions. However, the mere
presence of plasmalemmal constituents in elements of the Golgi complex
as revealed by cytochemical localization or by biochemical analysis of
tissue fractions does not distinguish between constituents that are
newly synthesized and those that are retrieved from the plasma membrane
by endocytosis.

Involvement of the Golgi Complex during Endocytosis ("Membrane Recycling") in Secretory Cells

As discussed in the preceding Sections, the Golgi complex is the receiving compartment for peripheral vesicles deriving from transitional
elements of the RER (proximal membrane interactions) and the donating
compartment during the formation of secretion granules to which in
turn the luminal cell surface is the receiving membrane (distal membrane interaction). It has been postulated that membranes translocated
during the secretory process are specifically removed from their receiving compartments and reutilized during the intracellular transport
of secretory proteins (Palade 1975; Jamieson and Palade 1977). However,
only the distal membrane interaction (Golgi→secretion granules→cell
surface) has been studied in this respect because during exocytosis
the granule membrane deriving in part from the Golgi complex is inserted into the luminal plasmalemma and, therefore, accessible to suitable
tracers.

Both morphological and biochemical observations have led to the postulate of re-utilization of granule membrane following exocytosis:

1. During exocytosis portions of the secretion granule membrane are
   inserted into the plasmalemma. For example, in the guinea pig exocrine pancreas cell, about 900 μ^2 of granule membrane gain continuity with the apical plasmalemma with a surface area of about 30 μ^2
   (see Table 1; Jamieson and Palade 1977). Under normal conditions,
   however, the surface area of the luminal membrane remains surprisingly constant (Palade 1959).

2. Following massive stimulation of rat parotid gland, the luminal surface area is temporarily increased followed by restoration to its
   normal size (Amsterdam et al. 1969; Cope and Williams 1973). Con-
   comitantly, smooth-surfaced apical vesicles presumably carrying a
   certain amount of luminal plasmalemma appear during the phase of
   restoration (Amsterdam et al. 1969).

3. Biochemical observations have clearly shown that the granule membrane proteins in bovine adrenal medulla (Winkler et al. 1972), in
   guinea pig exocrine pancreas (Meldolesi 1974), and in rat parotid
   gland (Wallach et al. 1975) are synthesized at a rate about tenfold
   slower than that of the exportable proteins.

Until recently, it has been unknown whether intact membrane fragments
are internalized as originally proposed by Palade (1959) or whether
membranes are degraded and reutilized indirectly as molecular consti-
uents (Hokin 1968). Membrane retrieval has been studied previously in
a variety of cell systems; however, the tracers were observed exclusively in lysosomes (Masur et al. 1972; Abrahams and Holtzman 1973;
Holtzman et al. 1973; Geuze and Kramer 1974; Kalina and Rabinovitch
1975; Oliver and Hand 1977). Our approach to this problem has been to
expose selectively the apical, the basolateral or the entire plasmalemma of different secretory cells to suitable electron-opaque tracers.
The tracers used were tolerated well by the cells and were either mar-

kers of the content because they do not bind to the plasmalemma or
"membrane markers"[1] because of their mild charge interaction with the
plasma membrane.

All tracers are internalized by coated pits regardless of whether the
tracer binds to the plasma membrane. Both neutral dextran (as a content
marker) and cationized ferritin (as a membrane marker) are found initially in coated pits (insets in Figs. 11 and 14). In acinar cells of
the rat parotid and lacrimal gland (Herzog and Farquhar 1977), in exocrine pancreas cells (Herzog and Reggio 1980), and in thyroid follicle
cells (Herzog and Miller 1979) the internalization of the tracer in
coated pits occurs preferentially along the luminal plasmalemma, whereas the participation of the basolateral plasmalemma appears negligible.
At least part of the coat is shed following vesicle formation because
most of the tracer-carrying apical vesicles are smooth-surfaced (Fig.
11). The vesicles are rapidly transported to various compartments. In
lacrimal and parotid glands and the exocrine pancreas (Figs. 11 and 12),
the stacked Golgi cisternae are reached first by the tracer already
2 min after application (Table 3). In the exocrine pancreas, condensing
vacuoles are reached at the same time, thus pointing to the functional
unity of both compartments (Fig. 12) (for structural similarities between both compartments see De Camilli et al. 1976). Later, at 15 min,
mature secretion granules contain the tracer (Fig. 12, inset) indicating indirectly that some luminal plasma membrane retrieved by endocytosis is reutilized for the formation of secretion granule membrane.
In addition, luminal content which normally contains discharged secretory products is packaged together with newly synthesized exportable
proteins indicating that "membrane recycling" may not be limited to
the re-use of membrane but may include also re-uptake of small quantities of luminal content. Lysosomes are reached later, if at all, at
30 min, by the tracer. Hence, in exocrine gland cells the main route
of incoming vesicles leads directly to compartments along the secretory pathway.

This is in contrast to the vesicle traffic in thyroid follicle cells
(Figs. 13-15). Here, regardless of the charge (Table 3), lysosomes are
reached first by the tracer (within 5 min) which accumulates there with
time (Fig. 14, inset). Later, at 30 min, cationized ferritin is found
also in stacked Golgi cisternae preferentially at their lateral rims,
thus indicating that portions of luminal plasma membrane retrieved
during endocytosis are inserted into stacked Golgi cisternae. Native
(anionic) ferritin which has a pI of 4.7 similar to thyroglobulin and
which does not bind to the plasma membrane remains within the lysosomal
matrix. These results with tracers mirror thyroid function when thyroglobulin is taken up from the lumen for lysosomal degradation and lib-
eration of thyroid hormones. The membrane of the vesicles, however,
which normally carry thyroglobulin, may detach from the lysosomes and
move to the stacked Golgi cisternae for its possible re-utilization
during packaging of thyroglobulin (Fig. 16) (Herzog and Miller 1979).

Traffic of internalized plasma membrane to Golgi cisternae has been
observed in a number of other secretory cell types (Farquhar 1978b;
Gonatas et al. 1977; Ottosen et al. 1980; Orci et al. 1978). Apparently,

---

[1]Cationized ferritin is a reliable marker for anionic binding sites on the plasma
membrane. During internalization, however, it may detach from the membrane if other
anionic binding sites become available (for example, in the lysosomal matrix; see
Fig. 14). At present, stable (covalently linked) and non-perturbing membrane markers
which would not detach from the membrane are not available.

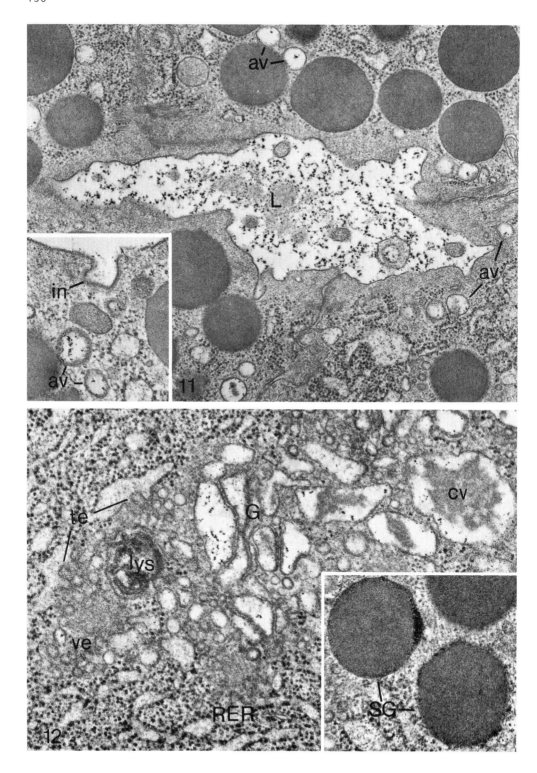

endocytic vesicles do not fuse with cisternae of RER, thus underlining the specificity of the membrane interactions. Furthermore, tracers which occupy the stacked cisternae of the Golgi complex are not transferred further across the interface between RER and Golgi cisternae because they are never found in transitional elements (see Fig. 12). From this observation it cannot be deduced, however, that a recycling of peripheral Golgi vesicles back to transitional elements does not occur.

Little is known concerning the factors which may regulate the traffic of endocytic vesicles (see Table 3). The net charge of the tracers may determine their intracellular route (Farquhar 1978b; Herzog and Miller 1979). Other factors may result from the specific function of a given cell type (compare, for example, the exocrine pancreas which is oriented towards the export of secretory proteins with the thyroid in which the lysosomal digestion of internalized thyroglobulin dominates; see Table 3).

Table 3. Pathways of endocytosis in secretory cells

Time points at which compartments are reached by endocytic vesicles						
Cell type	Tracer	GC	CV	SG	Lysosomes	References
Rat parotid	Dextran	5	60	?	30[a]	Herzog and Farquhar 1977
Rat exocrine pancreas	Dextran HRP	2 –	2 –	15 –	– 5	Herzog and Reggio 1980
Pig thyroid	NF CF	– 30	– ?	– ?	5 5	Herzog and Miller 1979

Abbreviations: GC, Golgi cisternae; CV, condensing vacuoles; SG, mature secretion granules; HRP, horseradish peroxidase; NF, native ferritin; CF, cationized ferritin.

Pathways of endocytosis in secretory cells may vary according to the cell type and the nature (composition and charge) of the tracer (compare rat exocrine pancreas with pig thyroid). In rat parotid gland, the transfer of the tracer to lysosomes was minor under in vivo conditions, but was increased in isolated acini incubated in vitro. This indicates that the physiologic state of the cell may also influence the traffic of endocytic vesicles ([a]).

Figs. 11 and 12. Studies on "Membrane Recycling" in rat exocrine pancreas using a content marker. Dextran infused into the pancreatic duct fills the acinar lumen (L) and is taken up by smooth surfaced apical vesicles (av, Fig. 11) which derive from coated pits (in) along the luminal plasma membrane (inset Fig. 11). 2-4 min after infusion, tracer particles become visible in Golgi cisternae (G) and condensing vacuoles (cv) but are absent from RER including transitional elements (te), peripheral Golgi vesicles (ve) and lysosomes (lys) (Fig. 12). After 15 min, dextran particles become also visible in mature secretion granules (SG, inset Fig. 12). Fig. 11, x 29,000; inset Fig. 11, x 41,000. Fig. 12, x 42,000; inset Fig. 12, x 42,000. (Herzog and Reggio, 1980; with permission of Wissenschaftliche Verlagsgesellschaft)

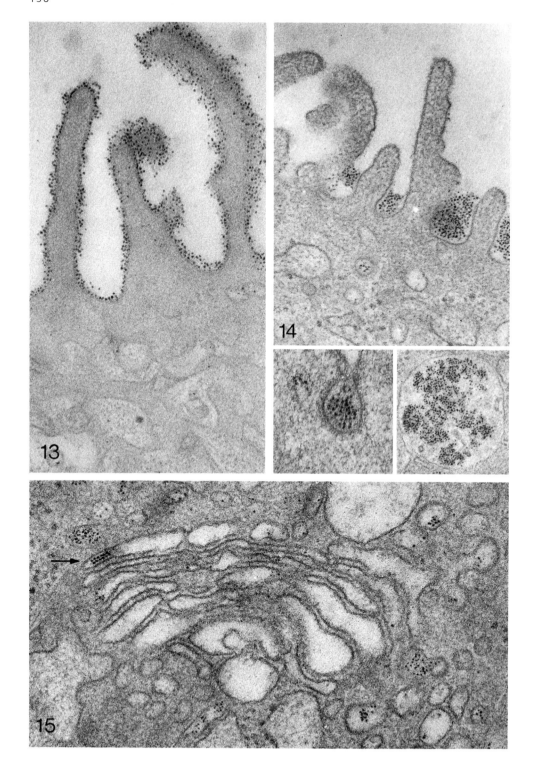

Figs. 13-15. Studies on "Membrane Recycling" in thyroid follicle cells using a membrane marker. Isolated and ruptured follicles are exposed to cationized ferritin which binds to the entire luminal plasma membrane (after glutaraldehyde fixation, Fig. 13. In unfixed follicles, cationized ferritin is found preferentially in coated pits (*left inset*, Fig. 14) located at the bases of microvilli (Fig. 14). Upon stimulation with TSH, vesicles are formed which carry cationized ferritin first to lysosomes (within 5 min) (*right inset*, Fig. 14). Later, at 30 min, cationized ferritin becomes also visible in stacked Golgi cisternae (*arrow*, Fig. 15) where the inserted membrane may be reutilized for granule formation. Fig. 13, x 80,000; Fig. 14, x 62,000; *left inset* Fig. 14, x 125,000; *right inset* Fig. 14, x 50,000; Fig. 15, x 70,000. (Herzog and Miller 1979; with permission of Wissenschaftliche Verlagsgesellschaft)

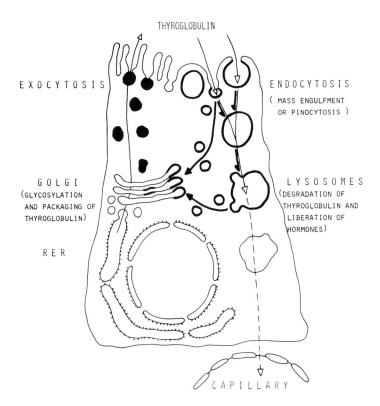

Fig. 16. Diagram showing the bidirectional transport of thyroglobulin (*thin arrows*) and the role of the Golgi complex in the concomitant membrane interactions (*thick arrows*). In the first phase, newly synthesized thyroglobulin is transported to secretion granules and released by exocytosis into the follicle lumen. During the second phase, thyroglobulin is internalized by mass engulfment or by pinocytosis and transferred to lysosomes where thyroid hormones are liberated from thyroglobulin and released into the bloodstream. The membranes retrieved during endocytosis are marked with thick contours. A direct route of membrane patches from the luminal surface to the Golgi cisternae as it is established in some exocrine glands (see also Table 3 and Figs. 11 and 12) cannot be excluded. The main pathway, however, leads to lysosomes from which part of the internalized membrane may pinch off and move forward to be inserted into the membranes of stacked Golgi cisternae (see also Figs. 13-15) for possible re-utilization during granule formation

In all gland cells studied, the Golgi complex appears to play an important role as a membrane reservoir (Palade 1959) which contains, at least in part, recycled plasma membrane. At present, we do not know whether the membrane inserted into Golgi cisternae is identical with the membrane removed from the luminal cell surface. Furthermore, it is unknown whether membrane patches inserted into Golgi membranes retain their specificity or intermix with the latter.

Summary

In polarized secretory epithelial cells, the Golgi apparatus is oriented with its cis- or entry-face towards the rough endoplasmic reticulum (RER) and with its trans- or exit-face towards the secretion granules. This cis-trans-polarity is expressed also by the structural and cytochemical heterogeneity of Golgi elements. The Golgi complex functions in the condensation and in chemical transformations of secretory proteins such as transfer of sulfate groups, partial proteolytic processing and terminal glycosylation. At all stages during the intracellular transport, secretory proteins are segregated by a membrane. Hence, membrane components and secretory content follow a common pathway entering the Golgi apparatus at its cis-face via small peripheral vesicles derived from transitional elements of the RER. Although postulated, it is still unknown whether peripheral vesicles return as empty containers to transitional elements. On the trans-face of the Golgi apparatus, condensing vacuoles are formed under participation of Golgi membranes. Tracer studies with a variety of cell types provide evidence that membranes of secretion granules inserted into the plasmalemma during exocytosis are retrieved by endocytosis and inserted into the membranes of stacked Golgi cisternae. This would indicate that the Golgi apparatus is a membrane reservoir consisting in part of recovered plasmalemma which may be reutilized for the formation of secretion granules. In addition, some lysosomal enzymes are transported through the Golgi apparatus which, therefore, may play a role also in the routing of membranes and of secretory content to various intracellular destinations.

Acknowledgements. I am grateful to Dr. F. Miller for discussions. I thank Ms. S. Fuchs for photographic work, and Ms. B. Laur and Ms. R. Schmittdiel for editorial help. Supported by Deutsche Forschungsgemeinschaft and by NATO.

References

Abrahams SJ, Holtzman E (1973) Secretion and endocytosis in insulin-stimulated rat adrenal medulla cells. J Cell Biol 56:540-558

Amsterdam A, Jamieson JD (1974) Studies on dispersed pancreatic exocrine cells. I. Dissociation technique and morphologic characteristics of separated cells. J Cell Biol 63:1037-1056

Amsterdam A, Ohad I, Schramm M (1969) Dynamic changes in the ultrastructure of the acinar cell of the rat parotid gland during the secretory cycle. J Cell Biol 41: 753-773

Bainton DF, Farquhar MG (1966) Origin of granules in polyleukocytes. Two types derived from opposite faces of the Golgi complex in developing granulocytes. J Cell Biol 28:277-301

Bainton DF, Farquhar MG (1968) Differences in enzyme content of azurophil and specific granules of polymorphonuclear leukocytes. II. Cytochemistry and electron microcscopy of bone marrow cells. J Cell Biol 39:299-317

Bainton DF, Farquhar MG (1970) Segregation and packaging of granule enzymes in eosinophil leukocytes. J Cell Biol 45:54-73

Berg NB, Young RW (1971) Sulfate metabolism in pancreatic acinar cells. J Cell Biol 50:469-483

Bergeron JJM, Evans WH, Geschwind II (1973) Insulin binding to rat liver Golgi fractions. J Cell Biol 59:771-775

Bergeron JJM, Borts D, Cruz J (1978) Passage of serum-destined proteins through the Golgi apparatus of rat liver. An examination of heavy and light Golgi fractions. J Cell Biol 76:87-97

Blobel G, Dobberstein B (1975) Transfer of proteins across membranes. I. Presence of proteolytically processed and unprocessed nascent immunoglobulin light chains on membrane-bound ribosomes of murine myeloma. J Cell Biol 67:835-851

Blobel G, Walter P, Chang CN, Goldman BM, Erickson AH, Lingappa VR (1979) Translocation of proteins across membranes: The signal hypothesis and beyond. In: Hopkins CR, Duncan CJ (eds) Secretory mechanisms. Cambridge University Press, Cambridge, pp 9-36

Bolender RP (1974) Stereological analysis of the guinea pig pancreas. I. Analytical model and quantitative description of nonstimulated pancreatic exocrine cells. J Cell Biol 61:269-287

Bretz R, Bretz H, Palade GE (1980) Distribution of terminal glycosyltransferases in hepatic Golgi fractions. J Cell Biol 84:87-101

Brown RM, Willison JHM (1977) Golgi apparatus and plasma membrane involvement in secretion and cell surface deposition, with special emphasis on cellulose biogenesis. In: Brinkley BR, Porter KR (eds) International cell biology. Rockefeller University Press, New York, pp 267-283

Camilli De P, Peluchetti D, Meldolesi J (1976) Dynamic changes of the luminal plasmalemma in stimulated parotid acinar cells. A freeze-fracture study. J Cell Biol 70:59-74

Castle JD, Jamieson JD, Palade GE (1972) Radioautographic analysis of the secretory process in the parotid acinar cell of the rabbit. J Cell Biol 53:290-311

Cheetham RD, Morré DJ, Friend DS (1971) Isolation of a Golgi apparatus - rich fraction from rat liver. J Cell Biol 49:899-905

Cheng H, Farquhar MG (1976) Presence of adenylate cyclase activity in Golgi and other fractions from rat liver. II. Cytochemical localization within Golgi and ER membranes. J Cell Biol 70:671-684

Chu LLH, MacGregor RR, Cohn DV (1977) Energy-dependent intracellular translocation of proparathormone. J Cell Biol 72:1-10

Claude A (1970) Growth and differentiation of cytoplasmic membranes in the course of lipoprotein granule synthesis in the hepatic cell. J Cell Biol 47:745-766

Clemente F, Meldolesi J (1975) Calcium and pancreatic secretion. I. Subcellular distribution of calcium and magnesium in the exocrine pancreas of the guinea pig. J Cell Biol 65:88-102

Cope GH, Williams MA (1973) Quantitative analysis of the constituent membranes of parotid acinar cells and of the changes evident after induced exocytosis. Z Zellforsch 145:311-330

Dalton AJ, Felix MD (1954) A study of the Golgi substance and ergoplasm in a series of mammalian cell types. Proc 8th Int Congr Cell Biol Leiden, The Netherlands, pp 274-293

Duve De C (1975) Exploring cells with a centrifuge. Science 189:186-194

Ehrenreich JH, Bergeron JM, Siekevitz P, Palade GE (1973) Golgi fractions prepared from rat liver homogenates. I. Isolation procedure and morphological characterization. J Cell Biol 59:45-72

Evans WH, Gurd JW (1971) Biosynthesis of liver membranes. Incorporation of 3H-leucine into proteins and of 14C-glucosamine into proteins and lipids of liver microsomal and plasma membranes. Biochem J 125:615-624

Farquhar MG (1971) Processing of secretory products by cells of the anterior pituitary gland. In: Heller H, Lederis K (eds) Subcellular organization and function in endocrine tissues. Cambridge University Press, Cambridge, pp 79-122

Farquhar MG (1978a) Traffic of products and membranes through the Golgi complex. In: Silverstein S (ed) Transport of macromolecules in cellular systems. Dahlem-Konferenzen, Berlin, pp 341-362

Farquhar MG (1978b) Recovery of surface membrane in anterior pituitary cells. Variations in traffic detected with anionic and cationic ferritin. J Cell Biol 78:R35-R42

142

Farquhar MG, Bergeron JJM, Palade GE (1974) Cytochemistry of Golgi fractions prepared from rat liver. J Cell Biol 60:8-25

Farquhar MG, Reid JJ, Daniel LW (1978) Intracellular transport and packaging of prolactin: A quantitative electron microscope autoradiographic study of mammotrophs dissociated from rat pituitaries. Endocrinologia, Bucharest, 102:296-311

Fleischer B, Fleischer S, Ozawa H (1969) Isolation and characterization of Golgi membranes from bovine liver. J Cell Biol 43:59-79

Franke WW, Morré DJ, Deumling B, Cheetham RD, Kartenbeck J, Jarasch ED, Zentgraf HW (1971) Synthesis and turnover of membrane proteins in rat liver: an examination of the membrane flow hypothesis. Z Naturforsch 266:1031-1039

Geuze JJ, Kramer MF (1974) Function of coated membranes and multivesicular bodies during membrane regulation in stimulated exocrine pancreas cells. Cell Tissue Res 156:1-20

Geuze JJ, Slot JW, Tokuyasu KT (1979) Immunocytochemical localization of amylase and chymotrypsinogen in the exocrine pancreatic cell with special attention to the Golgi complex. J Cell Biol 82:697-707

Golgi C (1898a) Sur la structure des cellules nerveuses. Arch Ital Biol 30:60-71

Golgi C (1898b) Sur la structure des cellules nerveuses des ganglions spinaux. Arch Ital Biol 30:278-286

Gonatas NK, Kim SU, Stieber A, Avrameas S (1977) Internalization of lectins in neuronal GERL. J Cell Biol 73:1-13

Graham RC, Karnovsky MJ (1966) The early stages of absorption of injected horseradish peroxidase in the proximal tubules of mouse kidney: ultrastructural cytochemistry by a new technique. J Histochem Cytochem 14:291-302

Herzog V (1979) The secretory process as studied by the localization of endogenous peroxidase. J Histochem Cytochem 27:1360-1362

Herzog V, Farquhar MG (1977) Luminal membrane retrieved after exocytosis reaches most Golgi cisternae in secretory cells. Proc Natl Acad Sci USA 74:5073-5077

Herzog V, Miller F (1970) Die Lokalisation endogener Peroxidase in der Glandula parotis der Ratte. Z Zellforsch 107:403-420

Herzog V, Miller F (1972) The localization of endogenous peroxidase in the lacrimal gland of the rat during postnatal development. Electron microscope cytochemical and biochemical studies. J Cell Biol 53:662-680

Herzog V, Miller F (1979) Membrane retrieval in epithelial cells of isolated thyroid follicles. Eur J Cell Biol 19:203-215

Herzog V, Reggio H (1980) Pathways of endocytosis from luminal plasma membrane in rat exocrine pancreas. Eur J Cell Biol 21:141-180

Hickman S, Neufeld EF (1972) A hypothesis for I-cell disease: defective hydrolases that do not enter lysosomes. Biochem Biophys Res Commun 49:992-999

Hokin LE (1968) Dynamic aspects of phospholipids during protein secretion. Int Rev Cytol 23:187-208

Holtzman E, Teichberg S, Abrahams SJ, Citkowitz E, Grain SM, Kavai N, Peterson ER (1973) Notes on synaptic vesicles and related structures, endoplasmic reticulum, lysosomes and peroxisomes in nervous tissue and the adrenal medulla. J Histochem Cytochem 21:349-385

Hopkins CR, Farquhar MG (1973) Hormone secretion by cells dissociated from rat anterior pituitaries. J Cell Biol 59:276-303

Howell SL, Whitfield M (1973) Synthesis and secretion of growth hormone in the rat anterior pituitary. I. The intracellular pathway, its time course and energy requirements. J Cell Sci 12:1-21

Howell KE, Ito A, Palade GE (1978) Endoplasmic reticulum marker enzymes in Golgi fractions - what does it mean? J Cell Biol 79:581-589

Jamieson JD (1978) Intracellular transport and discharge of secretory proteins: present status and future perspectives. In: Silverstein SC (ed) Transport of macromolecules in cellular systems. Dahlem Konferenzen, Berlin, pp 273-288

Jamieson JD, Palade GE (1967a) Intracellular transport of secretory proteins in the pancreatic exocrine cell. I. Role of the peripheral elements of the Golgi complex. J Cell Biol 34:577-596

Jamieson JD, Palade GE (1967b) Intracellular transport of secretory proteins in the pancreatic exocrine cell. II. Transport to condensing vacuoles and zymogen granules. J Cell Biol 34:597-615

Jamieson JD, Palade GE (1971a) Condensing vacuole conversion and zymogen granule discharge in pancreatic exocrine cells: metabolic studies. J Cell Biol 48:503-522

Jamieson JD, Palade GE (1971b) Synthesis, intracellular transport, and discharge of secretory proteins in stimulated pancreatic exocrine cells. J Cell Biol 50:135-158

Jamieson JD, Palade GE (1977) Production of secretory proteins in animal cells. In: Brinkley BR, Porter KR (eds) International cell biology. Rockefeller University Press, New York, pp 308-317

Kalina M, Rabinovitch R (1975) Exocytosis couples to endocytosis of ferritin in parotid acinar cells from isoprenalin stimulated rats. Cell Tissue Res 163:373-382

Kaplan A, Fischer D, Achord D, Sly W (1977) Phosphohexosyl recognition is a general characteristic of pinocytosis of lysosomal glycosidases by human fibroblasts. J Clin Invest 60:1088-1093

Katona E (1976) Incorporation of inorganic sulfate in rat liver Golgi. Eur J Biochem 63:583-590

Katz FN, Rothman JE, Lingappa VR, Blobel G, Lodish HF (1977) Membrane assembly in vitro: Synthesis, glycosylation, and asymmetric insertion of a transmembrane protein. Proc Natl Acad Sci USA 74:3278-3282

Keenan TW, Morré DJ, Basu S (1974) Ganglioside biosynthesis. Concentration of glycosphingolipid glycosyltransferases in Golgi apparatus from rat liver. J Biol Chem 249:310-315

Kern HF, Bieger W, Völkl A, Rohr G, Adler G (1979) Regulation of intracellular transport of exportable proteins in the rat exocrine pancreas. In: Hopkins CR, Duncan CJ (eds) Secretory mechanisms. Cambridge University Press, Cambridge, pp 79-99

Kraehenbuhl JP, Racine L, Jamieson JD (1977) Immunocytochemical localization of secretory proteins in bovine pancreatic exocrine cells. J Cell Biol 72:406-423

Kruse NT, Bornstein P (1975) Metabolic requirements for the intracellular movement and secretion of collagen. J Biol Chem 250:4841-4847

Leblond CP, Bennett G (1977) Role of the Golgi apparatus in terminal glycosylation. In: Brinkley BR, Porter KR (eds) International cell biology. Rockefeller University Press, New York, pp 326-336

Leduc EH, Avrameas S, Bouteille M (1968) Ultrastructural localization of antibody in differentiating plasma cells. J Exp Med 127:109-118

Masur SK, Holtzman E, Walter R (1972) Hormone-stimulated exocytosis in the toad urinary bladder. J Cell Biol 52:211-219

Matsukawa S, Hosoya T (1979) Process of iodination of thyroglobulin and its maturation. II. Properties and distribution of thyroglobulin labeled in vitro or in vivo with radioiodine, ^{3}H-tyrosine, or ^{3}H-galactose in rat thyroid glands. J Biochem, Tokyo, 86:199-212

Meldolesi J (1974) Dynamics of cytoplasmic membranes in guinea pig pancreatic acinar cells. I. Synthesis and turnover of membrane proteins. J Cell Biol 61:1-13

Meldolesi J, Borgese N, De Camilli P, Ceccarelli B (1978) Cytoplasmic membranes and the secretory process. In: Poste G, Nicolson GN (eds) Membrane fusion. Elsevier-North-Holland, Amsterdam, New York, pp 509-627

Michaels J, Leblond CP (1976) Transport of glycoprotein from Golgi apparatus to cell surface by means of "carrier" vesicles, as shown by radioautography of mouse colonic epithelium after injection of ^{3}H fucose. J Microsc Biol Cell 25:243-248

Miller F, Herzog V (1969) Die Lokalisation von Peroxidase und saurer Phosphatase in eosinophilen Leukocyten während der Reifung. Elektronenmikroskopisch-cytochemische Untersuchungen am Knochenmark von Ratte und Kaninchen. Z Zellforsch Mikrosk Anat 97:84-110

Morré JD (1971) Isolation of Golgi apparatus. Methods Enzymol 22:130-148

Morré DJ (1977) Membrane differentiation and the control of secretion: a comparison of plant and animal Golgi apparatus. In: Brinkley BR, Porter KR (eds) International cell biology. Rockefeller University Press, New York, pp. 293-303

Morré, DJ, Hamilton RL, Mollenhauer HH, Mahley RW, Cunningham WP, Cheetham RD, Le Quire US (1970) Isolation of a Golgi apparatus - rich fraction from a rat liver. I. Method and morphology. J Cell Biol 44:484-491

Morré DJ, Keenan TW, Huang CM (1974) Membrane flow and differentiation: Origin of Golgi apparatus membranes from endoplasmic reticulum. In: Ceccarelli B, Clementi F, Meldolesi J (eds) Advances in cytopharmacology, vol II. Raven Press, New York, pp 107-125

Morré JD, Kartenbeck J, Franke WW (1979) Membrane flow and interconversions among endomembranes. Biochim Biophys Acta 559:71-152

Neutra M, Leblond C (1966) Synthesis of the carbohydrate of mucus in the Golgi complex as shown by electron microscope autoradiography of goblet cells from rats injected with glucose-H3. J Cell Biol 30:119-136

Novikoff AB (1976) The endoplasmic reticulum: A cytochemists' view (A review). Proc Natl Acad Sci 73:2781-2787

Novikoff AB, Mori M, Quintana N, Yam A (1977) Studies of the secretory process in the mammalian exocrine pancreas. I. The condensing vacuoles. J Cell Biol 75:148-165

Oliver C, Hand RA (1977) Uptake and fate of luminally administered horseradish peroxidase in resting and isoproterenol-stimulated rat parotid acinar cells. J Cell Biol 76:207-220

Orci L, Perrelet A, Gorden P (1978) Less-understood aspects of the morphology of insulin secretion and binding. Recent Prog Horm Res 34:95-121

Ottosen P, Courtoy P, Farquhar MG (1980) Pathways followed by membrane recovered from the surface of plasma cells and myeloma cells. J Exp Med 152:1-19

Palade GE (1959) Functional changes in the structure of cell components. In: Hayashi T (ed) Subcellular particles. Ronald Press, New York, pp 64-83

Palade GE (1975) Intracellular aspects of the process of protein secretion. Science 189:347-358

Palade GE, Fletcher M (1977) Reversible alterations in the morphology of the Golgi complex induced by the arrest of secretory transport. J Cell Biol 75:371a

Patzelt C, Chan SJ, Duguid J, Hortin G, Keim P, Heinrikson RL, Steiner DF (1978) Biosynthesis of polypeptide hormones in intact and cell-free systems. In: Magnussen S et al (eds) Regulatory proteolytic enzymes and their inhibitors, vol 47. Proc Symp A 6. Pergamon Press, New York, pp 69-78

Rachubinski RA, Verma DPS, Bergeron JJM (1980) Synthesis of rat liver microsomal cytochrome b_5 by free ribosomes. J Cell Biol 84:705-716

Rambourg A, Hernandez W, Leblond CP (1969) Detection of periodic acid-reactive carbohydrate in Golgi saccules. J Cell Biol 40:395-414

Rambourg A, Clermont Y, Marraud A (1974) Three-dimensional structure of the osmium-impregnated Golgi apparatus as seen in the high voltage electron microscope. Am J Anat 140:27-46

Rawitch AB, Taurog A, Chernoff SB, Dorris ML (1979) Hog thyroid peroxidase: Physical, chemical and catalytic properties of the highly purified enzyme. Arch Biochem Biophys 194:244-257

Reggio H, Dagorn JC (1978) Ionic interactions between bovine chymotrypsinogen A and chondroitin sulfate A.B.C. A possible model for molecular aggregation in zymogen granules. J Cell Biol 78:951-957

Reggio H, Palade GE (1978) Sulfated compounds in the zymogen granules of the guinea pig pancreas. J Cell Biol 77:288-314

Rothman JE, Lodish HF (1977) Synchronized transmembrane insertion and glycosylation of a nascent membrane protein. Nature, London, 269:775-780

Salpeter MM, Bachmann L (1972) Autoradiography. In: Hayat MA (ed) Principles and techniques of electron microscopy, vol II. Van Nostrand Reinhold Col, New York, pp 221-278

Schachter H (1974) Glycosylation of glycoproteins during intracellular transport of secretory products. In: Ceccarelli B, Clementi F, Meldolesi J (eds) Advances in cytopharmacology, vol II. Raven Press Publ, New York, and North-Holland Publishing Company, Amsterdam, pp 207-218

Scheele GA, Palade GE, Tartakoff AM (1978) Cell fractionation studies on the guinea pig pancreas. Redistribution of exocrine proteins during tissue homogenization. J Cell Biol 78:110-130

Schimke RT (1975) Turnover of membrane proteins in animal cells. In: Korn ED (ed) Methods in membrane biology, vol III. Plenum Press, New York, pp 201-236

Siekevitz P (1972) Biological membranes: the dynamics of their organization. Annu Rev Physiol 34:117-140

Sly WW, Stahl P (1978) Receptor-mediated uptake of lysosomal enzymes. In: Silverstein SC (ed) Transport of macromolecules in cellular systems. Dahlem Konferenzen, Berlin, pp 229-244

Smith CE (1980) Ultrastructural localization of nicotinamide adenine dinucleotide phosphatase (NADP ase)activity to the intermediate saccules of the Golgi apparatus in rat incisor ameloblasts. J Histochem Cytochem 28:16-26

Stein EA, Hsin J, Fischer EH (1964) Alpha-amylases as calcium-metalloenzymes. I. Preparation of calcium-free apoamylases by chelation and electrodialvsis. Biochemistry 3:56-61

Steiner DF, Kemmler W, Clark JL, Oyer PE, Rubenstein AH (1972) The biosynthesis of insulin. In: Steiner DF, Freinkel N (eds) Handbook of physiology. Endocrinology I. Williams and Wilkins, Baltimore, pp 175-198

Tager HS, Patzelt C, Chan SJ, Quinn PS, Steiner DF (1979) The biosynthesis and conversion of islet cell hormone precursors. A brief review. Biol Cell 36:127-136

Tartakoff AM, Jamieson JD (1974) Subcellular fractionation of the pancreas. Methods Enzymol 31:41-59

Tartakoff AM, Vassalli P (1978) Comparative studies of intracellular transport of secretory proteins. J Cell Biol 79:694-707

Tartakoff AM, Greene LJ, Palade GE (1975) Studies on the guinea pig pancreas. Fractionation and partial characterization of exocrine proteins. J Biol Chem 249: 7420-7431

Tice LW, Wollman SH (1974) Ultrastructural localization of peroxidase on pseudopods and other structures of the typical thyroid epithelial cell. Endocrinologia, Bucharest, 94:1555-1567

Toyoshima S, Fukuda M, Osawa T (1972) Chemical nature of the receptor site for various phytomitogens. Biochemistry 11:4000-4005

Tweto J, Doyle D (1977) Turnover of proteins in the eukaryotic cell surface. In: Poste G, Nicolson GL (eds) Cell surface reviews, vol IV. Elsevier-North-Holland, Amsterdam, pp 137-164

Wallach D, Schramm M (1971) Calcium and the exportable protein in rat parotid gland - Parallel subcellular distribution and concomitant secretion. Eur J Biochem 21: 433-437

Wallach D, Kirshner N, Schramm M (1975) Non-parallel transport of membrane proteins and content proteins during assembly of the secretory granule in rat parotid gland. Biochim Biophys Acta 375:87-105

Whur P, Herscovics A, Leblond CP (1969) Radioautographic visualization of the incorporation of galactose-^{3}H and mannose-^{3}H by rat thyroids in vitro in relation to the stages of thyroglobulin synthesis. J Cell Biol 43:289-311

Winkler H, Schöpf JAL, Hörtnagl H, Hörtnagl H (1972) Bovine adrenal medulla: Subcellular distribution of newly synthesized catecholamines, nucleotides, and chromogranins. Naunyn-Schmiedebergs Arch Exp Pathol Pharmakol 273:43-61

Young RW (1973) The role of the Golgi complex in sulfate metabolism. J Cell Biol 57: 175-189

The Role of Free and Membrane-Bound Polysomes in Organelle Biogenesis

G. Kreibich, S. Bar-Nun, M. Czako-Graham, W. Mok, E. Nack, Y. Okada, M. G. Rosenfeld, and D. D. Sabatini[1]

Introduction

It has recently become clear that information contained within the nascent polypeptide chain of specific proteins determines that their syntheses proceed on polysomes which become associated with membranes of the endoplasmic reticulum (ER) (Blobel and Sabatini 1971; Milstein et al. 1972; Blobel and Dobberstein 1975a,b). Studies with secretory (for review see Campbell and Blobel 1976; Blobel 1977; Habener et al. 1978), lysosomal (Popov et al. 1978; Erickson and Blobel 1979), cellular (Bar-Nun et al. 1980), and viral membrane proteins (Toneguzzo and Gosh 1977, 1978; Katz et al. 1977; Bonatti and Blobel 1979) indicate that amino terminal regions of the nascent polypeptides, which may vary in length from 15 (Gaye et al. 1977) to 29 amino acids (Sherwood et al. 1979) and are rich in hydrophobic residues, play an essential role in directing the ribosomes to specific sites on the ER membranes. Polypeptides of secretory proteins are then transferred cotranslationally through the membranes into the ER lumen and from there to the Golgi apparatus before their discharge by exocytosis (Palade 1975). A similar transfer through the ER membrane takes place during the synthesis of the acid hydrolases which are ultimately packaged into lysosomes (Popov et al. 1978; Erickson and Blobel 1979). On the other hand, membrane proteins synthesized on bound polysomes (Bar-Nun et al. 1980; Okada et al. 1979; Chyn et al. 1979; Schechter et al. 1979; Lingappa et al. 1979; Dobberstein et al. 1979; Sabban et al. 1979; Rothman and Lodish 1977) are not completely transferred across the ER membrane, but upon completion of the polypeptide remain inserted into the phospholipid bilayer.

Following binding of the ribosome and insertion of the nascent polypeptide into the ER membrane, the amino terminal segments which serve as signals triggering this process are generally, but not always, removed from the nascent chain by a membrane-associated peptidase that appears to be located on the luminal side of the membrane (c.f. Blobel 1977; Jackson and Blobel 1977; Kreibich et al. 1980). During their syntheses on bound polysomes, the polypeptides may also be subjected to other structural modifications such as glycosylation (Lennarz 1975; Czichi and Lennarz 1977; Tabas et al. 1978; Hanover and Lennarz 1980) or hydroxylation (Olsen et al. 1973; Peterkofsky and Assad 1979) which are carried out by enzymatic systems located near the ribosome-membrane junction. Subsequent processing of the polypeptides during their transit through specific compartments can occur by further proteolysis (Russell and Geller 1975; Habener and Kronenberg 1978), changes which affect the oligosaccharide core (Hubbard and Robbins 1979; Grinna and Robbins 1979; Kornfeld et al. 1979; Elting et al. 1980; c.f. Leblond and Bennett 1977), or by modifications which affect specific amino acid residues such as the attachment of fatty acids (Schmidt and Schle-

[1]New York University School of Medicine, Department of Cell Biology, 550 First Avenue, New York, N.Y. 10016, USA

148

singer 1979, 1980), phosphorylation (Natowicz et al. 1979), or sulfation (Reggio and Palade 1978). It is likely that some of these modifications are important to ensure the correct subcellular distribution of the protein and in some cases are required for packaging of the products into storage compartments. The segregation of the specific enzymatic systems which carry out the co- and posttranslational changes characteristic of specific secretory membrane and lysosomal polypeptides requires that these products follow a well-defined subcellular route (Palade 1975; Sabatini and Kreibich 1976). This is ensured by the presence of ribosome binding sites and specific receptors for signal peptides in the rough ER (RER) membranes (Borgese et al. 1974; Czakó Graham et al. 1979; Kreibich et al. 1980).

Our recent work has been addressed to an elucidation of the role that specific ER membrane components play in the early stages of cotranslational transfer of polypeptides across ER membranes (Fig. 1), such as (a) the recognition of signal segments in nascent polypeptides (1C), (b) the binding of ribosomes to specific sites (1D), and (c) the modifications which affect growing polypeptides during insertion into the membrane and during vectorial discharge (1E and 1F).

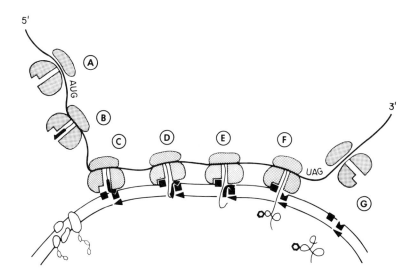

Fig. 1. Model depicting the major functional elements expected to participate in the assembly of membrane-bound polysomes. A likely sequence of events would be: (1) After initiation of translation (A) an amino terminal sequence (*thick line in B*) emerges from the 60S ribosomal subunit and recognizes a receptor (◣) on the RER membrane (C). (2) This interaction is reinforced by ionic linkages between the ribosome and its binding site on the membrane (■). (3) The growing nascent chain, initially in a loop configuration (D), crosses the ER membrane. (4) Cotranslational modifications such as proteolytic processing by the signal peptidase (◀) (E) or carbohydrate transfer to asparagine residues via a dolichol pyrophosphate oligosaccharide intermediate ensues (◖) (F). (5) After termination of protein synthesis, the secretory protein is segregated in the lumen of the ER. Inactive ribosomes are then detached (G), partially aided by a dissociation factor. (Mok et al. 1976)

Compositional Differences between Rough and Smooth Portions of the
Endoplasmic Reticulum

In parenchymal cells of the liver and other organs engaged in the synthesis of secretory proteins, a large portion of the ribosome population is found attached to membranes of the ER cisternae. In these specialized cells, of which the hepatocyte is a typical example, such ribosomes are topographically segregated to specific regions of the cytoplasm in which the rough cisternae may form characteristic stacks
(Fig. 2). On the other hand, the smooth endoplasmic reticulum (SER)

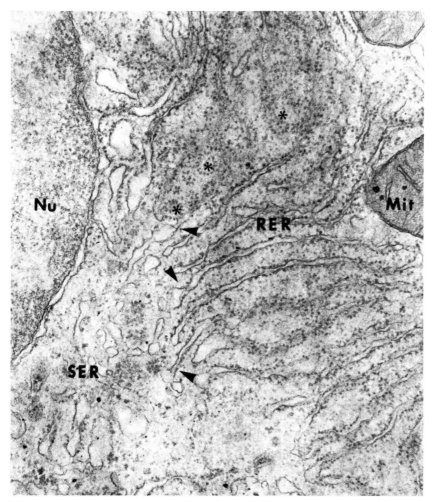

Fig. 2. Transition between rough and smooth ER membranes in rat liver hepatocyte.
The distinctly different morphological appearance of rough and smooth portions of
the ER is apparent. While RER membranes adopt a planar configuration within the
stacked cisternae, smooth membranes (*lower left*) form contorted tubules. *Arrow
heads* indicate the transition between the two types of ER membranes. *Stars* mark
the typical spiral and double row arrangement of membrane-bound polysomes observed
in grazing sections of the RER cisternae. Small pieces of liver tissue from a male
rat (150 g) starved for 18 h were fixed at 4°C in 2% glutaraldehyde (2.5 h), 1%
OsO_4 (2 h), and stained in block with 0.5% uranyl acetate (15 min). x 62,000

usually forms a tubular membrane system which varies in its extension with the physiologic state of the cell and frequently penetrates most other areas of the cytoplasm. Electron microscopic studies have provided numerous examples of continuities between the rough and smooth ER membrane systems (Bruni and Porter 1965; Kreibich et al. 1978c). Cell fractionation procedures are available for the isolation of membrane fragments derived from both portions of the ER in the form of rough and smooth microsomal vesicles (c.f., Adelman et al. 1973a). Biochemical studies have shown that aside from the presence of ribosomes, rough and smooth ER membranes from hepatocytes have similar protein and phospholipid compositions. This is reflected in the distribution of enzymatic markers and biochemical functions (Depierre and Dallner 1975; Amar-Costesec et al. 1974; Colbeau et al. 1971). It has been suggested that both types of ER membranes are related biogenetically and that regions of continuity between them allow for the direct flow of proteins synthesized on bound polysomes into the smooth ER compartment (c.f., Sabatini and Kreibich 1976). The continuity of the membranes, however, raises the question of how, in spite of the fluidity of ER membranes (Ojakian et al. 1977), the segregation of bound ribosomes to the rough cisternae is maintained.

Recently, biochemical and electrophoretic analysis has revealed the existence of compositional differences between the rough and smooth endoplasmic reticulum membranes which appear to be related to the function of the RER in protein synthesis (Kreibich et al. 1975, 1978a,b) (see Fig. 3). Rough microsomes (RM) contain two transmembrane glycoproteins (Rodriguez Boulan et al. 1978) of 65,000 and 63,000 daltons which are not found in smooth microsomes (SM). These proteins are solubilized when microsomes are treated with DOC, but are quantitatively

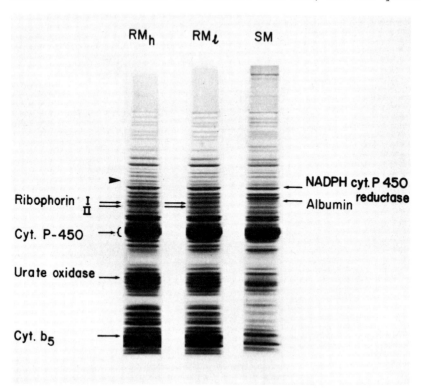

recovered in association with sedimentable ribosomes when neutral detergents are used to solubilize microsomal membranes. When intact microsomes were treated with low concentrations of the reversible bifunc-
tional crosslinking reagent methyl-4-mercaptobutyrimidate, the proteins
characteristic of RM were crosslinked to the ribosomes and were recovered with them after DOC treatment. For these reasons they were designated Ribophorins I and II (Kreibich et al. 1978a,b,c, 1980), a name
which suggests a possible role in mediating the attachment of ribosomes
to the membranes.

Densitometric tracings of SDS PAGE patterns of rough microsomes and
ribosomal sediments containing ribophorins have revealed a stoichiometric relationship between these proteins and the ribosome content
(1.4:1). Subfractionation of total microsomal fractions by isopycnic
sedimentation into heavy and light RM (RM_h and RM_1 in Fig. 3) and SM
has shown that the ribophorin content of membrane vesicles decreases
in parallel with the amount of ribosomes associated with the membranes.

We have found that although the apparent molecular weights of ribophorins vary slightly in different animal species, microsomes from different organs of the same animal contain seemingly identical ribopho-
rin-like proteins (Grebenau et al. 1977; Kreibich et al. 1978c, 1980).
These comparative analyses reveal therefore, that the presence of ribophorins is a characteristic feature of all RER membranes, as would
be expected if those proteins play a role in the functions associated
with the binding of a specific class of polysomes.

Ribophorins also appear to play a role in maintaining the characteristic structural configuration of RER, and their presence may account
for the morphological differences with SER (Kreibich et al. 1978c).
The proteins have a strong tendency to interact with each other and
in microsomal sediments obtained after treatment with nonionic detergents they form curved planar aggregates which have the appearance of
membrane remnants. A network of these proteins may therefore control
the distribution of ribosome binding sites in the plane of the ER membrane. The characteristic configurations of membrane-bound polysomes
(Palade 1964) in rosettes, spirals, and double rows, which is apparent
in grazing sections of RER membranes (see Fig. 2), suggests that mechanisms indeed exist for the local control of the configuration of
ribosome binding sites.

In Vitro Binding of Ribosomes to Sites on ER Membranes

It has been shown that, in addition to the link provided by the nascent
polypeptide chain, a direct ionic interaction between membrane receptors and ribosomes plays a role in maintaining the association of active polysomes with ER membranes (Adelman et al. 1973b). In vitro as-

Fig. 3. Analysis of microsomal fractions by SDS acrylamide gel electrophoresis.
Rat liver microsomal subfractions of varying isopyknic density (corresponding to
different ratios of RNA/protein) were isolated according to Kruppa and Sabatini
(1977) and used for analysis on SDS polyacrylamide gels (5%-13% gradient). RNA/
protein ratios of the subfractions were: heavy RM (RM_h): 0.29; light RM (RM_ℓ):
0.165; SM: 0.05. Various amounts of sample from the different microsomal subfractions,
containing in all cases approximately 250 μg of membrane protein, were loaded into
each slot. The two ribophorins and a band with a mobility corresponding to 86,000
daltons (indicated by an *arrow head*; see also Sharma et al. 1978) are characteristic
of RM and absent from SM. The position of other proteins characteristically found
in microsomal fractions are indicated by *arrows*

152

says have been developed which demonstrate that specific sites on rough microsomal membranes, exposed after the membranes are stripped of ribosomes, are capable of ribosome binding in media of low ionic strength, even in the absence of nascent polypeptide chains (Borgese et al. 1974). These membrane sites are unable to bind 50S subunits derived from prokaryotic ribosomes (Mok et al. 1976), but at nearly physiologic ionic strength they show a higher affinity for 60S than 40S ribosomal subunits (Borgese et al. 1974). Ribosome binding sites are inactivated by mild trypsinization, an observation which indicates the involvement of protein components in the binding process (Borgese et al. 1974; for review see Sabatini and Kreibich 1976).

Several investigators have measured the ribosome-binding capacity of SM and have shown it to be significantly smaller than that of stripped RM (RMstr). The number of ribosome binding sites in SM, however, is not negligible and in crude fractions may amount to 30% of that in RMstr. It should be noted, however, that the composition of the smooth microsome fractions varies substantially with the cell fractionation procedure used and may be affected by pretreatment of the animals by starvation (Yap et al. 1978) or administration of drugs (Stäubli et al. 1969). Therefore, actual levels of ribosome binding sites in SM fractions are variable and the contribution of membranes derived from the SER to binding is frequently difficult to assess. Purified plasma membranes, Golgi membranes, and mitochondrial outer membrane fractions have been shown to have negligible capacities for ribosome binding (Fig. 4, see also Mok et al. 1976). Hence, two main alternatives should be considered in interpreting the finding of binding sites in smooth microsome fractions: (a) these sites are contributed exclusively by membranes derived from portions of the RER which at the moment of fractionation bear small numbers of ribosomes and are recovered in the smooth fractions; (b) binding sites are indigenous to the SER and are uniformly distributed over these membranes. In an attempt to resolve

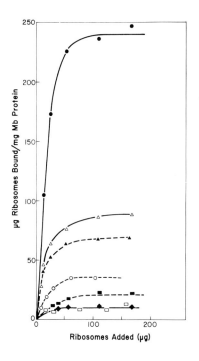

Fig. 4. Saturation binding of 80s ribosomes to different subcellular membrane fractions from rat liver. Various amounts of ^{3}H-80s ribosomes were incubated for 30 min at 3^0C in a total of 80 µl TKM (50 mM Tris HCl pH 7.5; 25 mM KCl; 5 mM MgCl$_2$) with either rough microsomal membranes stripped of ribosomes (0.10 mg protein; •), smooth microsomes (0.15 mg; ▲), smooth microsomal membranes stripped of ribosomes (0.12 mg; △), rough microsome (0.14 mg; ○), Golgi membranes (0.28 mg; ▪), mitochondria (0.6 mg; ▫), or plasma membranes (0.20 mg; ◆). The amount of ribosomes bound to the different subcellular membrane fractions is expressed in µg ribosome bound per mg membrane protein. For experimental details, see Borgese et al. (1974). RM, SM, and mitochondrial membranes were prepared according to Adelman et al. (1973a). The procedures of Dorling and Le Page (1973) and Ehrenreich et al. (1973) were used to prepare plasma membranes and Golgi membranes respectively

this question, we measured the binding capacity of smooth microsomes
prepared from rats treated with phenobarbital (PB) (Kreibich et al.
1978c), a drug which causes an extensive proliferation of the SER
(Stäubli et al. 1969) and allows the preparation of SM fractions with
a better-defined subcellular origin. We found that SM prepared from
PB-treated rats are almost completely devoid of ribosome-binding sites.
These observations suggest the existence of mechanisms which maintain
the segregation of ribosome-binding sites in spite of the continuity
of rough and smooth endoplasmic reticulum membranes.

Segregation of Proteins into the Lumen of the Endoplasmic Reticulum

The observation that ribophorins are found only in the rough portion
of the ER correlates with results from in vitro translation experiments.
The experiment shown in Fig. 5 compares the relative capacity of RM
and SM from HeLa cells to effect vectorial discharge and cotranslational processing of human placental lactogen. RM, when present during
translation (Fig. 5b,f), but not when added posttranslationally (Fig.
5c), are capable of this process, while SM are incompetent under both
conditions (Fig. 5d,e,g). This result is consistent with a model in
which ribosome receptors present only in the RER function in conjunction with membrane elements involved in nascent chain recognition, vectorial discharge, and cotranslational cleavage of the precursor (but
see Bielinska et al. 1979). It remains to be demonstrated, however,
whether all these elements are segregated solely to rough portions of
the ER.

Site of Synthesis of Membrane Proteins of the Microsomal Electron
Transport Chain

The relationship of an integral membrane protein with respect to the
phospholipid bilayer appears to be related to the mechanism of its insertion into the membrane. Proteins exposed only on cytoplasmic faces
of membranes may be synthesized on free polysomes and inserted into
the membrane posttranslationally. Synthesis on membrane-bound polysomes and cotranslational insertion, on the other hand, appears to be
the main mechanism that ensures the transmembrane disposition of proteins and the passage of luminally exposed proteins through the membrane (Sabatini and Kreibich 1976; but see Wickner 1979). Nascent polypeptides of membrane proteins synthesized on bound polysomes must
contain insertion signals analogous to those of secretory proteins
which initiate cotranslational insertion into the membrane. Because
membrane proteins synthesized on bound polysomes may be retained in
the ER or transferred to other compartments, such as the Golgi apparatus or the plasma membrane, mechanisms which operate posttranslationally must discriminate between the different classes of transmembrane
proteins and control their lateral diffusion along continuous phospholipid bilayers or their uptake into vesicles which shuttle between
compartments (Palade 1975; Sabatini and Kreibich 1976; Rothman et al.
1980).

Cytochrome P-450 is a major integral membrane protein of the ER in hepatocytes. It is found in both rough and smooth portions of this mem-
branous organelle and provides an excellent model to study cotranslational insertion of a protein into a membrane. The polypeptide appears
to have a transmembrane disposition since its accessibility to macromolecular labeling probes is enhanced when microsomal membranes are
made permeable by low detergent concentration (Rodriguez Boulan et al.
1978). In vitro translation experiments using free and membrane-bound

154

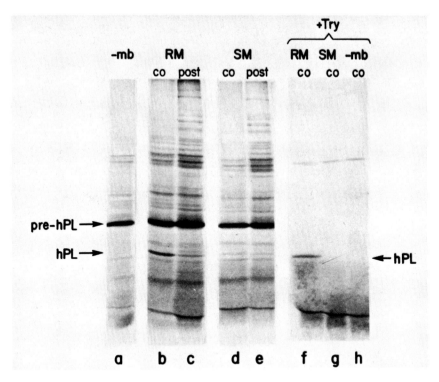

Fig. 5. Cotranslational processing and vectorial discharge of pre-hPL by HeLa cell rough microsomes. HeLa cells were washed twice in PBS and resuspended in 0.5 M sucrose containing 3 mM DTT. The cells were gently homogenized and centrifuged for 10 min at 10,000xg. The postmitochondrial supernatant was subfractioned according to the procedure of Morimoto, Feldman, and Gaetani (personal communication). The RM were resuspended in 50 mM Tris (pH 7.5) and 50 mM KCl and EDTA was added to a concentration of 3 μmol EDTA/10 OD_{260}. The microsomes were placed on a sucrose gradient (20%-30%) over a 1.6 M sucrose cushion containing the same ion concentration as the sample. After centrifugation (39 K-SW41-45 min), the stripped RM were collected from the 1.6 M sucrose cushion and the stripping procedure repeated. The microsomes were collected, diluted three times with 0.25 M sucrose and pelleted in an SW56 rotor for 20 min at 50 K. The microsomes were resuspended in a solution containing 20 mM Hepes (pH 7.4) and 3 mM DTT and were used in the translation assay. For further subfractionation, SM were resuspended in a buffer containing (50 mM Tris and 50 mM KCl TK), placed on a discontinuous sucrose gradient (0.8 M-10 M-1.15 M all in TK) and centrifuged at 40 K for 3 h (SW41 Spinco rotor). Microsomes were collected from the 1.0-1.5 M interface and stripped as described above for rough microsomes. Translation was carried out at 28°C in a reticulocyte lysate system (Palmiter et al. 1977) supplemented with rat liver tRNA (55 μg/ml) and spermidine (0.2 mM), using 1 μg placental mRNA and 25 μCi ^{35}S-methionine per 25 μl translation mixture. Membranes were present in the translation at a final concentration of 6 A_{260} units/ml. After translation, the mixture was incubated with proteolytic enzymes (80 μg each of trypsin and chymotrypsin per ml) for 3 h at 0°C followed by the addition of 30 units of Trasylol. Total translation products were analyzed by SDS/polyacrylamide gel electrophoresis (10%-15%) followed by autoradiography. The autoradiogram shows translation products synthesized in the absence of membranes (a), or in the presence of stripped RM (b), or SM (d). Posttranslational addition of either RM (c) or SM (e) did not lead to processing of pre-hPL to hPL ($arrows$). Lanes f, g, and h correspond to samples for b, d, and e, which after translation were incubated with proteolytic enzymes

polysomes showed that phenobarbital administration produced a substantial increase in levels of translatable liver cytochrome P-450 mRNA which was primarily associated with the bound polysomes (see Fig. 6) (Negishi et al. 1976). In vitro synthesis of cytochrome P-450 in the presence of RM demonstrated that the nascent protein was directly inserted into the ER membranes. Labeled and immunoprecipitable polypeptides could not be released from the sedimentable microsomes by treatment with low detergent concentrations which released albumin and other microsomal proteins from their lumen (Bar-Nun et al. 1980).

The electrophoretic mobility of cytochrome P-450 synthesized in vitro in the absence of membranes was indistinguishable from that of mature cytochrome P-450. This finding suggested that, in contrast to most secretory proteins (c.f., Kreibich et al. 1980; see also the steps described in Fig. 7A-C,E) and the envelope glycoprotein of VSV (Lingappa et al. 1978; see Fig. 7A,B,D,F), the primary translation product for cytochrome P-450 is not proteolytically processed during or after synthesis. This has been proven by sequence analysis of the amino-terminal portions of the in vitro product of cytochrome P-450 mRNA and the mature protein, which were found to be identical. Similarly ovalbumin (Palmiter et al. 1978), bovine retinal opsin (Schechter et al. 1979), and the PE_2 glycoprotein of Sindbis virus (Bonatti and Blobel 1979) are not proteolytically processed co- or posttranslationally.

The amino-terminal sequence of cytochrome P-450 resembles the signal segment of a typical presecretory protein, such as that of albumin, but differs in that the putative insertion signal remains as a permanent feature of the mature protein (Bar-Nun et al. 1980). It is likely that even after synthesis is completed, the hydrophobic amino-terminal segment that appears to initiate insertion of the polypeptide into the membrane remains embedded within the phospholipid bilayer (Fig. 7A,B, D,G).

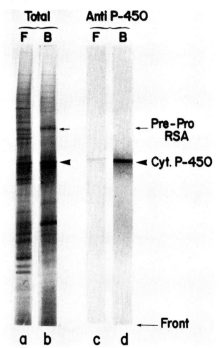

Total

F B

Anti P-450

F B

← Pre-Pro RSA

◄ Cyt. P-450

← Front

a b   c d

Fig. 6. Site of synthesis of cytochrome P-450. Free and membrane-bound polysomes were prepared from livers of starved (Ramsey and Steele 1976) or unstarved (Ramsey and Steele 1979) Long Evans rats pretreated with a single phenobarbital dose (100 mg/kg body weight) 12 h before being killed. In vitro translation was carried out at 37°C in a reticulocyte lysate system (Palmiter et al. 1977) supplemented with rat liver tRNA (55 ug/ml) and spermidine (0.2 mM), using 20.0 A_{260} units/ml of polysomes and ^{35}S-methionine as a label (500 µCi/ml; 100 Ci/mmol). In vitro synthesized cytochrome P-450 was immunoprecipitated with anti-cytochrome P-450 IgG (Bar-Nun et al. 1980). Total translation products and immunoprecipitates were analyzed by SDS/polyacrylamide gel electrophoresis followed by autoradiography. Lanes *a* and *b* show autoradiograms of total translation products of free *a* and bound *b* polysomes, while lanes *c* and *d* correspond to immunoprecipitates from these translation mixtures, respectively. The *arrowhead* and the *arrow* indicate the position of cytochrome P-450 and pre-proalbumin, respectively

156

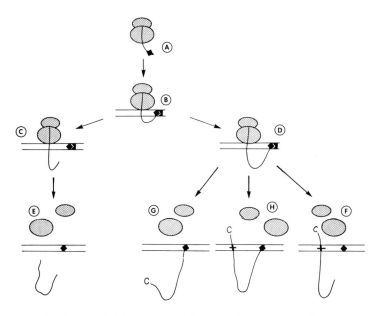

Fig. 7. Cotranslational insertion of microsomal membrane proteins into the ER. The amino terminal sequence serving as an insertion signal (◆;A) is recognized by a specific receptor in the RER (▶) (B) which initiates vectorial discharge (B,C,D). The final disposition of the newly synthesized peptide is determined by subsequent events such as cotranslational proteolytic processing (C or D) and the presence of stop transfer signals (H and F) that prevent completion of the discharge into the cisternal lumen

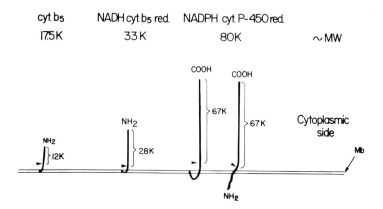

Fig. 8. Spatial relationshop of proteins of the microsomal electron transport chain to the ER membranes. The sites on which cytochrome b_5, NADH cytochrome b_5 reductase, and NADPH cytochrome P-450 reductase are cleaved when exogenous proteases are added to isolated rat liver microsomes are marked by arrowheads. The molecular weights of the intact membrane proteins as well as of the released fragments (indicated by *brackets*) are also indicated (for details see text)

Since a portion of the mature cytochrome P-450 molecule remains exposed on the cytoplasmic side of the microsomal membrane (Matsuura et al. 1978; Welton and Aust 1974; Nilsson et al. 1978; Rodriguez Boulan et al. 1978), one must postulate that a halt transfer signal also exists within the polypeptide which prevents its further transfer through the membrane (Fig. 7A,B,D,H). A second hydrophobic stretch, followed by a group of charged residues immediately adjacent to the amino-terminal insertion signal or located further inside the polypeptide chain, could serve this function. In this case a loop disposition within the membrane would be achieved.

We have also studied the biosynthesis of cytochrome b_5, NADH-cytochrome b_5 reductase, and NADPH-cytochrome P-450 reductase, which together with cytochrome P-450 constitute the two well-characterized microsomal electron transport chains (Strittmatter 1963; c.f., Sato and Omura 1978). These integral membrane proteins of 17,500, 33,000, and 80,000 daltons respectively, are exposed mainly on the cytoplasmic face of the ER membranes (Fig. 8). Characterization of segments cleaved by proteolysis indicated that in the case of cytochrome b_5 and its reductase, carboxy-terminal segments of these integral proteins serve as hydrophobic anchors which maintain an association with the membranes (Strittmatter et al. 1972; Rogers and Strittmatter 1974, 1975). The NADPH-cytochrome P-450 reductase, however, appears to be inserted into the membrane through a segment near its amino-terminal end (Black et al. 1979).

Contrary to previously reported results (Sargent and Vadlamudi 1968; Omura and Kuriyama 1971; Harano and Omura 1977; Ragnotti et al. 1969; Lowe and Hallinan 1973; Omura and Kuriyama 1971; Harano and Omura 1977),

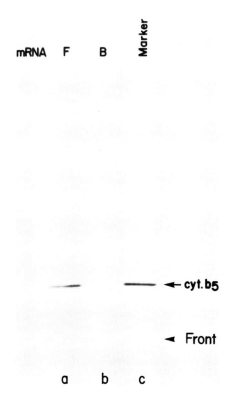

Fig. 9. Biosynthesis of microsomal cytochrome b_5 on free polysomes. Messenger RNA (0.02 OD_{260} each), isolated from free and membrane-bound rat liver polysomes prepared according to Ramsey and Steele (1976), was used to direct in vitro protein synthesis in 50 µl of a reticulocyte lysate (60%) (Palmiter et al. 1977) containing 25 µCi ^{35}S-methionine. Monospecific IgG against cytochrome b_5 was used for immunoprecipitation in the presence of SDS. Autoradiograms of immunoprecipitates from translation mixtures containing mRNA from free polysomes (*lane a*) or bound polysomes (*lane b*) were analyzed on SDS acrylamide gels (6%-13%). Cytochrome b_5, obtained by detergent extraction from rat liver microsomes (Imai 1976) and labeled with ^{125}I, was used as a marker (*lane c*)

recent experiments have shown that cytochrome b_5 (Fig. 9) and NADH-cytochrome b_5 reductase are synthesized exclusively on free polysomes (Okada et al. 1979; Rachubinski et al. 1980). In both cases the products of in vitro translation had apparent molecular weights in SDS gels which corresponded to those of the mature proteins extracted by detergents. This suggests that the insertion of these proteins into the membrane is not accompanied by proteolytic processing, which in other cases removes insertion signal segments. It is well known that cytochrome b_5, purified from microsomal membranes by solubilization with detergents, is an amphipathic molecule that contains a hydrophobic segment of about 40 amino acids near the carboxy-terminus (Ozols and Gerard 1977). The site of synthesis of cytochrome b_5 and the apparent lack of processing during the insertion of the polypeptide into the membrane suggests that the hydrophobic carboxy-terminal region, which is only exposed after termination of protein synthesis, also serves as a signal for posttranslational insertion into the ER membrane

A similar posttranslational mechanism seems to operate for NADH-cytochrome b_5 reductase, which is also synthesized on free polysomes (not shown; Okada et al. 1979; see also Borgese and Gaetani 1980). Although little information on the primary structure of this membrane protein is available, our in vitro translation experiments indicate that a hydrophcbic segment is also located near the carboxy-terminus of the polypeptide. The location of this segment explains the finding that an amino-terminal hydrophilic portion (28,000 daltons) of the NADH-cytochrome b_5 reductase molecule can be released from rat liver microsomes by treatment with cathepsin D (Spatz and Strittmatter 1973).

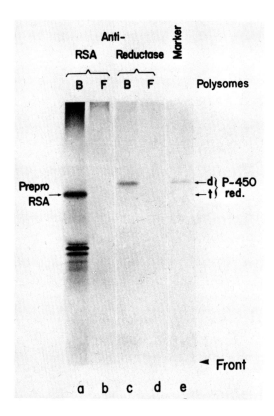

Fig. 10. Biosynthesis of NADPH cytochrome P-450 reductase on membrane-bound polysomes. Messenger RNA samples (0.1 OD_{260} each), isolated from free (lanes b and d) and membrane-bound (lanes a and c) rat liver polysomes, were used for translation (60 min at 25°C) in 100 μl of a 23S wheat germ system (Roman et al. 1976) containing 50 μCi ^{35}S-methionine. Indirect immunoprecipitation with anti-RSA (lanes a and b) and rabbit anti-P-450 reductase (lanes c and d) shows that P-450 reductase is made exclusively on membrane-bound polysomes. ^{125}I-labeled NADPH P-450 reductase was used as a marker (lane e)

In vitro translation experiments with free and bound rat liver polysomes, followed by immunoprecipitation and gel electrophoretic analysis, showed that, as is the case with cytochrome P-450 (Bar-Nun et al. 1980), the reductase of this hemoprotein is synthesized on membrane-bound polysomes (Okada et al. 1979; Gonzalez and Kasper 1980; see also Fig. 10). Because a large portion of the carboxy-terminal region of this membrane protein is exposed on the cytoplasmic side of the ER membrane (see Fig. 8), it may be expected that, in this case too, a stop transfer signal within the polypeptide serves to halt its transfer through the membrane shortly after its insertion has begun.

Site of Synthesis of Lysosomal and Peroxysomal Content Proteins

The topographic relationship of lysosomes to the ER and Golgi apparatus, the discharge of lysosomal hydrolases into phagosomes by a phenomenon akin to exocytosis, and the glycoprotein nature of lysosomal enzymes have all led to the generally accepted idea that lysosomal proteins, like secretory proteins, are synthesized on bound polysomes and vectorially segregated into the lumen of the ER (De Duve and Wattiaux 1960; Swank and Paigen 1973). Our studies on the biosynthesis of the lysosomal enzymes β-glucuronidase (Popov et al. 1978) and Cathepsin D demonstrate that these polypeptides are indeed products of bound polysomes and that the ER is involved in the pathway of distribution of these enzymes. As shown for β-glucuronidase, the primary translation

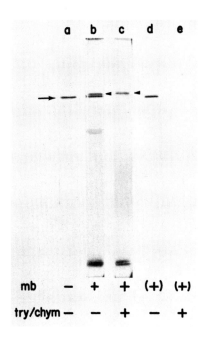

Fig. 11. Cotranslational processing and segregation of rat preputial β-glucuronidase. Messenger RNA (0.03 A_{260}), isolated from rat preputial glands according to Liu et al. (1979), was used for translation (25°C, 60 min) in a wheat germ S-23 supernatant containing 12.5 µCi ^{35}S-methionine in a final volume of 25 µl. Dog pancreas microsomes stripped of ribosomes were prepared as described by Shields and Blobel (1978) and were added to the translation mixture at a final concentration of a A_{260} units/ml either co- (*lanes* b and c) or posttranslationally (*lanes* d and e). After translation, both mixtures were incubated with proteolytic enzymes (80 µg each of trypsin and chymotrypsin per ml) (*lanes* d and e) for 3 h at 0°C followed by the addition of 30 units Trasylol. The in vitro synthesized β-glucuronidase was purified by indirect immunoprecipitation (Goldman and Blobel 1978) with the addition of 10 µg affinity purified IgG. Immunoprecipitates were analyzed by polyacrylamide gel (10%) electrophoresis followed by fluorography. *Lane* a shows immunoreactive translation product in the absence of added membranes (pre β-glucuronidase); *lane* b: dog pancreas microsomal membranes present cotranslationally; *lane* c: as in *lane* b, but followed by incubation with proteolytic enzymes; *lane* d: translation products isolated when immunoprecipitation was preceeded by a posttranslational incubation with dog pancreas microsomal membranes; *lane* e: as in *lane* d, but incubated with proteolytic enzymes before immunoprecipitation. The *arrowheads* point to the position of cotranslationally "processed" β-glucuronidase, while the *arrow* in track a indicates the position of native β-glucuronidase, the primary translation product

160

product (Fig. 11a) was modified when dog pancreas microsomal membranes were present cotranslationally yielding a product of slightly slower electrophoretic mobility (Fig. 11b). Partial sequence analysis suggests that the decrease in electrophoretic mobility results from the transfer of a core oligosaccharide to the proteolytically processed polypeptide. β-glucuronidase synthesized in the presence of microsomal membranes was cotranslationally segregated into the microsomal lumen and became inaccessible to added proteases (Fig. 11c).

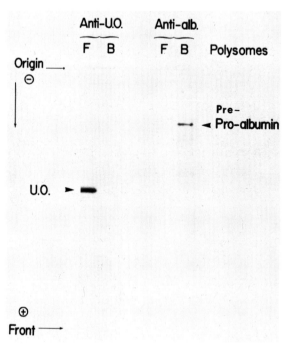

Fig. 12. Biosynthesis of urate oxidase, a peroxisomal enzyme marker, on free polysomes. Free (F) and membrane-bound (B) polysomes were prepared from livers of unstarved rats essentially according to Ramsey and Steele (1979). Polysomes were translated at a final concentration of 25 $A_{260/ml}$ in 200 µl of a reticulocyte lysate system. Immunoprecipitates were prepared essentially according to Goldman and Blobel (1978), using 30 µg of rabbit anti rat urate oxidase IgG, or as a control 30 µg of rabbit anti RSA IgG, followed by protein A Sepharose. Immunoprecipitates were analyzed by polyacrylamide gel electrophoresis (10%-15%) followed by fluorography. The results show that urate oxidase ($U.O.$) is made exclusively on free polysomes

Although it had been proposed that peroxysomal content proteins follow an intracellular pathway similar to that of lysosomal enzymes (Higashi and Peters, 1963a,b; De Duve and Baudhuin 1966; Novikoff and Shin 1969; Essner 1967; Reddy et al. 1974), biochemical studies have generally failed to show an early segregation of newly synthesized catalase into rough microsome vesicles (Redman et al. 1972; Lazarow and De Duve 1973 a,b). Recent experiments have shown that the peroxisomal enzymes catalase and urate oxidase (Goldman and Blobel 1978 and Fig. 12) are synthesized on free polysomes. This reveals a pathway of incorporation into the peroxisomes which must involve the posttranslational passage through an organellar membrane. The primary translation products for these proteins, which are not glycoproteins, have the same electrophoretic mobility as the native enzymes. This is consistent with a mechanism for posttranslational uptake which would not involve proteolytic modification. It remains to be demonstrated if the peroxisomal enzymes are incorporated directly into the mature organelle or if they are taken up into developing peroxisomes which may bud off from ER membranes.

Acknowledgements. This work was supported by Grants GM 21971, GM 20277, and AG 00378 from NIH. G.K. is the recipient of a Research Career Development Award (GM 00232) from NIH. M. C-G. was supported by a graduate training grant from NIH (GM 07238) and M.R. holds a postdoctoral fellowship from NIH (GM 07177).

References

Adelman MR, Blobel G, Sabatini DD (1973a) J Cell Biol 56:191-205

Adelman MR, Sabatini DD, Blobel G (1973b) J Cell Biol 56:206-229

Amar-Costesec A, Beaufay H, Wibo M, Thines-Sempoux D, Feytmans E, Robbi M, Berthet J (1974) J Cell Biol 61:201-212

Bar-Nun S, Kreibich G, Adesnik M, Alterman L, Negishi M, Sabatini DD (1980) Proc Natl Acad Sci USA 77:965-969

Bielinska M, Rogers G, Rucinsky T, Boime I (1979) Proc Natl Acad Sci USA 76:6152-6156

Birken S, Smith DL, Canfield RE, Boime I (1977) Biochem Biophys Res Commun 74:106-112

Black S, French J, Williams CH Jr, Coon MJ (1979) Biochem Biophys Res Commun 91:1528-1535

Blobel G (1977) FEBS Fed Eur Biochem Soc, 11th Meet, Copenhagen, vol 43. Academic Press Symposium A2, pp 99-108

Blobel G, Dobberstein B (1975a) J Cell Biol 67:835-851

Blobel G, Dobberstein B (1975b) J Cell Biol 67:852-862

Blobel G, Sabatini DD (1971) In: Manson LA (ed) Biomembranes, vol II. Plenum Publ Corp, New York, pp 193-195

Bonatti S, Blobel G (1979) J Biol Chem 254:12261-12264

Borgese N, Gaetani S (1980) FEBS Lett 112:216-220

Borgese D, Mok W, Kreibich G, Sabatini DD (1974) J Mol Biol 88:559-580

Bruni C, Porter KR (1965) Am J Pathol 46:691-756

Burstein Y, Schechter I (1977) Proc Natl Acad Sci USA 74:716-720

Campbell PN, Blobel G (1976) FEBS Lett 72:215-226

Chyn TL, Martonosi AN, Morimoto T, Sabatini DD (1979) Proc Natl Acad Sci USA 76:1241-1245

Colbeau A, Nachbaur J, Vignais PM (1971) Biochim Biophys Acta 249:462-492

Czakô-Graham M, Boime I, Sabatini DD, Kreibich G (1979) Fed Proc 38:2071

Czichi U, Lennarz WJ (1977) J Biol Chem 252:7901-7904

De Duve C, Baudhuin P (1966) Physiol Rev 46:323-358

De Duve C, Wattiaux R (1966) Annu Rev Physiol 28:435

De Pierre JW, Dallner G (1975) Biochem Biophys Acta 415:411-472

Dobberstein B, Garoff H, Warren G, Robinson PJ (1979) Cell 17:759-769

Dorling PR, Le Page RN (1973) Biochim Biophys Acta 318:33-44

Ehrenreich JH, Bergeron JJM, Siekevitz P, Palade GE (1973) J Cell Biol 59:45-72

Elting JJ, Chen WW, Lennarz WJ (1980) J Biol Chem 255:2325-2331

Erickson AH, Blobel G (1979) J Biol Chem 254:11771-11774

Essner E (1967) Lab Invest 17:71-87

Gaye P, Gautron JP, Mercier JC, Haze G (1977) Biochem Biophys Res Commun 79:903-911

Goldman BM, Blobel G (1978) Proc Natl Acad Sci USA 75:5066-5070

Gonzalez FJ, Kasper CB (1980) Biochemistry 19:1790-1796

Grebenau R, Sabatini DD, Kreibich G (1977) J Cell Biol 75:234a

Grinna LS, Robbins PW (1979) J Biol Chem 254:8814-8818

Habener JF, Kronenberg HM (1978) Fed Proc 37:2561-2566

Habener JF, Rosenblatt M, Kemper B, Kronenberg HM, Rich A, Potts JT Jr (1978) Proc Natl Acad Sci USA 75:2616-2620

Hanover JA, Lennarz WJ (1980) J Biol Chem 255:3600-3604

Harano T, Omura T (1977) J Biochem 82:1551-1557

Haugen DA, Coon MJ (1976) J Biol Chem 251:7929-7939

Higashi T, Peters T (1963a) J Biol Chem 238:3945-3951

Higashi T, Peters T (1963b) J Biol Chem 238:3952-3954

Himeno M, Ohhara H, Arakawa Y, Kato K (1975) J Biochem 77:427-438

Hubbard SC, Robbins PW (1979) J Biol Chem 254:4568-4576

Imai Y (1976) J Biochem 80:267-276

Jackson RC, Blobel G (1977) Proc Natl Acad Sci USA 74:5598-5602

Jarasch ED, Kartenbeck J, Bruder G, Fink A, Morrê DJ, Franke WW (1979) J Cell Biol 80:37-52

Katz FN, Rothman JE, Lingappa VR, Blobel G, Lodish HF (1977) Proc Natl Acad Sci USA 74:3278-3282

Kornfeld S, Gregory W, Chapman A (1979) J Biol Chem 254:11649-11654
Kreibich G, Ulrich B, Sabatini DD (1975) J Cell Biol 67:225a
Kreibich G, Ulrich BL, Sabatini DD (1978a) J Cell Biol 77:464-487
Kreibich G, Freienstein CM, Pereyra BN, Ulrich BL, Sabatini DD (1978b) J Cell Biol
    77:488-506
Kreibich G, Czakó-Graham M, Grebenau R, Mok W, Rodriguez-Boulan E, Sabatini DD
    (1978c) J Supramol Struct 8:279-302
Kreibich G, Czakó-Graham M, Grebenau RC, Sabatini DD (1980) Ann NY Acad Sci 343:
    17-33
Kruppa J, Sabatini DD (1977) J Cell Biol 74:414-427
Lazarow P, De Duve CH (1973a) J Cell Biol 59:491-506
Lazarow P, De Duve CH (1973b) J Cell Biol 59:507-524
Leblond CP, Bennett G (1977) In: Brinkley BB, Porter KR (eds) International cell
    biology. Rockefeller Univ Press, New York, pp 326-336
Lennarz WJ (1975) Science 188:986-991
Lingappa VR, Katz FN, Lodish HF, Blobel G (1978) J Biol Chem 253:8667-8670
Lingappa VR, Cunningham BA, Jazwinski SM, Hopp TP, Blobel G, Edelman GM (1979) Proc
    Natl Acad Sci USA 76:3651-3655
Liu CP, Slate DL, Gravel R, Ruddle FH (1979) Proc Natl Acad Sci USA 76:4503-4506
Lowe D, Hallinan T (1973) Biochem J 136:825-828
Matsuura S, Fujii-Kuriyama Y, Tashiro Y (1978) J Cell Biol 78:503-519
Milstein C, Brownlee GG, Harrison TM, Mathews MB (1972) Nature (London) New Biol
    239:117-120
Mok W, Freienstein C, Sabatini D, Kreibich G (1976) J Cell Biol 70(2):393a
Natowicz MR, Chi MMY, Lowry OH, Sly WS (1979) Proc Natl Acad Sci USA 76:4322-4326
Negishi M, Fujii-Kuriyama Y, Tashiro Y, Imai Y (1976) Biochem Biophys Res Commun
    71:1153-1160
Nilsson OS, De Pierre YW, Dallner G (1978) Biochim Biophys Acta 511:93-104
Novikoff AB, Shin WY (1964) J Microsc 3:187
Ojakian GK, Kreibich G, Sabatini DD (1977) J Cell Biol 72:530-551
Okada Y, Sabatini DD, Kreibich G (1979) J Cell Biol 83:437a
Olsen BR, Berg RA, Kishida Y, Prockop DJ (1973) Science 182:825-827
Omura T, Kuriyama Y (1971) J Biochem 69:651-658
Ozols J, Gerard C (1977) Proc Natl Acad Sci USA 74:3725-3729
Palade GE (1964) Proc Natl Acad Sci USA 52:613-634
Palade GE (1975) Science 189:347-358
Palmiter RD, Gagnon J, Ericsson LH, Walsh KA (1977) J Biol Chem 252:6386-6393
Palmiter RD, Gagnon J, Walsh KA (1978) Proc Natl Acad Sci USA 75:94-98
Pelham HRB, Jackson RJ (1976) Eur J Biochem 67:247-256
Peterkofsky B, Assad R (1979) J Biol Chem 254:4714-4720
Popov D, Alterman L, Sabatini DD, Kreibich G (1978) J Cell Biol 79:364a
Rachubinski RA, Verma DPS, Bergeron JJM (1980) J Cell Biol 84:705-716
Ragnotti G, Lawford GR, Campbell PN (1969) Biochem J 112:139-147
Ramsey JC, Steele WJ (1976) Biochemistry 15:1704-1711
Ramsey JC, Steele WJ (1979) Anal Biochem 92:305-313
Reddy JK, Azarnoff DI, Svoboda DV, Prasad JD (1974) J Cell Biol 61:344-358
Redman CM, Grab DJ, Irukulla R (1972) Arch Biochem Biophys 152:496-501
Reggio HA, Palade G (1978) J Cell Biol 77:288-314
Rodriguez Boulan E, Sabatini DD, Pereyra BN, Kreibich G (1978) J Cell Biol 78:894-
    909
Rogers MJ, Strittmatter P (1974) J Biol Chem 249:895-900
Rogers MJ, Strittmatter P (1975) J Biol Chem 250:5713-5718
Roman R, Brooker JD, Seal SN, Marcus A (1976) Nature (London) 260:359-360
Rothman JE, Lodish HF (1977) Nature (London) 269:775-780
Rothman JE, Bursytyn-Pettegrew H, Fine RW (1980) J Cell Biol 86:162-171
Russell JH, Geller DM (1975) J Biol Chem 250:3409-3413
Sabatini DD, Kreibich G (1976) In: Martonosi A (ed) The enzymes of biological mem-
    branes, vol II. Plenum Publ Corp, New York, pp 531-579
Sabatini DD, Tashiro Y, Palade GE (1966) J Mol Biol 19:503-524
Sabban E, Sabatini DD, Adesnik M, Marchesi V (1979) J Cell Biol 83:437a
Sargent JR, Vadlamudi BP (1968) Biochem J 107:839-849

Sato R, Omura T (1978) In: Cytochrome P-450. Sato R (ed) Kodansha, Tokyo and Academic Press, London, New York, pp 23-35, 138-163

Schechter I, Burstein Y, Zemell R, Ziv E, Kantor F, Papermaster D (1979) Proc Natl Acad Sci USA 76:2654-2658

Schmidt MFG, Schlesinger MJ (1979) Cell 17:813-819

Schmidt MFG, Schlesinger MJ (1980) J Biol Chem 255:3334-3339

Sharma RN, Behar-Bannelier M, Rolleston FS, Murray RK (1978) J Biol Chem 253:2033-2043

Sherwood LM, Burstein Y, Schechter I (1979) Proc Natl Acad Sci USA 76:3819-3823

Shields D, Blobel G (1978) J Biol Chem 253:3753-3756

Spatz L, Strittmatter P (1973) J Biol Chem 248:793

Stäubli W, Hess R, Weibel ER (1969) J Cell Biol 42:92-112

Strittmatter P (1963) The Enzymes. Boyer PD, Lardy H, Myrback K (eds) vol VIII. Academic Press, London, New York, p 113

Strittmatter P, Rogers MJ, Spatz L (1972) J Biol Chem 247:7188-7194

Swan D, Aviv H, Leder P (1972) Proc Natl Acad Sci USA 69:1967-1971

Swank RT, Paigen K (1973) J Mol Biol 77:371-389

Tabas I, Schlesinger S, Kornfeld S (1978) J Biol Chem 253:716-722

Toneguzzo F, Gosh HP (1977) Proc Natl Acad Sci USA 74:1516-1520

Toneguzzo F, Gosh HP (1978) Proc Natl Acad Sci USA 75:715-719

Welton AF, Aust SD (1974) Biochim Biophys Acta 373:197-210

Wickner W (1979) Annu Rev Biochem 48:23-45

Yap SH, Strair RK, Shafritz DA (1978) Biochem Biophys Res Commun 83:427-433

Transport of Membranes and Vesicle Contents During Exocytosis

M. Gratzl[1]

Introduction

In many electronmicroscopical studies, it has been observed that secretory vesicles release their contents into the extracellular fluid during exocytosis and their limiting membranes become inserted into the cell membrane. From these observations it has been deduced that secretory vesicle membranes may act as precursors of cell membrane components and thus may be involved in the biogenesis of the cell membrane.

There is abundant evidence that many membrane proteins, including those destined for the cell membrane, are synthesized by membrane-bound ribosomes at the endoplasmic reticulum (see Kreibich, this vol.). It has been suggested that cell membrane components then join the secretory pathway and are transferred via the Golgi apparatus and secretory vesicles to the cell membrane. However, considerable transport of membranes also exists from the cell membrane back to the cytoplasm (endocytosis) which seems to be coupled with the exocytotic pathway. Endocytotic vesicles end up in lysosomes or are reused in the Golgi apparatus to reform secretory vesicles (membrane recycling) (see Herzog, this vol.).

Little is known concerning the biosynthesis of individual cell membrane proteins within the cell, the kinetics of their intracellular transport to the cell membrane (is it coupled to exocytosis?), their removal from the cell membrane (endocytosis) and, if not degraded within lysosomes, their reuse in secretory vesicles. All of these mechanisms are involved in the biogenesis of the cell membrane and in the turnover of its components.

Since secretory vesicles are everted during exocytosis, those membrane proteins localized at the extracellular side of the cell membrane should also be found at the intracisternal side of secretory vesicles. The first part of this contribution will discuss whether this postulate is fulfilled. The second part will deal with the insertion mechanism itself, the initiation of membrane fusion and the respective roles of membrane lipids and proteins.

Presence of Cell Membrane Components on the Intracisternal Side of Secretory Vesicles

5'-nucleotidase is considered to be a cell membrane marker and is widely used as such. However, 5'-nucleotidase is not restricted in its dis-

[1]Department of Physiological Chemistry, University of Saarland, 665 Homburg/Saar, FRG

tribution to the cell membrane, as is usually assumed, but has also been found to be present in the endoplasmic reticulum (Widnell 1972) where it is synthesized (Bergeron et al. 1975), and in secretory vesicles from rat liver (Farquhar et al. 1974; Little and Widnell 1975).

5'-nucleotidase activity in isolated secretory vesicle fractions is latent. Thus, the enzyme was inaccessible to both antibody and concanavalin A and an increase of activity was found when detergent was included in the assay, suggesting an intravesicular location of this enzyme. This was confirmed by cytochemical procedures with which the reaction product of 5'-nucleotidases was localized on the inside of secretory vesicles (Farquhar et al. 1974; Little and Widnell 1975). With cell membranes, however, reaction product is localized on the extracellular side of the membrane both in isolated cell fractions and in situ (cf. ref. Little and Widnell 1975). These studies are in agreement with the postulated exocytotic insertion of the secretory vesicle membrane 5'-nucleotidase into the cell membrane, since the inner aspect of the secretory vesicle membranes becomes the outer aspect of the cell membrane when secretory vesicles are everted during exocytosis.

Although the primary biological action of insulin is probably exerted at the cell surface, receptors for this hormone have also been found within several intracellular membranes including secretory vesicles (Bergeron et al. 1973). Freeze-thawing markedly augmented the binding of insulin as well as that of growth hormone in secretory vesicle fractions (Bergeron et al. 1978) This behavior is compatible with the location of the hormone-binding sites on the cisternal face of secretory vesicles.

In chromaffin cells, acetylcholinesterase has been histochemically demonstrated on the cell membrane as well as in the cisternae of endoplasmic reticulum. The Golgi apparatus, where the secretory vesicles are formed, very rarely showed acetylcholinesterase activity. In secretory vesicles themselves, however, no reaction product has been found (Somogyi et al. 1975). Despite this, release of acetylcholinesterase and catecholamines into the perfusate has been observed when chromaffin cells were stimulated to secrete with depolarizing concentrations of K^+ or carbachol. The fact that the presence of Ca^{2+} in the external medium is necessary before acetylcholinesterase is released provided evidence that the release is by the processes of exocytosis (Chubb and Smith 1975).

Recently, we have isolated highly purified secretory vesicles from bovine adrenal medulla by differential and density gradient centrifugation on iso-osmolal gradients using Percoll^{TM} (Gratzl et al. 1980). These vesicles gradually released their content (e.g., adrenalin) when incubated in media of osmolalities <400 mosm/kg (Fig. 1). Under these conditions acetylcholinesterase activities increased to values comparable to values found in assays with Triton X 100 (0.12% final) included (Fig. 2). The marked increase in enzyme activity (10-15 times), resulting from the osmotic lysis or the permeability changes when detergent was added seems to establish that acetylcholinesterase is localized on the inside of adrenal medullary secretory vesicles. The origin of secretory vesicle acetylcholinesterase is difficult to evaluate. The presence of this enzyme in the endoplasmic reticulum of chromaffin cells suggests that the enzyme in secretory vesicles may be, at least in part, newly synthesized by these cells. On the other hand, in adrenal medulla, fusion of small coated vesicles with prosecretory vesicles has been observed (Bendeczky and Smith 1972). Coated vesicles in this cell have also been detected along exocytotic profiles at the cell membrane (cf. ref. Winkler 1977). Therefore, acetylcholinesterase synthesized and

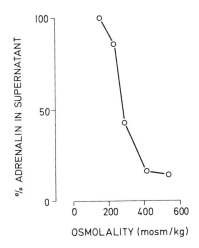

Fig. 1. Stability of secretory vesicles isolated from bovine adrenal medulla in media of different osmolality. Vesicles were incubated for 30 min at 37°C in 20 mM MOPS (pH 7.0), 5 mM EGTA and sucrose to obtain the osmolality indicated at the abscissa. Vesicles were separated from the media by centrifugation (12,000 g for 10 min) and the adrenalin released into the supernatant was determined (Gratzl et al., unpubl.)

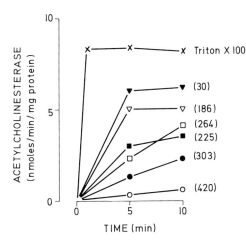

Fig. 2. Latency of acetylcholinesterase of adrenal medullary secretory vesicles. Isolated secretory vesicles were lyzed at 20°C in media of different osmolality as described in Fig. 1 or with Triton X 100 (0.12% final concentration) and the specific activity of this enzyme was determined (Ellman et al. 1961) at the times indicated at the abscissa. The values given in *brackets* represent the osmolalities of the incubation media (Gratzl et al., unpubl.)

released by the adrenal medullary cells or other cells in this tissue might have been taken up by coated vesicles during exocytosis and then transferred to secretory vesicles.

The cell membrane components found on the intracisternal side of secretory vesicles, as well as the biochemical evidence for this localization, are summarized in Table 1.

Recently, another "cell membrane enzyme" has been found in subfractionation studies in secretory vesicles. The activity of this enzyme, adenylate cyclase, in secretory vesicle membranes was even higher than in cell membranes (Cheng and Farquhar 1976a). But, however, its sidedness was exactly opposite to that of 5'-nucleotidase, hormone receptors or acetylcholinesterase. In both secretory vesicles and cell membranes it faces the cytoplasmic side; a fact which is compatible with the possibility that adenylate cyclase is also inserted exocytotically into the cell membrane (Cheng and Farquhar 1976b).

168

Table 1. Cell membrane components on the intercisternal side of secretory vesicles

Component	Evidences
5'-Nucleotidase	inaccessible to antibody and Con A, activation of enzyme activity by detergent, cytochemical localization
Hormone receptors (Insulin, Growth Hormone)	enhancement of hormone binding by freeze thawing
Acetylcholinesterase	activation of enzyme activity by detergent, activation of enzyme activity by hypotonic treatment

In stimulated cells, secretory vesicles fuse with the cell membrane as well as with each other, a process termed "compound exocytosis" which has been observed in many secretory cells (Fig. 3, cf. ref. Dahl et al. 1979; Gratzl et al. 1980). It is reasonable to assume that the membrane components responsible for the specific attachment and fusion of the membranes are localized on the interacting surfaces of the membranes; namely on the cytoplasmic surfaces of both secretory vesicle membranes and cell membranes and are, therefore, similarly arranged as adenylate cyclase.

The elucidation of the molecular mechanism of membrane fusion during exocytosis has been hampered by the lack of suitable systems for studying this process. This is mainly due to the difficulties involved in the isolation of both interacting membranes in a reasonable state of purity and, especially cell membranes, in an appropriate orientation. The fact that secretory vesicles fuse together in stimulated cells indicates that the components required for membrane fusion are present in secretory vesicle membranes. This points to the possibility of studying this process in vitro using isolated secretory vesicles. In such experiments substances, suggested to trigger exocytosis in stimulated cells, can be tested for their ability to induce membrane fusion. Furthermore, questions concerning the role of membrane lipids and proteins in this process can probably be answered.

Fusion of Secretory Vesicles in Vitro

Secretory vesicles isolated from liver (Gratzl and Dahl 1976, 1978), pancreatic islets (Dahl and Gratzl 1976; Gratzl et al. 1980), neurohypophyses (Gratzl et al. 1977) and adrenal medulla (Dahl et al. 1979;

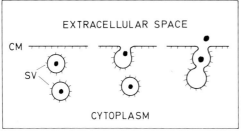

Fig. 3. Schematic representation of secretory vesicles (*SV*) close to the cell membrane (*CM*) (*left*). The cytoplasmic surfaces of the membranes are marked with *strokes*. During exocytosis (*middle*) secretory vesicles fuse with the cell membrane and discharge their content into the extracellular space. The inner surface of the secretory vesicle membrane becomes the outer surface of the cell membrane. Compound exocytosis (*right*) is characterized by fusion of secretory vesicles with each other and the cell membrane

Gratzl et al. 1980) have been subject of fusion studies on the subcellular level.

As seen in freeze-fractured suspensions, secretory vesicles in buffered sucrose media containing EGTA are dispersed. Upon addition of divalent cations (10^{-4}M final concentration) to the medium, vesicles become attached to each other. If the media were supplemented with Ca^{2+}, in addition, fused vesicles could be detected. Fused rat liver secretory vesicles are shown in Fig. 4 in freeze-fracture electronmicrographs. The continuity of the membranes of "twinned vesicles" is indicated by the continuous cleavage plane in both membrane faces exposed by freeze-fracturing. Interaction of vesicle contents after exposure of rat liver secretory vesicles to Ca^{2+} was demonstrated by the mixing of vesicles containing labeled proalbumin but inactivated converting enzyme with unlabeled, active vesicles. Addition of 10^{-4}M Ca^{2+} increased the conversion of proalbumin into albumin within the vesicles and provides quite strong evidence for the induction of fusion between the two types of vesicles (Quinn and Judah 1978).

The number of fused secretory vesicles increased with the Ca^{2+} concentration in the medium. If the percentage of fused vesicles is plotted as a function of the free Ca^{2+}-concentration the curve shown in Fig. 5 is obtained. The percentage of fused vesicles increases sigmoidally between 10^{-7}M and 10^{-4}M Ca^{2+} and reaches a plateau. None of the other divalent cations tested was able to induce fusion of secretory vesicles in concentrations lower than 1 mM. Simultaneous addition of Mg^{2+} and Ca^{2+} to secretory vesicles resulted in lower percentages of fused vesicles than was observed in the presence of 10^{-4}M Ca^{2+} alone (Table 2). Pretreatment of rat liver secretory vesicles with proteolytic en-

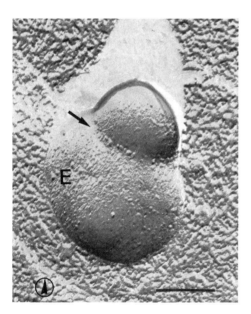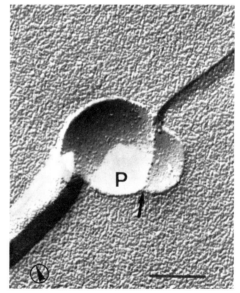

Fig. 4. Secretory vesicles isolated from rat liver in media containing 2×10^{-5}M Ca^{2+}. Twinned vesicles with a continuous cleavage plane in the membrane EF-face (*left*) as well as the membrane PF-face (*right*). *Encircled arrowhead* indicates direction of shadowing. Scale: 0.2 μm. Fracture faces are denoted according to the nomenclature introduced (Branton et al. 1975). (Gratzl and Dahl 1976)

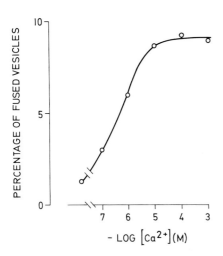

Percentage of fused vesicles as function of the Ca^{2+}-concentration. The experiments were evaluated by counting 500 vesicles for each Ca^{2+}-concentration (Gratzl and Dahl 1976, 1978)

Table 2. Cation specificity of the fusion of rat liver secretory vesicles

Cations	Percentage of fused vesicles
-	1.3
10^{-4} M Ca^{2+}	10.2
10^{-4} M Mg^{2+}	1.5
10^{-4} M Sr^{2+}	1.8
10^{-4} M Ba^{2+}	1.7
10^{-4} M La^{3+}	1.4
10^{-4} M Mn^{2+}	2.6
10^{-4} M Ca^{2+} + 10^{-4} M Mg^{2+}	7.0
10^{-4} M Ca^{2+} + 10^{-3} M Mg^{2+}	5.0

The experiments were evaluated by counting 400 vesicles for each incubation (Gratzl and Dahl 1976, 1978).

zymes, neuraminidase, or glutaraldehyde abolished fusion induced by Ca^{2+} <1 mM (Gratzl and Dahl 1978).

Fusion of isolated secretory vesicles in vitro, triggered by low [µM] concentrations of Ca^{2+}, the ineffectiveness of other divalent cations in replacing Ca^{2+}, and the inhibition of Ca^{2+}-induced fusion by Mg^{2+} are all properties common to the fusion of secretory vesicles isolated from different tissues (Gratzl and Dahl 1976, 1978; Dahl and Gratzl 1976; Gratzl et al. 1977, 1980; Gratzl et al. 1977; Dahl et al. 1979; Gratzl et al. 1980). The different efficacy of the alkaline earths and the Ca^{2+}/Mg^{2+} antagonism may well provide a clue to the biochemical nature of the Ca^{2+}-sensitive apparatus in secretory vesicle membranes.

Ca^{2+}-specificity for fusion (albeit at higher concentrations) and sensitivity to proteolytic attack was also observed with isolated cell membranes from myoblasts (Schudt et al. 1976; Dahl et al. 1978). Ca^{2+}-specific fusion of these membranes was not only abolished by enzymatic pretreatment or prefixation with glutaraldehyde but was also reduced when cell membranes were isolated from cultures supplemented with cycloheximide (Dahl et al. 1978). All of these experiments support the suggestion that membrane proteins participate in the Ca^{2+}-specific fusion of biological membranes.

Trypsinized secretory vesicles from adrenal medulla, which are unable to fuse with μM Ca^{2+}, could only be fused when exposed to very high concentrations of Ca^{2+} (>2.5 mM). However, at these concentrations Mg^{2+} and other divalent cations were equally effective (Dahl et al. 1979; Gratzl et al. 1980). Apparently, upon enzymatic attack of the membrane proteins, secretory vesicles not only lose their ability to fuse with low [Ca^{2+}], but also their specificity for this cation.

To establish further the role of membrane proteins or lipids in the fusion of isolated secretory vesicles, liposomes have been prepared from the extracted lipids of adrenal medullary secretory vesicle membranes. These liposomes, as trypsinized secretory vesicles, fuse with Ca^{2+} or other divalent cations in concentrations higher than 2.5 mM. Moreover, if Ca^{2+} and Mg^{2+} are added to the liposomes, the effect of both ions is additive (Dahl et al. 1979; Gratzl et al. 1980). Thus, the ionic requirements of liposome fusion are similar for liposomes prepared from either the membrane lipids of secretory vesicles or pure phospholipids (cf. ref. Papahadjopoulos 1978).

Comparison of Secretory Vesicle Fusion in Vitro and Secretion by Exocytosis of Intact Cells

This chapter will discuss whether fusion of secretory vesicles in vitro matches the properties of exocytosis by intact cells.

The intracellular concentration of Ca^{2+} in resting cells is low ($\leqslant$10^{-7}M) (Baker et al. 1976; Di Polo et al. 1976) but increases during stimulation. Release of transmitter is directly correlated with a rise in the intracellular Ca^{2+}-concentration (Llinás and Nicholson 1975). This was shown in the giant synapse, where the intracellular concentration of Ca^{2+} was directly monitored with injected Ca^{2+} indicators. Similarly, the intracellular Ca^{2+}-concentration of sea urchin eggs, activated to release cortical vesicle contents by exocytosis, increased to a mean value of 2.5-4.5 μM (Steinhardt et al. 1977) and in the medaka egg in the space beneath the cell membrane to 30 μM (Gilkey et al. 1978).

To find out whether the observed increase in intracellular [Ca^{2+}] parallels exocytosis or controls exocytosis directly, intracellular concentration of Ca^{2+} was increased by microinjection. Actually Ca^{2+}, injected into the presynaptic nerve terminal of the giant synapse induced transmitter release, while Mg^{2+} and Mn^{2+} were ineffective. Mg^{2+} and Mn^{2+} led to a slight reduction in the amount of transmitter released by Ca^{2+} (Miledi 1973). Also Ca^{2+}, but not Mg^{2+} injected into mast cells, elicited extrusion of secretory granules or resulted in exocytosis of cortical vesicles of amphibian oocytes (Kanno et al. 1973; Hollinger and Schütz 1976).

The concentration of intracellular Ca^{2+} can also be modified in cells rendered permeable to small molecular weight substances by high voltage discharges (Baker and Knight 1978). Such "leaky" adrenal medullary cells release less than 1% of total intracellular catecholamine when EGTA is present in the incubation medium. Addition of μM Ca^{2+} induces release of catecholamines, but not lactate dehydrogenase from the cytoplasm. Raising the Mg^{2+}-concentration to 2-50 mM reduces the amount of catecholamines released by Ca^{2+} (Baker and Knight 1978).

By comparing exocytosis by intact cells and fusion of secretory vesicles in vitro, it is obvious that both processes occur under similar conditions. Secretory vesicle fusion is low at the Ca^{2+}-concentration found in resting cells but increases with $[Ca^{2+}]$ found in stimulated cells. Exocytosis by intact cells can be triggered by increasing experimentally the intracellular Ca^{2+}-concentrations. Mg^{2+} and other divalent cations are ineffective in inducing both exocytosis and secretory vesicle fusion. Moreover, Mg^{2+} inhibits exocytosis as well as the secretory vesicle fusion activated by Ca^{2+}. Thus, it appears that Ca^{2+} is able to act specifically as the final trigger of exocytosis by initiating membrane fusion.

The fusion of liposomes prepared from the membrane lipids of secretory vesicles and trypsinized secretory vesicles does not retain the characteristic ionic requirements of the fusion of intact secretory vesicles. Fusion of the former requires >mM concentration of Ca^{2+}, whilst Mg^{2+} and other divalent cations can replace and supplement the Ca^{2+}-effect. Therefore, it can be concluded that membrane proteins of secretory vesicles account for the characteristic properties of secretory vesicle fusion. The precise role of membrane proteins in secretory vesicle fusion and exocytosis by intact cells is not known yet. They may bind Ca^{2+} specifically, act as recognition sites between the interacting membranes and/or be involved in the membrane rearrangements taking place during membrane fusion. It has recently been demonstrated that binding of Ca^{2+} to high affinity sites on secretory vesicle membranes parallels Ca^{2+}-induced fusion (Dahl et al. 1979). Further characterization and identification of the membrane components with which Ca^{2+} interacts promises progress towards the molecular mechanism of exocytotic membrane fusion.

Acknowledgments. Studies from the author's laboratory were supported by the Deutsche Forschungsgemeinschaft, Sonderforschungsbereich 38, "Membranforschung".

References

Baker PD, Knight DE (1978) Calcium-dependent exocytosis in bovine adrenal medullary cells with leaky plasma membranes. Nature (London) 276:620-622

Baker PF, Hodgkin AL, Ridgway EB (1971) Depolarization and calcium entry in squid giant axons. J Physiol 218:709-755

Benedeczky I, Smith AD (1972) Ultrastructural studies on the adrenal medulla of golden hamster. Origin and fate of secretory granules. Z Zellforsch Mikrosk Anat 124:367-386

Bergeron JJM, Evans WH, Geschwind II (1973) Insulin binding to rat liver Golgi membranes. J Cell Biol 59:771-776

Bergeron JJM, Berridge MV, Evans WH (1975) Biogenesis of plasmalemmal glycoproteins. Intracellular site of synthesis of mouse liver plasmalemmal 5'-nucleotidase as determined by the sub-cellular location of messengers RNA coding for 5'-nucleotidase. Biochim Biophys Acta 407:325-337

Bergeron JJM, Posner BI, Josefsberg Z, Sikstrom R (1978) Intracellular polypeptide hormone receptors. J Biol Chem 253:4058-4066

Branton D, Bullivant S, Gilula N, Karnovsky M, Moor H, Mühlethaler K, Northcote N, Packer L, Satir B, Satir P, Speth V, Staehlin L, Weinstein R (1975) Freeze-etching nomenclature. Science 190:54-56

Cheng H, Farquhar MG (1976a) Presence of adenylate cyclase activity in Golgi and other fractions from rat liver. I. Biochemical determination. J Cell Biol 70: 660-670

Cheng H, Farquhar MG (1976b) Presence of adenylate cyclase activity in Golgi and other fractions from rat liver. II. Cytochemical localization within Golgi and ER membranes. J Cell Biol 70:670-684

Chubb IW, Smith AD (1975) Release of acetylcholinesterase into the perfusate from the ox adrenal gland. Proc R Soc London Ser B 191:263-269

Dahl G, Gratzl M (1976) Calcium-induced fusion of isolated secretory vesicles from the islet of Langerhans. Cytobiologie 12:344-355

Dahl G, Schudt C, Gratzl M (1978) Fusion of isolated myoblast plasma membranes. An approach to the mechanism. Biochim Biophys Acta 514:105-116

Dahl G, Ekerdt R, Gratzl M (1979) Models for exocytotic membrane fusion. Symp Soc Exp Biol 33:349-368

Dipolo R, Requena J, Brinley FJ, Mullins LJ, Scarpa A, Tiffert T (1976) Ionized calcium concentrations in squid axons. J Gen Physiol 67:433-467

Ellman GL, Courtney KD, Andres V, Featherstone RM (1961) A new and rapid colorimetric determination of acetylcholinesterase activity. Biochem Pharmacol 7:88-95

Farquhar MG, Bergeron JJM, Palade GE (1974) Cytochemistry of Golgi fractions prepared from rat liver. J Cell Biol 60:8-25

Gilkey JC, Jaffe LF, Ridgway WB, Reynolds G (1978) A free calcium wave traverses the activating egg of the medaka Oryzias latipes. J Cell Biol 76:448-466

Gratzl M, Dahl G (1976) Ca^{2+}-induced fusion of Golgi-derived secretory vesicles isolated from rat liver. FEBS Lett 62:142-145

Gratzl M, Dahl G (1978) Fusion of secretory vesicles isolated from rat liver. J Membr Biol 40:343-364

Gratzl M, Dahl G, Russell JT, Thorn NA (1977) Fusion of neurohypophyseal membranes in vitro. Biochim Biophys Acta 470:45-57

Gratzl M, Ekerdt R, Dahl G (1980a) The role of Ca^{2+} as trigger for membrane fusion. Horm Metab Res, Suppl 10:144-149

Gratzl M, Krieger-Brauer H, Ekerdt R (1980b) The presence of latent acetylcholinesterase in purified secretory vesicles of adrenal medulla. Eur J Cell Biol 22:186

Gratzl M, Schudt C, Ekerdt R, Dahl G (1980c) Fusion of isolated biological membranes. A tool to investigate basic processes of exocytosis and cell-cell fusion. In: Bittar EE (ed) Membrane structure and function, vol III. John Wiley, New York, pp 59-92

Hollinger TG, Schütz AW (1976) "Cleavage" and cortical granule breakdown in Rana Pipiens oocytes induced by direct microinjection of calcium. J Cell Biol 71:395-401

Kanno T, Cochrane DE, Douglas WW (1973) Exocytosis (secretory granule extrusion) induced by injection of calcium into mast cells. Can J Physiol Pharmacol 51:1001-1004

Little JS, Widnell CC (1975) Evidence for translocation of 5'-nucleotidase across hepatic membranes in vivo. Proc Natl Acad Sci USA 72:4013-4017

Llinás R, Nicholson C (1975) Calcium role in depolarization-secretion coupling: An aequorin study in squid giant synapse. Proc Natl Acad Sci USA 72:187-190

Miledi R (1973) Transmitter release induced by injection of calcium ions into nerve terminals. Proc R Soc London Ser B 183:421-425

Papahadjopoulos D (1978) Calcium induced phase changes and fusion in natural and model membranes. In: Poste G, Nicolson GL (eds) Cell surface reviews, vol V. North Holland Publ Com Amsterdam, New York, Oxford, pp 766-790

Quinn PS, Judah JD (1978) Calcium-dependent Golgi vesicle fusion and cathepsin B in the conversion of proalbumin into albumin in rat liver. Biochem J 172:301-309

Schudt C, Dahl G, Gratzl M (1976) Calcium-induced fusion of plasma membranes isolated from myoblasts grown in culture. Cytobiologie 13:211-223

Somogyi P, Chubb IW, Smith AD (1975) A possible structural basis for the extracellular release of ecetylcholinesterase. Proc R Soc London Ser B 191:271-283

Steinhardt R, Zucker R, Schatten G (1977) Intracellular calcium release at fertilization in the sea urchin egg. Dev Biol 58:185-196

Widnell CC (1972) Cytochemical localization of 5'-nucleotidase in subcellular fractions isolated from rat liver. I. The origin 5'-nucleotidase activity in micro somes. J Cell Biol 52:542-558
Winkler H (1977) The biogenesis of adrenal chromaffin granules. Neuroscience 2: 657-683

Assembly and Turnover of the Subsynaptic Membrane

H. Betz and H. Rehm[1]

Introduction

Synapses are the basic structural elements for information transfer between excitable cells in the nervous system, and their properties and
specific arrangements confer on the brain its enormous capacity for sensory, motor and higher brain functions. During the past decade, the development of chemical synapses has attracted the interest of numerous
molecular biologists, and currently many attempts are being made to
elucidate the different processes underlying synaptogenesis, such as
outgrowth of neurites, selective innervation of the proper target organs, differentiation of pre- and post-synaptic elements or maintenance
of functionally "correct" and elimination of redundant "incorrect" synaptic contacts. In the present paper, we discuss one particular aspect
of synapse formation, the assembly of the postsynaptic membrane, with
special reference to two questions which are closely related to the
central topics of this conference:

1. How is the expression of postsynaptic proteins regulated, and

2. what mechanisms are involved in the synaptic localization of these
   proteins?

Most of the following data concern the nicotinic acetylcholine receptor (AChR), an integral protein of the subsynaptic membrane of vertebrate neuromuscular junctions and some other cholinergic synapses in
the peripheral and central nervous system. It appears, however, reasonable to speculate that mechanisms of regulation similar to those
demonstrated for AChR may also be implicated in the ontogenesis of
synapses other than cholinergic.

Structure and Function of a Typical Vertebrate Synapse, the Neuromus-
cular Junction

Hitherto the best-characterized synapse in vertebrates is that between
motor neurons and striated muscle fibers, the neuromuscular junction
or motor endplate (for review see Fambrough 1976; Changeux 1979). As
illustrated in Fig. 1, this synapse exhibits elaborate specializations
of both its pre- and postsynaptic cells as well as in the extracellular
material associated with the synaptic cleft, the synaptic basal lamina.
The presynaptic nerve terminal contains numerous vesicles of 30 to 60
nm diameter. These are filled with the neurotransmitter acetylcholine
(ACh) and are particularly clustered in the vicinity of specialized
sites of transmitter release, the so-called active zones. Underneath

[1]Max-Planck-Institut für Psychiatrie, Abt. Neurochemie, 8033 Martinsried, FRG

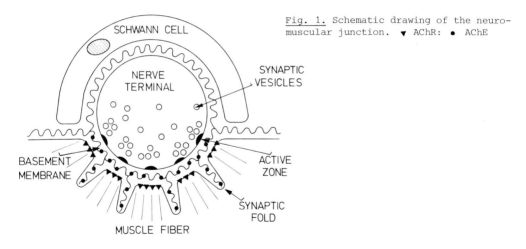

Fig. 1. Schematic drawing of the neuro-muscular junction. ▼ AChR: ● AChE

the nerve terminal, the plasma membrane of the postsynaptic muscle cell is considerably thickened and thrown in regular folds which, at their crests, contain receptors for ACh (AChR) at a very high density ($\geq 10^4$ per μm^2). In fact, AChR is the major protein there and is present in a semicrystalline arrangement. Between the folds, numerous bundles of "subneural" filaments 5 nm in diameter are observed which anchor the postsynaptic membrane to the muscle cell cytoskeleton. Within the synaptic cleft, the enzyme acetylcholinesterase (AChE) is highly concentrated in the basal lamina of the muscle cell, the total number of enzyme molecules being equal to that of AChR.

Depolarization of the nerve terminal by an invading action potential causes several hundreds of synaptic vesicles to be fused with the presynaptic membrane and their ACh to be released into the synaptic cleft. The transmitter diffuses through the cleft and binds to the AChR, thereby inducing the opening of transmembrane channels that are permeable to sodium and potassium ions. This depolarizes the muscle cell and ultimately leads to contraction. During the short time required for binding of ACh to the receptor, only a small amount of transmitter is thought to be removed by AChE via direct binding. ACh dissociating from the receptors, however, is rapidly hydrolyzed by the enzyme, and thus neurotransmission is terminated in about one millisecond.

Properties and Distribution of AChR on Embryonic and Adult Muscle Cells

During the past ten years, the structure and the function of AChR have been investigated in great detail. Due to the availability of highly selective ligands, α-toxins from snake venoms which form extremely stable complexes with AChR, and a natural source very rich in this protein, fish electric organ, AChR could be obtained in large quantities in a highly purified form. Thus it was the first receptor for hormones and neurotransmitters to be amenable to biochemical and biophysical analysis. Some major properties of AChR are listed in Table 1. The picture that currently emerges from various studies is that AChR is a highly complex oligomeric membrane protein with different and multiple binding sites for ACh agonists and antagonists, α-toxins and noncompetitive inhibitors of ACh-mediated ion translocation (for review see Heidmann and Changeux 1978).

177

Table 1. Properties of AChR

AChR

a) is an integral membrane glycoprotein which can be solubilized by nondenaturing detergents;

b) spans the lipid bilayer of the subsynaptic membrane;

c) is a multimeric protein (MW about 250 K) composed of 4 different subunits $(\alpha_2 \beta \gamma \delta)$;

d) contains one type of polypeptide (MW 40 to 50 K) which is labeled by affinity reagents for the acetylcholine recognition site;

e) contains additional polypeptides (50 to 65 K) with binding sites for substances interfering with ACh-mediated ion translocation (histrionicotoxin and local anesthetics);

f) has two binding sites for α-bungarotoxin and other snake α-toxins.

Electrophysiological mapping of ACh sensitivity and labeling with radio-active snake α-toxins have been widely used to quantify and localize AChR on embryonic and adult muscle cell membranes (Fambrough 1979). These studies revealed that AChR undergoes a developmental regulation and redistribution on the myofiber surface as shown in Fig. 2. The fusion of mononucleated myoblasts to form myotubes induces the synthesis of AChR, the newly synthesized receptors being randomly incorporated into the membrane of the developing myotubes. Upon innervation, receptors cluster under the contacting nerve terminal, and the embryonic "extrasynaptic" receptors disappear from the remainder of the cell surface. In the adult innervated fiber, AChR is highly concentrated in the subsynaptic part of the muscle membrane, its density there being at least 1,000 times higher than in the extrasynaptic regions and 1 to 2 orders of magnitude larger than on noninnervated embryonic myotubes.

Because of their rather different distribution, the properties of extra- and subsynaptic AChR's have been compared in great detail. Some of the data obtained by different groups are listed in Table 2. In addition to their unequal distribution, extra- and subsynaptic receptors can be distinguished by a different affinity for the antagonist d-tubocurarine and a different isoelectric point. Also, extrasynaptic receptors exhibit lateral mobility in the plane of the membrane, whereas clustered (sub-synaptic) receptors are essentially immobile. Upon activation, the ion channels of extrasynaptic AChR stay open for a longer time period than those of the subsynaptic receptors. Antisera of patients suffering from myasthenia gravis, an autoimmune disease against muscle AChR, detect more determinants on extra- than on subsynaptic receptor. Finally, both

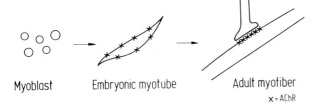

Myoblast   Embryonic myotube   Adult myofiber

x = AChR

Fig. 2. Cell surface distribution of AChR during muscle development. Explanation see text

Table 2. Comparison of extra- and subsynaptic AChR's. For references see
Fambrough 1979

	Extrasynaptic (embryonic)	Subsynaptic (adult)
Distribution	Diffuse $(10^2$ to 10^3 per $\mu m^2)$	Clustered (ca. 10^4 per $\mu m^2)$
Pharmacology (d-tubocurarine)	App $K_{D\ extra}$ $(5.5 \times 10^{-7}$ M) $>$	App $K_{D\ sub}$ $(4.5 \times 10^{-8}$ M)
Isoelectric point	5.3	5.1
Diffusion coefficient	Large $(5 \times 10^{-11}$ $cm^2/s)$	Small $(\ll 10^{-12}$ $cm^2/s)$
Channel opening time	3 to 5 ms	0.7 to 1 ms
Antigenic determinants detected by myasthenia gravis antibodies	++++	++
Metabolic half-life	15 to 30 h	5 to 15 days

types of AChR have different metabolic properties in vivo: extrasynaptic receptors are subjected to continuous turnover with a half-life of about 20 h, whereas subsynaptic receptors in the adult are very stable and exhibit a half-life of at least 5 days, presumably weeks.

From Fig. 2 and Table 2 it is evident that motor endplate formation, as far as AChR is concerned, involves at least the following events:

1. The induction of AChR synthesis and its incorporation into the plasma membrane of the muscle cell;

2. the aggregation of AChR in the developing subsynaptic membrane;

3. the removal of extrasynaptic AChR sites;

4. the metabolic stabilization of subsynaptic AChR; and

5. alterations in the physicochemical properties of AChR.

Recent experiments have shown that these events occur at different stages of development. Apparently, endplate formation is a complex multi-step process.

Regulation and Synaptic Localization of AChR during Endplate Formation

Several lines of evidence suggest that the various changes in AChR properties and cell surface distribution occurring during endplate formation are related to the innervation and contractile activation of the muscle cell. In order to probe for the role of the nerve terminal in the regulation of AChR metabolism and distribution, two main experimental paradigms have been used: (1) the blockade of neuromuscu-

lar transmission and muscle activity in the developing embryo by injecting ACh antagonists, i.e., curare or its analogs, and (2) the modulation of AChR expression and aggregation in muscle cell cultures by
adding dissociated motor neurons, nerve extracts, drugs etc. The cell
culture approach also has been very useful to unravel the basic metabolism of extrasynaptic AChR in noninnervated myotubes (Fambrough 1979).
A brief summary of these results is given in Fig. 3 in the form of a
diagram which represents the dynamic flow or "life cycle" of AChR in
the muscle cell. Basically, the synthesis and insertion into the plasma
membrane of AChR, its clustering and its internalization and subsequent
degradation follow the pathways used by any other membrane protein.
A particular feature of AChR, however, is that the first two processes
are known to be modified by nerve-derived signals.

Induction of AChR Clustering by Neural Aggregation Factors

In a series of elegant experiments Anderson et al. (1977) demonstrated
that the addition of motor neurons to cultured *Xenopus* myocytes whose
AChR had previously been labeled with fluorescent α-bungarotoxin causes
an accumulation of fluorescence under sites of nerve-muscle contact.
This suggests that motor neurons can induce a redistribution of receptors on the surface of muscle cells. Furthermore, extracts from spinal

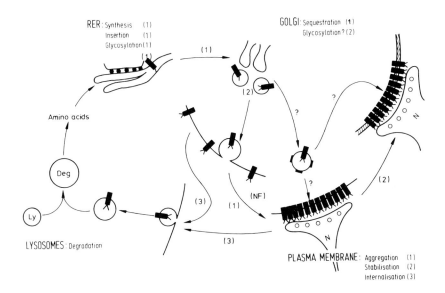

Fig. 3. "Life cycle of AChR". At the rough endoplasmic reticulum (*RER*), AChR is synthesized, inserted into the ER membrane and glycosylated (*1*). After its transport
to the Golgi region, it is sequestered in vesicles (*1*) and presumably further glycosylated (*2*). AChR-containing vesicles then fuse with the noninnervated plasma membrane. In the plane of the plasma membrane, AChR is laterally mobile. Upon innervation in vivo or spontaneously in culture, AChRs aggregate into clusters (*1*). Diffuse and clustered (subsynaptic) AChR is internalized into vesicles (*3*) which fuse
with primary lysosomes and thus initiate degradation of the AChR protein. In the
subsynaptic plasma membrane, AChR becomes resistant to degradation at later stages
of development (*2*). Golgi-derived AChR-containing coated vesicles may provide a pathway for the directional incorporation of AChR into the developing subsynaptic membrane (Bursztajn and Fischbach 1979)

cord, brain or nerve cell lines have been shown to contain proteinaceous factors which enhance the aggregation of AChR on cultured myotubes (Podleski et al. 1978; Christian et al. 1978). One major active principle has recently been identified as a low MW peptide (1,600 daltons; Jessels et al. 1979). Its mechanism of action, whether serving as a lectin-like glue between different receptor molecules or rendering the AChR aggregation-competent by some direct or indirect structural modification, is unknown.

Regulation of AChR Metabolism by Muscle Activity

In 1960, Miledi demonstrated that denervation of adult muscle increases the acetylcholine sensitivity in extrasynaptic regions of the myofiber. Later on, Lømo et al. (1976) found that this increase in ACh sensitivity can be prevented by electrical stimulation of the muscle, thus, muscle activity is an important factor in the regulation of extrasynaptic AChR in adult muscle. "Neurotrophic" factors other than ACh have, however, also been implicated (Fambrough 1979).

Neurally evoked muscle activity also occurs in the embryo at very early stages of development, signs of motility being detectable in the embryonic chick at 4 days of incubation. Here, such activity is essential for the restriction of ACh sensitivity to the developing endplate region. As shown in Fig. 4, the total AChR content of a typical fast muscle of the chick, the breast muscle, reaches a maximum around in ovo at 14 days, i.e., at a time where typical AChE-positive endplates become detectable by histochemical staining. Thereafter, the AChR content of the muscle decreases until the end of embryonic development (Betz et al. 1977; Burden 1977a). By autoradiographic techniques, this decrease in receptor number can be shown to be due to the degradation of excess extrasynaptic AChR not used in the assembly of the subsynaptic membrane (Betz et al. 1977; Burden 1977a). Labeling of AChR in vivo by {^{125}I} α-bungarotoxin injection allows the determination of the half-life of the receptor protein. As summarized in Table 3, the latter does not change during chick embryo development, but only several weeks after hatching. (The situation is, however, different in the rat.) The fall in AChR after the 14th day thus results from a repression of AChR

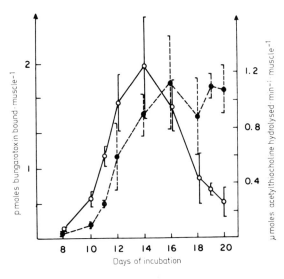

Fig. 4. AChR and AChE contents of chick embryo breast muscle during in ovo development. o AChR; ● AChE

Table 3. Half-life of AChR during the development of chick and rat muscle. Data from Devreotes and Fambrough 1975; Kao and Drachman 1977; Berg and Hall 1974; Burden 1977a,b; Betz et al. 1980; see also Fambrough 1979

Species and types of muscle	$t_{1/2}$ in hours
Chicken	
Cultured embryonic myotubes	16-28
Pectoral muscle and PLD, 8 to 20 days in ovo	25-3o
Pectoral muscle and PLD, 10 days after hatching	25-30
Pectoral muscle and PLD, 3 to 5 weeks after hatching	65 to >120
Rat	
Cultured embryonic myotubes	16-22
+ myasthenia gravis serum	<4-11
Adult diaphragm,	
extrasynaptic regions	7-19
junctional regions	>120

synthesis. Comparative studies with different muscles of the chick and paralysis experiments revealed that this repression depends on the extent of neurally evoked muscle activity (Bourgeois et al. 1978; Betz et al. 1980; Burden 1977a). In conclusion, after the formation of functional neuromuscular contacts the size of the pool of extrasynaptic AChR is controlled by neurotransmitter release from the innervating motor neuron terminal.

In contrast to AChR synthesis, the degradation of the receptor is insensitive to muscle activity, but accelerated by AChR antibodies (Table 3). The mechanisms underlying the stabilization of subsynaptic AChR in maturing endplates (3 weeks after hatching in the chick, around birth in the rat; see Table 3) are unknown. Covalent modifications of AChR have been implicated in this process (Changeux and Danchin 1976). In membrane preparations from fish electric organ, AChR can be phosphorylated in vitro by an endogenous protein kinase (Gordon et al. 1977; Teichberg et al. 1977). The physiological relevance of this phosphorylation is, however, unsolved. Alternatively, changes in synaptic architecture during endplate maturation may render the entire postsynaptic membrane resistant to degradation.

A Hypothesis for the Control of AChR Synthesis by Muscle Activity

Organ and monolayer cultures of avian and mammalian muscle cells have proven very useful in studies on the control of AChR by muscle activity. Upon morphological maturation, chick myotubes in vitro express AChR, develop action potentials and exhibit spontaneous contractile activity. Paralysis of the cells by tetrodotoxin, a blocker of voltage-dependent sodium channels, leads to an increased incorporation of AChR into the plasma membrane (Shainberg and Burstein 1976; Betz and Changeux 1979; see Table 4). In contrast, depolarization by electrical stimulation

Table 4. Accumulation of ^{125}I-α-bungarotoxin binding sites in 7-day-old muscle cultures after saturation of pre-existing AChR sites with unlabeled α-bungarotoxin. By this protocol, only newly synthesized AChR is detected. Modified according to Betz and Changeux 1979

Culture conditions	Concentration (M)	^{125}I-α-bungarotoxin bound (cpm ± SEM per dish)
Control	-	15.023 ± 1.412
Veratridine	1.5×10^{-6}	8.688 ± 1.318
Tetrodotoxin (TTX)	10^{-6}	25.799 ± 1.624
Dibutyryl cGMP	2.5×10^{-4}	7.772 ± 339
Dibutyryl cAMP	2.5×10^{-4}	18.863 ± 568
TTX + dibutyryl cGMP	As above	8.473 ± 393
TTX + dibutyryl cAMP	As above	33.260 ± 1.676
Veratridine + dbcGMP	As above	8.001 ± 588
Veratridine + dbcAMP	As above	11.155 ± 847

or veratridine, a drug which activates the voltage-dependent sodium channels, lowers the number of surface receptors below controls. In other words, in mature myotubes in vitro AChR is subject to regulation by activity as it is in vivo. Interestingly, dibutyryl cGMP has been found to inhibit AChR synthesis in a similar fashion as veratridine and to abolish the increase in AChR caused by tetrodotoxin (Table 4; Betz and Changeux 1979). Dibutyryl cAMP, by contrast, increases AChR and protein synthesis in myotubes. Since repression of AChR by muscle activity is known to be dependent on Ca^{2+} (Prives et al. 1976), it is postulated that cyclic nucleotides act as intracellular messengers of AChR regulation (Fig. 5): cyclic AMP is thought to be involved in the induction of receptor synthesis, whereas Ca^{2+} entering during excitation mediates its repression by a cGMP-dependent mechanism. Recent results on cGMP levels in carbachol-treated or electrically stimulated adult muscle support this model (Garber et al. 1978; Beam at al. 1977; Nestler et al. 1978).

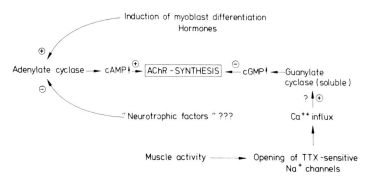

Fig. 5. Hypothetical scheme on the regulation of extrasynaptic AChR synthesis

Regulation and Synaptic Localization of Muscle AChE

Though AChE is present in the synaptic cleft in amounts equal to that
of AChR and these proteins are suspected of having complementary functions in neurotransmission, they are regulated differently during motor-
endplate formation: (1) the total activity of AChE in embryonic muscle
or cultured myotubes follows a developmental pattern different from
that of AChR (Fig. 4; Prives et al. 1976; Betz et al. 1977); (2) denervation decreases total muscle AChE, but paralysis has no effect on
its accumulation (Hall 1973; Bourgeois et al. 1978; Betz et al. 1980).
All the currently available results are compatible with the assumption
that the expresssion of total AChE is relatively insensitive to muscle
activity[1] but stimulated by neurotrophic factors released from the
nerve terminal (Younkin et al. 1979). Soluble neuronal proteins (MW
25 or 84 k daltons) supporting myotube differentiation and/or AChE
accumulation have been described and, in one case, isolated in pure
form (Davey et al. 1979; Markelonis and Oh 1979). A precise analysis
of the effects of these proteins is, however, complicated by the fact
that at least three different molecular forms of AChE are present in
innervated muscle which can be distinguished by their sedimentation
coefficients (Hall 1973; Vigny et al. 1976). The heavy form of the
enzyme (19.5S in chick, 16S in rat) is sensitive to collagenase and
is specifically associated with the endplate region, whereas the
lighter forms are synthesized and secreted by denervated muscle or
cultured myotubes, too (Wilson et al. 1973; Rotundo and Fambrough 1979).
The biochemical relationship between the different species of AChE is
unknown. For the chick, kinetic data suggest that all forms have a
common catalytic subunit. Interestingly, heavy AChE is also found in
neurons and rapidly transported down the innervating axons (Di Giamberardino and Couraud 1979). Some of the heavy synaptic AChE thus might
be neurogenic.

In contrast to what is observed for the expression of total AChE, the
accumulation of sites of esterase activity at newly formed synapses
depends on muscle activity: both in the embryo and in nerve-muscle
cocultures, paralysis by AChR antagonists or tetrodotoxin prevents
the appearance of histochemically detectable AChE patches and of the
endplate-specific heavy form of the enzyme (Giacobini et al. 1973;
Bourgeois et al. 1978; Betz et al. 1980; Rubin et al. 1980). In vitro,
the effects of paralysis can be reversed by the addition of dibutyryl
cGMP. Apparently, the control of the developmental localization of
AChE within the endplate involves mechanisms similar to that postulated
for the activity-mediated repression of AChR synthesis (see Fig. 4).

Neuronal α-Bungarotoxin Receptors

Various regions of the central and peripheral nervous system have been
shown to contain receptors for α-bungarotoxin with pharmacologically
typical nicotinic binding characteristics. Although the nature of these
receptors is still a matter of debate, they may be tentatively identified as nicotinic AChR. In the retina of the chick embryo, these receptors are concentrated at certain synapses in the inner plexiform layer
(Vogel and Nirenberg 1976). In tissue culture, a high percentage of
retinal neurons expresses binding sites for the toxin which are evenly

---

[1]Controversal findings, i.e., a reduction of AChE accumulation by electrical stimulation, have also been reported (Walker and Wilson 1975)

184

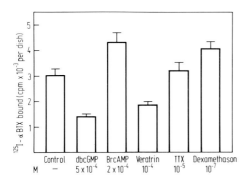

Fig. 6. Accumulation of α-bungarotoxin binding sites in chick embryo retina cultures. The different drugs were added from day 2 to 6 in vitro

distributed over the cell surface. Recently we observed that the accumulation of these receptors in culture can be modulated by depolarizing agents, cyclic nucleotide derivatives and steroid hormones. As shown in Fig. 6, 8-bromo-cAMP or dexamethasone increase the number of receptors per dish, whereas veratridine or dibutyryl cGMP reduce it. These observations suggest that putative neurotransmitter receptors in the central nervous system may be regulated by neuronal activity, the latter's action being coupled to cyclic nucleotides.

Conclusions and Perspectives

Present evidence suggests that, in the case of AChR, formation of the subsynaptic membrane involves modulations both of the metabolism of this protein and of the association between receptor molecules or between receptors and other constituents of the developing synapse. As summarized in Table 5, both processes are independently controlled by "orthograde" signals released from the presynaptic cell, the neurotransmitter itself and coreleased "trophic" proteins acting as hormonal messengers for this regulation. The same or similar signals also serve as inducers of the synaptic localization and of the synthesis of AChE.

In analogy to these orthograde signals, retrograde signals from muscle to nerve are required for synaptogenesis. At present, it is not under-

Table 5. Cell-cell signaling during motor endplate development

I.	*Orthograde signals from nerve to muscle:*		
	a) Acetylcholine	-	Repression of extrasynaptic AChR by muscle activity
		-	Induction of the endplate specific form of AChE by muscle activity
	b) "Trophic" proteins	-	Aggregation of AChR in the subsynaptic membrane
		-	Induction of total muscle AChE
II.	*Retrograde signals from muscle to nerve:*		
	a) "Trophic" proteins	-	Induction of choline acetyltransferase
	b) Factors postulated to be controlled by muscle activity	-	Regulation of the size and of the assembly of the nerve terminal

stood how the development of the presynaptic compartment is matched
with that of the postsynaptic membrane. One protein generally implicated in such retrograde regulations is nerve growth factor, a hormone
which markedly affects the survival and differentiation of sensory and
sympathetic neurons (Levi-Montalcini and Angeletti 1968). Currently,
studies on the assembly of the presynaptic membrane are hampered by
the lack of specific markers for this membrane. β-bungarotoxin, a neu-
rotoxin affecting transmitter release from cholinergic synapses (Strong
et al. 1976), might provide a useful probe for such investigations.
Recently, a specific binding of this toxin to retinal membranes has
been demonstrated (Rehm and Betz, in prep.).

Acknowledgments. One of us, H.B., is greatly indebted to Drs. J.P. Bourgeois and
J.P. Changeux for their collaboration, generous support and hospitality during part
of the work described in this publication. We also thank Ms. U. Müller for technical
assistance, Drs. D. Edgar and F. Pfeiffer for a critical reading, and Mrs. C. Bauer-
eiß and E. Eichler for help during the preparation of the manuscript.
Supported by the Deutsche Forschungsgemeinschaft (Be 718/1,2,4,5).

References

Anderson MJ, Cohen MW, Zorychta E (1977) Effects of innervation on the distribution
  of acetylcholine receptors on cultured muscle cells. J Physiol 268:731-756
Beam KG, Nestler EJ, Greengard P (1977) Increased cyclic GMP levels associated with
  contraction in muscle fibers of the giant barnacle. Nature (London) 267:534-535
Berg DK, Hall ZW (1974) Fate of α-bungarotoxin bound to acetylcholine receptors of
  normal and denervated muscle. Science 184:473-475
Betz H, Changeux JP (1979) Regulation of muscle acetylcholine receptor synthesis
  in vitro by cyclic nucleotide derivatives. Nature (London) 278:749-752
Betz H, Bourgeois JP, Changeux JP (1977) Evidence for degradation of the acetylcholine (nicotinic) receptor in skeletal muscle during the development of the
  chick embryo. FEBS Lett 77:219-224
Betz H, Bourgeois JP, Changeux JP (1980) Evolution of cholinergic proteins in developing slow and fast muscles of chick embryo. J Physiol (London) 302:197-218
Bourgeois JP, Betz H, Changeux JP (1978) Effets de la paralysie chronique de l'em-
  bryon de poulet par le flaxedil sur le développement de la jonction neuromuscu-
  laire. C R Acad Sci Paris Ser D 286:773-776
Burden S (1977a) Development of the neuromuscular junction in the chick embryo:
  the number, distribution, and stability of acetylcholine receptors. Dev Biol
  57:317-329
Burden S (1977b) Acetylcholine receptors at the neuromuscular junction: developmental
  change in receptor turnover. Dev Biol 61:79-85
Bursztain S, Fischbach GD (1979) Coated vesicles in cultured myotubes contain acetylcholine receptors. Proc Soc Neurosci 9:1613
Changeux JP (1979) Molecular interactions in adult and developing neuromuscular
  junctions. In: Schmitt FO, Worden FG (eds) The neurosciences, 4th Study Program.
  MIT press, Cambridge, pp 749-778
Changeux JP, Danchin A (1976) Selective stabilization of developing synapses as a
  mechanism for the specification of neuronal networks. Nature (London) 264:705-712
Christian CN, Daniels MP, Sugiyama H, Vogel Z, Jacques L, Nelson PC (1978) A factor
  from neurons increases the number of acetylcholine receptor aggregates on cultured
  muscle cells. Proc Natl Acad Sci USA 75:4011-4015
Davey B, Younkin LH, Younkin SG (1979) Neural control of skeletal muscle cholinester-
  ase: a study using organ-cultured rat muscle. J Physiol 289:501-515
Devreotes PN, Fambrough DM (1975) Acetylcholine receptor turnover in membranes of
  developing muscle fibers. J Cell Biol 65:335-358
Garber AJ, Entman ML, Birnbaumer L (1978) Cholinergic stimulation of skeletal muscle
  alanine and glutamine formation and release. Evidence for mediation by a nicotinic
  cholinergic receptor and guanosine 3':5'-monophosphate. J Biol Chem 253:7924-7930

Giacobini G, Filogamo G, Weber M, Boquet P, Changeux JP (1973) Effects of a snake α-neurotoxin on the development of innervated skeletal muscles in chick embryo. Proc Natl Acad Sci USA 70:1708-1712

Giamberardino Di L, Couraud JY (1978) Rapid accumulation of high molecular weight acetylcholinesterase in transected sciatic nerve. Nature (London) 271:170-172

Gordon AS, Davis CG, Diamond I (1977) Phosphorylation of membrane proteins at a cholinergic synapse. Proc Natl Acad Sci USA 74:263-267

Hall ZW (1973) Multiple forms of acetylcholinesterase and their distribution in endplate and non-endplate regions of rat diaphragm muscle. J Neurobiol 4:343-361

Heidmann T, Changeux JP (1978) Structural and functional properties of the acetylcholine receptor protein in its purified and membrane-bound states. Annu Rev Biochem 47:317-357

Jessels TM, Siegel RE, Fischbach GD (1979) Induction of acetylcholine receptors on cultured skeletal muscle by a factor extracted from brain and spinal cord. Proc Natl Acad Sci USA 76:5397-5401

Kao I, Drachman DB (1977) Myasthenic immunoglobulin accelerates acetylcholine receptor degradation. Science 196:527-529

Levi-Montalcini R, Angeletti PU (1968) Nerve growth factor. Physiol Rev 48:534-569

Lømo T, Westgaard RH (1976) Control of ACh sensitivity in rat muscle fibers. Cold Spring Harbor Symp. Quant Biol 40:263-274

Markelonis G, Oh TH (1979) A sciatic nerve protein has a trophic effect on development and maintenance of skeletal muscle cells in culture. Proc Natl Acad Sci USA 76:2470-2474

Miledi R (1960) The ACh sensitivity of frog muscle fibers after complete or partial denervation. J Physiol 151:1-23

Nestler EJ, Beam KG, Greengard P (1978) Nicotinic cholinergic stimulation increases cyclic GMP levels in vertebrate skeletal muscle. Nature (London) 27:451-453

Podleski TR, Axelrod D, Ravdin P, Greenberg I, Johnson MM, Salpeter MM (1978) Nerve extract induces increase and redistribution of acetylcholine receptors on cloned muscle cells. Proc Natl Acad Sci USA 75:2035-2039

Prives J, Silman I, Amsterdam A (1976) Appearance and disappearance of acetylcholine receptor during differentiation of chick skeletal muscle in vitro. Cell 7:543-550

Rotundo RL, Fambrough DM (1979) Molecular forms of chicken embryo acetylcholinesterase in vitro and in vivo. Isolation and characterization. J Biol Chem 254:4790-4799

Rubin LL, Schuetze SM, Weill CL, Fischbach GD (1980) Regulation of acetylcholinesterase appearance at neuromuscular junctions in vitro. Nature (London) 283:264-267

Shainberg A, Burstein M (1976) Decrease of acetylcholine receptor synthesis in muscle cultures by electrical stimulation. Nature (London) 264:368-369

Strong PN, Goerke J, Oberg SG, Kelly RB (1976) β-Bungarotoxin, a pre-synaptic toxin with enzymatic activity. Proc Natl Acad Sci USA 73:178-182

Teichberg V, Sobel A, Changeux JP (1977) In vitro phosphorylation of the acetylcholine receptor. Nature (London) 267:540-542

Vigny M, Koenig J, Rieger F (1976) The motor endplate specific form of acetylcholinesterase: appearance during embryogenesis and re-innervation of rat muscle. J Neurochem 27:1347-1353

Vogel Z, Maloney GJ, Ling A, Daniels MP (1977) Identification of synaptic acetylcholine receptor sites in retina with peroxidase-labeled α-bungarotoxin. Proc Natl Acad Sci USA 74:3268-3272

Walker CR, Wilson BW (1975) Control of acetylcholinesterase by contractile activity of cultured muscle cells. Nature (London) 256:215-216

Wilson BW, Nieberg PS, Walker CR, Linkhart TA, Fry DM (1973) Production and release of acetylcholinesterase by cultured chick embryo muscle. Dev Biol 33:285-299

Biogenesis of Peroxisomes
and the Peroxisome Reticulum Hypothesis

P. B. Lazarow, H. Shio, and M. Robbi[1]

In considering the biogenesis of peroxisomes, there are five types of experimental observations to take into account. We will briefly review each, and then attempt to formulate a consistent explanation of them.

Data

Biochemical Homogeneity of Peroxisomes

Peroxisomes display a remarkable biochemical homogeneity. Differences in enzyme distributions have not been observed as a function of the size or density of the organelle (Beaufay et al. 1964; Leighton et al. 1968; Poole et al. 1970; Lazarow and De Duve 1973b). About the same proportion (~30%) of each of the various peroxisomal enzymes is found in soluble form in homogenates, most likely due to breakage of some peroxisomes during homogenization. The one apparent exception to this homogeneity is urate oxidase; this is due to the fact that it is located in the peroxisome's insoluble core structure. Figure 1 illustrates the homogeneity of peroxisomes as a function of density. The four peroxisomal enzymes measured have very similar distributions, except for urate oxidase, whose slight excess of activity at the bottom of the gradient is accounted for by free peroxisomal cores (Leighton et al. 1968).

Peroxisomes also appear to be homogeneous as a function of age. Poole et al. (1970) looked for growth of peroxisomes and found none, within a time span of 3 h to 1 week: young peroxisomes were indistinguishable from old ones. Lazarow and De Duve (1973b) found that the distribution of newly synthesized, radioactive catalase was the same as total enzymatically active catalase 1 h after labeling (Fig. 2). Moreover, this included the newly made precursor of catalase that had not yet been assembled into active tetramers.

Uniform Turnover of Peroxisomal Proteins

Peroxisomal proteins appear to have the same half-life (Poole et al. 1969). The measured $t_{1/2}$ with leucine as labeled precursor is 3.5 days. Poole (1971) has shown that when these data are corrected for leucine reutilization, the half-life drops to 1.5 days, in good agreement with the value obtained for catalase turnover by two other independent methods (Price et al. 1962).

---

[1]The Rockefeller University, New York, NY 10021, USA

188

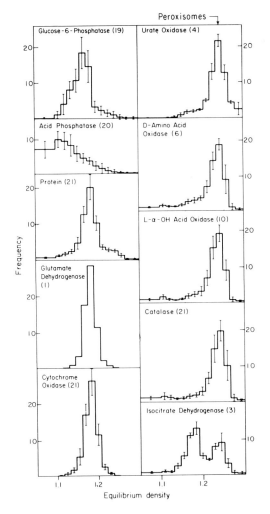

Fig. 1. Subfractionation of small mitochondrial fraction. A small mito- chondrial "lambda" fraction was pre- pared by differential centrifugation, and then subfractionated by isopycnic centrifugation in a linear sucrose gradient. On average, the lambda frac- tion contained 39% of the peroxisomes, 26% of the mitochondria and 19% of the lysosomes, but only 8.1% of the protein and 2.4% of the microsomes of the homo- genate. Marker enzymes: glucose-6-phos- phatase (microsomes), acid phosphatase (lysosomes), glutamate dehydrogenase and cytochrome oxidase (mitochondria). Isocitrate dehydrogenase is located partly in mitochondria and partly in peroxisomes. The *numbers in parentheses* are the number of experiments. (Leighton et al. 1968)

Morphological Observations

In reviewing the substantial literature of electron microscopic studies of peroxisomes, as well as our own observations, several things stand out that may be relevant to the problem of biogenesis. As first noted by Novikoff and Shin (1964), there is a tendency for peroxisomes to appear in clusters, as illustrated in Fig. 3A. Occasionally clusters give the impression of beads on a string (Fig. 3B). Clustering has been described by Legg and Wood (1970), Essner (1970), Reddy and Svoboda (1971), Hicks and Fahimi (1977), and Dougherty and Lee (1967).

Peroxisomes sometimes appear to be interconnected with each other (Legg and Wood 1970; Rigatuso et al. 1970; Reddy and Svoboda 1971, 1973). Several examples of dumb-bell shaped peroxisomes are illustrated in Fig. 4. Although these images are rare in normal liver, they appear more frequently at times of peroxisome proliferation, either due to treatment with clofibrate (Reddy and Svoboda 1971, Figs. 7, 12, 15) or other hypolipidemic drugs (Leighton et al. 1975, Fig. 15) or to partial hepatectomy (Rigatuso et al. 1970, Figs. 4, 5). Additional examples of

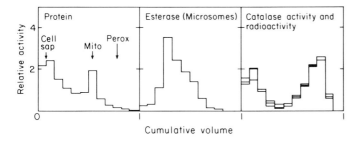

Fig. 2. Distribution of young (1-h-old) and total catalase. One hour after the in-
traportal injection of ^{3}H-leucine into a rat, a liver postnuclear supernatant was
prepared and centrifuged into a sucrose gradient. The position of the starting layer
was the left-most two fractions. The smaller peroxisomes did not quite reach their
equilibrium position near the bottom of the gradient. In addition, about 1/3 of the
catalase activity remained at the top of the gradient due to peroxisomes broken
during homogenization. The *right-hand box* contains three plots: catalase enzymatic
activity; radioactive catalase including precursors (about 36% of the total cpm)
isolated immunochemically; radioactive catalase, excluding precursors, isolated
immunochemically after a partial chemical purification procedure in which the precursors are lost. In each box the total activity (or radioactivity) is normalized
to 1. (Figure redrawn from data included in Lazarow and De Duve 1973b; the protein
and esterase distributions were among the averaged data of their Fig. 4)

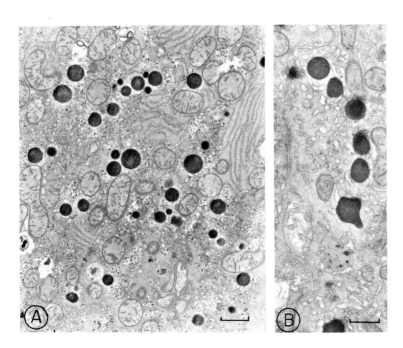

Fig. 3A,B. Clusters of peroxisomes. The peroxisomes have been stained by incubating
rat liver in the alkaline diaminobenzidine medium for catalase (Novikoff et al.
1972); electron opaque reaction product fills the peroxisomes. (A) Peroxisomes of
various sizes grouped in several clusters. (B) A cluster giving the impression of
beads on a string. Glutaraldehyde fixation, lead citrate stain. *Bar* 1 μm

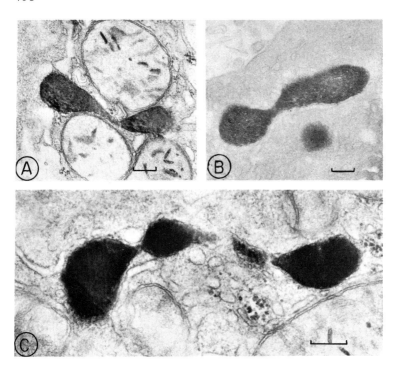

Fig. 4A-C. Interconnected peroxisomes. Rat liver stained for catalase showing dumb-bell shapes. Glutaraldehyde fixation. (A and C) lead citrate stain. (B) unstained. *Bar* 200 nm

interconnected or dumb-bell shaped peroxisomes in normal tissues are: Hruban and Rechcigl (1969) Fig. 61, liver; Beard (1972) Fig. 3 inset, adrenal cortex; Black and Bogart (1973) Fig. 15, fetal adrenal cortex; Gulyas and Yuan (1975) Figs. 2, 7, corpus luteum; Hicks and Fahimi (1977) Figs. 2, 7, 8, myocardium; Sternlieb and Quintana (1977) Fig. 3, human liver; and Arnold et al. (1979) Fig. 1, kidney.

Peroxisomes sometimes have tails (Fig. 5). Figure 5A shows a peroxisome whose membrane unambiguously runs into a narrow, apparently tubular structure (electron micrograph courtesy of Dr. George Palade). The tubular element apparently ends with a slight enlargement, but may in fact continue above or below the plane of this section. Other examples of peroxisomal tails may be seen in the studies of Novikoff and Shin (1964) Figs. 6b, 10a,b; Tsukada et al. (1968) Figs. 7, 8; Fahimi (1969) Fig. 6; Hruban and Rechcigl (1969) Figs. 13, 16; Essner (1970) Fig. 17; Reddy and Svoboda (1971) Figs. 18-20; Novikoff et al. (1973) Figs. 22-25; Black and Bogart (1973) Fig. 15; Reddy et al. (1974) Figs. 10, 11; Jézéquel (1975) Fig. 15; and Sternlieb and Quintana (1977) Fig. 4. These tails have been interpreted by some workers as connections to the endoplasmic reticulum. However, this is a matter of interpretation: they could equally well be connections to other peroxisomes.

Some tails on peroxisomes show loops or other unusual structures, which have been described as "gastruloid cisternae" by Hruban et al. (1974).

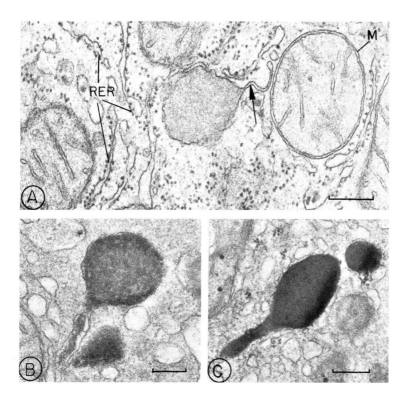

Fig. 5A-C. Peroxisomes with tails. (A) The tail (*arrow*) is an apparently tubular stretch of smooth unit membrane clearly in continuity with the peroxisome membrane. Nearby are rough endoplasmic reticulum (*RER*) and mitochondria (*M*). (Micrograph courtesy of Dr. G.E. Palade). (B,C) Tails stain positively for catalase. Glutaraldehyde fixation, uranyl acetate and lead citrate stain. *Bar* 200 nm

Peroxisomes are sometimes observed in close proximity to endoplasmic reticulum (ER), especially after clofibrate treatment (Fig. 6). Many studies have described *continuities* between peroxisomes and ER: Rhodin (1963), Hruban et al. (1963), Novikoff and Shin (1964), Svoboda and Azarnoff (1966), Essner (1967), Tsukada et al. (1968), and Sternlieb and Quintana (1977). Novikoff and Shin (1964) and Novikoff et al. (1973) have suggested that many or all peroxisomes might be attached to the ER.

On the other hand, some investigations have not observed these continuities (Rigatuso et al. 1970; Legg and Wood 1970; Fahimi et al. 1976). One good criterion for proof of luminal continuity between two structures is the ability to follow the unit membrane from one structure through the connection to the other structure. Many images of connections between peroxisomes and adjacent ER fail to satisfy this criterion, because the membranes become fuzzy or invisible in the region of apparent continuity. Tangential sections of overlapping structures would be expected to artifactually produce some such images.

Novikoff et al. (1973) have clearly demonstrated that tilting the electron microscopic ultrathin section in the electron beam by means of a goniometer stage permits one to bring membranes parallel to the beam so that they are sharply visible. Under these conditions, peroxisomes

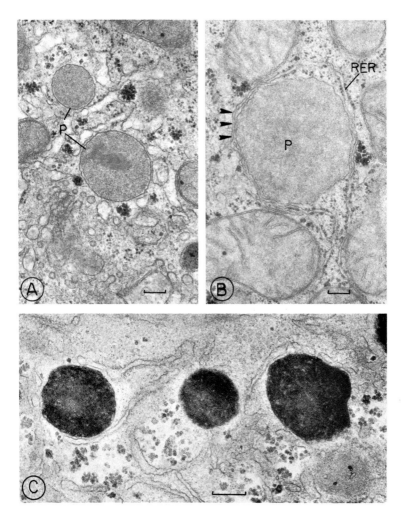

Fig. 6A-C. Proximity of endoplasmic reticulum to peroxisomes (*P*). (A) Normal rat liver. Glutaraldehyde fixation, uranyl acetate and lead citrate stain. (B) Clofibrate-treated rat liver. The large peroxisome is encircled by ER; where the ER is close to the peroxisome, ribosomes are absent (*arrowheads*). No continuities are observed between the ER and the peroxisome. Paraformaldehyde/glutaraldehyde fixation, uranyl acetate and lead citrate stain. (C) Peroxisomes of normal rat liver heavily stained with diaminobenzidine. The ER is free of reaction product, even when adjacent to peroxisomes. Paraformaldehyde/glutaraldehyde fixation, lead citrate stain. *Bar* 200 nm

show no continuities to the ER (their Figs. 6 and 7), whereas at other angles of tilt, where the membranes are not clearly visible, images suggestive of continuities are seen.

In one reported case, serial sectioning did not show apparent peroxisome-ER continuities except where the membranes were tangentially sectioned (Fahimi et al. 1976).

The shape of peroxisomes is not invariant. Although they are usually round or ovoid, a number of drugs produce highly elongated, or oddly shaped peroxisomes (Hruban et al. 1966, 1974).

Biochemical Differences between Peroxisomes and Endoplasmic Reticulum

Within the resolution of modern cell fractionation methods, there appears to be very little overlap between the biochemical composition of peroxisomes and endoplasmic reticulum. When peroxisomes are purified according to Leighton et al. (1968), they are nearly free of glucose-6-phosphatase, a characteristic marker enzyme for the endoplasmic reticulum (Fig. 1).

Some catalase activity is present in the microsomal fraction prepared by differential centrifugation, consisting mainly of vesicles derived from the endoplasmic reticulum, with Golgi elements, pieces of plasma membrane and fragments of the outer mitochondrial membrane. Subfractionation of the microsomal fraction by equilibrium density centrifugation (Beaufay et al. 1974) largely separates the catalase from the various other cell structures. As illustrated in Fig. 7, the catalase is located at an equilibrium density of about 1.23 (the usual equilibrium density of peroxisomes) near the bottom of the gradient. This demonstrates that the catalase present in microsomal fractions is not located within the microsomal vesicles themselves, but rather is in small peroxisomes which contaminate the microsomes.

These biochemical differences between peroxisomes and endoplasmic reticulum may also be observed by electron microscopic cytochemistry. Figures 3-6 illustrate preparations in which catalase-containing structures were stained with the alkaline diaminobenzidine procedure specific for catalase (Novikoff et al. 1972). The peroxisomes are intensely stained. We looked for, but did not observe staining in the endoplasmic reticulum, not even in ER vesicles near or touching peroxisomes (Figs. 4B, 4C, and 6C). The absence of DAB reaction product from ER cisternae

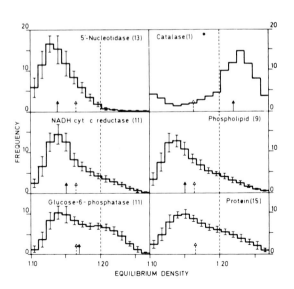

Fig. 7. Subfractionation of a microsomal fraction. A microsomal (P) fraction was prepared by differential centrifugation and then subfractionated by isopycnic centrifugation. On average the P fraction contained 19% of the protein, 49% of the phospholipid, 73% of the glucose-6-phosphatase, 59% of the NADH cytochrome c reductase, and 46% of the 5'-nucleotidase, but only 5.9% of the catalase of the homogenate. The *numbers in parentheses* are the number of experiments. The *solid arrows* mark the medians of the distributions. The *open arrows* mark the median of the protein distribution, which is repeated on each graph to facilitate comparisons. The *dotted line* at density 1.20 is present for the same purpose. (Redrawn from Beaufay et al. 1974, Figs. 2 and 7)

may also be noted in the micrographs of many other investigators: Novikoff and Goldfischer (1969) Figs. 1, 2; Essner (1969) Figs. 19-22;
Fahimi (1969) Figs. 3-7; Legg and Wood (1970) Figs. 5-8; and Hruban et
al. (1972).

Figure 8A illustrates a preparation on which the cytochemical procedure
for glucose-6-phosphatase (Leskes et al. 1971) was carried out. The
phosphate split from glucose-6-phosphate by the membrane-bound enzyme
diffuses until it is precipitated by lead, forming an electron-dense
precipitate. The lead phosphate reaction product is found in patches
in the cisternae of the ER, but is not observed in other structures,
including peroxisomes. Figure 8A includes ER cisternae touching peroxisomes: reaction product fills the ER but has not diffused into the
peroxisomes.

Figure 8B shows a preparation in which both the glucose-6-phosphatase
and catalase cytochemical reactions were carried out. The peroxisomes
are darker than the mitochondria, due to the diaminobenzidine, but the
intense glucose-6-phosphatase reaction product clearly does not extend
into the peroxisomes.

In addition, Lazarow et al. (1979) have recently found that the polypeptide compositions of the peroxisomal and ER membranes are highly
dissimilar.

Donaldson et al. (1972) claimed the presence of several ER enzymes in
peroxisomes. However, the presence of ER contaminants in their peroxisomal fraction, documented by the amount of glucose-6-phosphatase activity present, is more than sufficient to account for their observations.

One exception, in which a protein has been demonstrated both in peroxisomes and in ER, is cytochrome b_5, which is located in several
intracellular membranes, including very small amounts in peroxisomes
(Fowler et al. 1976; Remacle et al. 1976).

Any model of peroxisome biogenesis must take into account *diffusion*,
both of soluble proteins, and of membrane proteins within the plane of
the membrane. It is clear from the biochemical results that there is
little overlap in the biochemical composition of peroxisomes and ER.
It is clear from the cytochemical results that the reaction product
does not diffuse from one structure into the other. These facts are
not trivially reconciled with the idea of luminal connections between
peroxisomes and ER.

Studies on Biogenesis

Experiments in Vivo

Lazarow and De Duve (1971, 1973a) found that catalase, the principal
enzyme of rat liver peroxisomes, is synthesized in vivo as a precursor, specifically an apomonomer. At the age of 8 min, it is found in
soluble form upon fractionation of the liver (Lazarow and De Duve,
1973b) suggesting that it is located in the cell sap in vivo. However,
we must consider the possibility that the catalase precursor could
have leaked from the ER cisternae during homogenization of the liver
(when the ER cisternae rupture and reseal to form the microsomal vesicles). For example, Scheele et al. (1978) found that some leakage of
organelle contents occurred when they homogenized guinea pig pancreas;
subsequent reabsorption of these leaked proteins increased with the

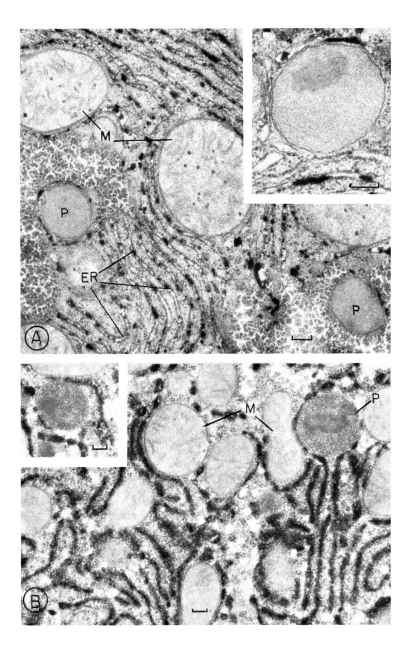

Fig. 8A,B. Lack of diffusion of cytochemical reaction product from ER to peroxisomes. (A) Glucose-6-phosphatase cytochemistry according to Leskes et al. (1971): the lead phosphate reaction product is found within the endoplasmic reticulum, including ER adjacent to two peroxisomes (P). *Inset* stained ER next to lead phosphate-free peroxisome with core. (B) Both glucose-6-phosphatase and catalase cytochemistry. The ER is full of the electron-dense lead phosphate, while the peroxisomes show only a weak diaminobenzidine reaction. (A and B) Liver from clofibrate-treated rat. Glutaraldehyde fixed, uranyl acetate and lead citrate stained. *Bar* 200 nm

alkalinity of their isoelectric points. The following experiment, however, makes it very unlikely that the catalase precursor leaked from
the ER.

A rat's liver was fractionated according to Lazarow and De Duve (1973b)
after pulse labeling with ^{35}S-methionine. Newly synthesized catalase
and albumin were isolated immunochemically from each of the fractions,
and subjected to SDS gel electrophoresis and fluorography. As shown in
Fig. 9, the labeled albumin was found entirely in microsomes, whereas
labeled catalase was found in soluble form at the top of the gradient.
Rat liver catalase and albumin have isoelectric points of about 6 (unpubl. observations) and 4.6 (Anderson et al. 1959), respectively. Therefore, if anything, catalase would be more likely to stick to the ER.
The absence of soluble labeled albumin agrees with the results of
Peters (1962).

Redman et al. (1972) independently found newly synthesized catalase in
the high-speed supernatant, not in microsomes.

The catalase precursor enters peroxisomes post-translationally, with a
half-time of approximately 14 min (Lazarow and De Duve 1973b), and does
not undergo detectable proteolytic cleavage as it does so (Fig. 10)
(Robbi and Lazarow 1978). Inside the organelle it acquires heme and
aggregates to form the active tetrameric enzyme (Lazarow and De Duve
1973b). This pathway is summarized in Fig. 11.

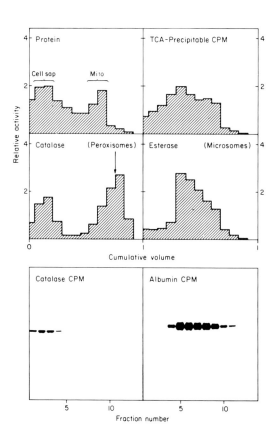

Fig. 9. Initial locations of
catalase and albumin in rat
liver. A rat was pulse-labeled
with ^{35}S-methionine, and a liver
postnuclear supernatant fractionated by centrifugation into
a sucrose gradient (as in Lazarow
and De Duve 1973b). After measure-
ment of the marker enzymes, catalase and albumin were isolated
immunochemically from aliquots of
each fraction. The immunoprecipitates were subjected to SDS gel
electrophoresis followed by fluorography. The *lower 2 panels* are
photographs of the fluorograms.
Centrifugation was from left to
right. The position of the mitochondria was determined by measuring cytochrome oxidase (not
illustrated)

Fig. 10. Comparison of sizes of extra-peroxisomal catalase precursor synthesized in vivo and the subunit of mature peroxisomal catalase. *Lanes 8m* 8-min-old precursor immuno-precipitated from high speed supernatant. *1d* 1-day-old catalase immunoprecipitated from purified peroxisomes. *50:50* equal mixture of the two. Fluorogram of SDS gradient gel; ^{35}S-methionine labeling. The lower band in the *1d* lanes is due to radioactive peroxisomal cores that have sedimented with the immunoprecipitates

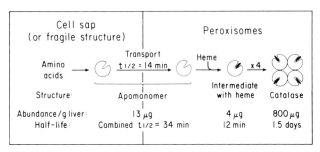

Fig. 11. Summary of catalase biogenesis in vivo in rat liver. (Based on Lazarow and De Duve 1973a,b). The representation of the intermediate with heme as a monomer is hypothetical for rat liver, but is supported by the findings of Zimniak et al. (1975) on *S. cerevisiae*

Zimniak et al. (1975, 1976) have studied the biosynthesis of catalase in *Saccharomyces cerevisiae* and concluded that it is synthesized as an apomonomer, converted to a monomer, and then to an active tetramer. Thus yeast and rat liver catalase apparently follow similar biochemical pathways.

Experiments in Vitro

Robbi and Lazarow (1978) translated liver mRNAs in the wheat germ and reticulocyte lysate cell-free protein-synthesizing systems and immuno-precipitated the catalase mRNA translation product (Fig. 12). In each case, it was found to have the same mobility on SDS polyacrylamide gradient slab gels as the subunit of mature peroxisomal catalase (Fig. 13). This differs from all secretory proteins [except ovalbumin (Palmiter et al. 1978)], which are synthesized in vitro as larger precursors with hydrophobic amino-terminal extensions (e.g., see the sympo-

Fig. 12. Translation of rat liver mRNAs in the wheat germ protein-synthesizing system. Fluorogram of SDS gradient gel; ^{35}S-methionine labeling. 1 Total products in 0.2 μl of translation mixture; 2 catalase immunoprecipitated from 130 μl of translation mixture. (From experiment of Robbi and Lazarow 1978)

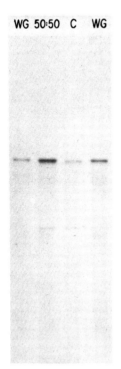

Fig. 13. Comparison of sizes of wheat germ translation product of catalase mRNA (WG) and subunit of 1-day-old peroxisomal catalase (C). 50:50 is an equal mixture of the two. Fluorogram of SDS gradient gel; ^{35}S-methionine labeling

sium proceedings edited by Zimmerman et al. 1980). This result argues
that catalase is not segregated into the ER, consistent with the in
vivo results.

Goldman and Blobel (1978) have also reported that catalase mRNA is
translated by the reticulocyte lysate with the same size as the subunit of the enzyme, and have extended this observation to urate oxidase. In addition, they have shown that catalase mRNA is found in the
free ribosomal RNA fraction, and the cell-free product is not cotranslationally segregated into dog pancreas microsomes. This confirms that
newly made catalase is not found in soluble form in vivo due to leakage from the ER.

In summary, the biochemical data prove that catalase biosynthesis differs in fundamental ways from secretory protein biosynthesis, and they
do not support the concept that peroxisomes are attached to, or bud
from, the ER.

Biogenesis of Glyoxysomes

The plant seed organelle, the glyoxysome, shown by Breidenbach and
Beevers (1967) to contain all of the glyoxalate cycle enzymes, is closely related to the peroxisome. Both contain oxidases and catalase (De
Duve et al. 1960; Breidenbach et al. 1968) and both catalyze the β-oxidation of fatty acids (Cooper and Beevers 1969; Lazarow 1978). Therefore one might expect them to have similar mechanisms of biogenesis.
Indeed, it has been suggested that glyoxysomes may be converted into
peroxisomes during plant development (Trelease et al. 1971; Burke and
Trelease 1975; Schopfer et al. 1976).

Several studies suggest a role for the ER in glyoxysome biogenesis.
Gonzalez and Beevers (1976) reported the presence of malate synthetase
and citrate synthetase enzymatic activities in microsomal fractions
isolated from castor bean cotyledons actively forming glyoxysomes.
Lord and Bowden (1978) claimed that newly synthesized radiolabeled
synthetase was first found in microsomes and then transferred to castor
bean glyoxysomes. However, the specificity of their immunochemical isolation procedure was not documented. Vigil (1970) has described apparent connections between glyoxysomes and ER.

On the other hand, recent work from Kindl's laboratory sheds a different light on these questions. Köller and Kindl (1978) found that in
germinating cucumber cotyledons, malate synthetase activity is indeed
associated with microsomal fractions under certain centrifugation conditions. However, upon longer centrifugation, the malate synthetase
could be largely separated from the microsomal marker enzymes. Kindl
et al. (1980) studied the biosynthesis of three glyoxysomal enzymes,
catalase, isocitrate lyase and malate synthetase, in germinating cucumber seedlings. The newly synthesized radioactive precursors, isolated immunochemically, were in each case found first in soluble form and
later in the glyoxysomes. Only very slight amounts were found in the
microsomal fraction. In the case of malate synthetase, the immunoprecipitates were subjected to SDS gel electrophoresis, and the malate
synthetase was then cut out of the gel and counted, leaving little
doubt about the accuracy of the results. These observations suggest
that the biogenesis of peroxisome and glyoxysome proteins may be similar.

One apparent difference comes from the work of Walk and Hock (1978),
who have reported that watermelon cotyledon glyoxysomal malate dehydrogenase is synthesized in vitro as a precursor 5,000 daltons larger than
the purified enzyme.

Models of Peroxisome Biogenesis

A number of models of peroxisome biogenesis have been proposed during the past 14 years, of which we will discuss four.

Model 1. Budding from the ER

On the basis of the morphological observations (Hruban et al. 1963; Rhodin 1963; Novikoff and Shin 1964) and biochemical data (Higashi and Peters 1963) available at the time, De Duve and Baudhuin (1966) proposed in their review that peroxisomes might form by budding from the endoplasmic reticulum. It was envisioned that peroxisomal proteins would be synthesized by bound ribosomes, be vectorially discharged within the lumen of the ER like secretory proteins, traverse the ER cisternae and accumulate within outpouchings, which, when full, could pinch off, becoming peroxisomes. This attractive model engendered a series of experiments to test various of its features (Poole et al. 1969, 1970; Lazarow and De Duve 1971, 1973a,b) and to the formulation and testing of three specific versions of the model (reviewed by De Duve 1973, and De Duve et al. 1974). This model is still widely accepted today, despite the fact that by 1973, it was clear that none of the biochemical experiments had provided any support for the model, and some of the results were inconsistent with it (specifically the finding of newly made catalase outside the microsomes).

Model 2. Synthesis on Free Polysomes and Subsequent Uptake

In 1973, Lazarow and De Duve (1973b) suggested that the simplest interpretation of their results was that catalase (and presumably other peroxisomal proteins) were synthesized on free polysomes, traversed the cell sap, and entered peroxisomes. It was noted that it was not easy to reconcile the uptake of a 60,000 dalton polypeptide with the current ideas about the permeability of phospholipid bilayer membranes.

This model was discussed recently in some detail by Lazarow (1980), who pointed out that the uptake of proteins into the organelle (against a strong concentration gradient) could in principle occur by two very different mechanisms: (A) selective uptake, e.g., by a receptor-mediated mechanism, or (B) selective retention, i.e., a form of thermodynamic self-assembly. It is now clear that some proteins enter chloroplasts and mitochondria post-translationally (e.g., Chua and Schmidt 1978; Highfield and Ellis 1978; Maccecchini et al. 1979), but whether this occurs by either of these mechanisms is still unknown.

Model 3. Synthesis by Polysomes Bound to a Fragile Precursor Structure

Lazarow and De Duve (1973b) also proposed an alternative hypothesis, which superficially resembled Model 1. This was the cotranslational segregation of catalase (and other peroxisomal proteins) within a special, fragile membrane vesicle. Filling of this vesicle and aggregation of the urate oxidase into a core could result in a peroxisome. Alternatively, this vesicle could convey its contents to a peroxisome. Rupture of this fragile structure upon homogenization of the liver would result in finding the newly made catalase in soluble form.

Such a precursor vesicle could in principle be derived from the ER, but its membrane would have to undergo substantial modification in order to account for its selective binding of ribosomes synthesizing peroxisomal

proteins and for its fragility. The vesicle need not *look* different
from the ER. This idea, though not ruled out by the available data,
now appears less likely in view of the finding that the composition
of the peroxisomal membrane differs very greatly from that of the ER
(Lazarow et al. 1979).

Model 4. Both Budding and Uptake

Goldman and Blobel (1978) suggested that the membrane of the peroxisome
buds from the ER while the soluble proteins enter the bud by post-translational uptake. They state that newly synthesized peroxisomal proteins
will only enter forming buds, not mature peroxisomes. Moreover, they
argue that the cell-free translation product of catalase mRNA is the
same size as the subunit of the mature enzyme due to the artifactual
cleavage of a hypothetical "signal sequence" by a putative peroxisomal
"signal peptidase" postulated to be present in both the reticulocyte
lysate and wheat germ cell-free protein-synthesizing systems.

The Peroxisome Reticulum Hypothesis

In order to explain in a consistent fashion the experimental data reviewed above, let us consider the possibility that peroxisomes may be
interconnected with each other, but *not* with the endoplasmic reticulum
(Fig. 14). Peroxisomes could undergo fission and fusion, and/or they
could be connected, permanently or transiently, by membrane-bounded
channels. This idea will be elaborated as we consider its implications
for the five types of data discussed above.

1. Morphology. This hypothesis implies that the tails that are observed
on peroxisomes by many investigators are not connections to the ER,
but rather connections to other peroxisomes. These tails would be expected to lack ER enzymes, and to contain proteins characteristic of
the peroxisome membrane. Thus the "peroxisome reticulum" would consist
of both larger, approximately spherical structures, and narrow tubular
elements. This would account for the observed clustering of peroxisomes
and for the dumb-bell shaped images, as well as for the tails. The close
associations that occur between peroxisomes and bonafide ER would facilitate biochemical collaboration between the two organelles, but would
not involve luminal continuities.

2. Biochemical Homogeneity. Interconnections among peroxisomes would permit
the interchange of their contents due to diffusion, which would result
in the observed biochemical homogeneity.

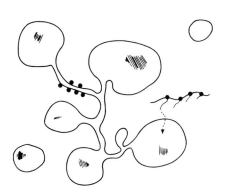

Fig. 14. Schematic drawing of the peroxisome reticulum, as envisioned by the
authors. There is no connection to the
endoplasmic reticulum. The *dotted arrow*
indicating uptake of peroxisomal proteins
from the cytosol is hypothetical; it corresponds to Model 2 in the text. The presence of ribosomes on a narrow tubular element of the system is a strictly hypothetical alternative, and corresponds to one
version of Model 3. It must be emphasized
that most peroxisomes need *not* be interconnected at any *one* time. We envision
that they are connected in *space and time*
through fission and fusion and/or interconnections

3. Turnover. It has been suggested that peroxisomes may be destroyed by autophagy (De Duve and Baudhuin 1966). Mixing of peroxisomal contents through interconnections or fission/fusion, plus occasional autophagic destruction, would account for the observed apparent synchrony in turnover of the peroxisomal proteins, as was pointed out by Poole et al. (1970) and by De Duve (1973).

4. Biochemical Differences between Peroxisomes and ER. If peroxisomes are never connected to the ER, then there is no difficulty in explaining the fact that these two organelles appear to have very little in common biochemically, and have different polypeptides in their membranes.

5. Biogenesis. Peroxisomes need not be constructed de novo, but rather additional peroxisome constituents could be added to the preexisting mass of peroxisomes. Division of peroxisomes in two (as suggested by Legg and Wood 1970), or budding from the peroxisome reticulum, would produce new ones. Peroxisomal constituents to be added include soluble, core and membrane proteins, as well as membrane phospholipids.

At the present time, the most likely means of adding proteins would appear to be Model 2 (above), namely synthesis on free polysomes, followed by post-translational uptake. Membrane as well as soluble proteins could be added by this means. Cytochrome b_5 establishes a precedent for the post-translational insertion of a membrane protein. It is synthesized on free polysomes according to Rachubinski et al. (1980) and is capable of binding to membranes in vitro (Remacle 1978). Moreover, the observations of Kindl et al. (1980) on malate synthetase synthesis (see above) suggest that this glyoxysomal membrane protein follows this route. However, it is not entirely certain whether it is an integral membrane protein.

The present data do not exclude the possibility of cotranslational segregation into a fragile element of a peroxisome reticulum (model 3). Ribosomes are not observed on rat liver peroxisomes (even after injection of cycloheximide into rats; Shio and Lazarow, unpubl. observations). However, ribosomes could conceivably bind to tubular structures connected to peroxisomes. Such ribosome-studded structures might look like ER morphologically, but biochemically and functionally would be different. A precedent for the binding of ribosomes to an organelle membrane is found in the observation of ribosomes on the outer membrane of mitochondria in yeast spheroplasts (Kellems and Butow 1972; Kellems et al. 1974, 1975).

Rupture of these fragile elements upon homogenization of the liver would account for finding newly synthesized peroxisomal proteins in soluble form. Ribosome-laden, unsealed membrane fragments might end up in the free ribosome fraction, as has been pointed out previously (Lazarow 1980).

Summary

We have reviewed five types of data that bear on peroxisome biogenesis: biochemical homogeneity of peroxisomes, uniform turnover of peroxisome proteins, peroxisome morphology, biochemical differences between peroxisomes and endoplasmic reticulum, and experiments on the synthesis of peroxisomal proteins. In addition, four models that have been proposed for peroxisome biogenesis were summarized.

Consideration of the data led us to point out that the classical model of peroxisome budding from the ER, although widely accepted, is not supported by the biochemical evidence. We suggest, as a hypothesis to be tested, that peroxisomes may be interconnected permanently and/or transiently, into a "peroxisome reticulum" distinct from the ER. Peroxisome formation would occur by addition of peroxisome constituents to this pre-existing system, rather than by de novo creation. Peroxisomal proteins could be biosynthesized and added to peroxisomes by either Model 2 or Model 3, previously proposed, although the former appears more likely at present.

These concepts are capable of explaining most of the published data. Whether or not they are correct remains to future experiments to determine.

Acknowledgments. The research of the authors was supported by NIH grants AM-19394 and HL-20909, and by NSF grants PCM77-11151 and PCM76-16657. We thank Dr. G.E. Palade for the electron micrograph of Fig. 5A.

References

Anderson NG, Canning RE, Anderson ML, Shellhamer RH (1959) Studies on isolated cell components. XI. Rat blood proteins. Exp Cell Res 16:292-308

Arnold G, Liscum L, ltzman E (1979) Ultrastructural localization of D-amino acid oxidase in microperoxisomes of the rat nervous system. J Histochem Cytochem 27: 735-745

Beard ME (1972) Identification of peroxisomes in the rat adrenal cortex. J Histochem Cytochem 20:173-179

Beaufay H, Jacques P, Baudhuin P, Sellinger OZ, Berthet J, De Duve C (1964) Tissue fractionation studies. 18. Resolution of mitochondrial fractions from rat liver into three distinct populations of cytoplasmic particles by means of density equilibration in various gradients. Biochem J 92:184-205

Beaufay H, Amar-Costesec A, Thines-Sempoux D, Wibo M, Robbi M, Berthet J (1974) Analytical study of microsomes and isolated subcellular membranes from rat liver. III. Subfractionation of the microsomal fraction by isopycnic and differential centrifugation in density gradients. J Cell Biol 61:213-321

Black VH, Bogart BI (1973) Peroxisomes in inner adrenocortical cells of fetal and adult guinea pigs. J Cell Biol 57:345-358

Breidenbach RW, Beevers H (1967) Association of the glyoxylate cycle enzymes in a novel subcellular particle from castor bean endosperm. Biochem Biophys Res Commun 27:462-469

Breidenbach RW, Kahn A, Beevers H (1968) Characterization of glyoxysomes from castor bean endosperm. Plant Physiol 43:705-713

Burke JJ, Trelease RN (1975) Cytochemical demonstration of malate synthase and glycolate oxidase in microbodies of cucumber cotyledons. Plant Physiol 56:710-717

Chua NH, Schmidt GW (1978) Post-translational transport into intact chloroplasts of a precursor to the small subunit of ribulose-1,5-bisphosphate carboxylase. Proc Natl Acad Sci USA 75:6110-6114

Cooper TG, Beevers H (1969) β-Oxidation in glyoxysomes from castor bean endosperm. J Biol Chem 244:3514-3520

Donaldson RP, Tolbert NE, Schnarrenberger C (1972) A comparison of microbody membranes with microsomes and mitochondria from plant and animal tissue. Arch Biochem Biophys 152:199-215

Dougherty WJ, Lee MM (1967) Light and electron microscope studies of smooth endoplasmic reticulum in dividing rat hepatic cells. J Ultrastruct Res 19:200-220

de Duve C (1973) Biochemical studies on the occurrence, biogenesis and life history of mammalian peroxisomes. J Histochem Cytochem 21:941-948

de Duve C, Baudhuin P (1966) Peroxisomes (microbodies and related particles). Physiol Rev 46:323-357

de Duve C, Beaufay H, Jacques P, Rahman-Li Y, Sellinger OZ, Wattiaux R, De Coninck S (1960) Intracellular localization of catalase and of some oxidase in rat liver. Biochim Biophys Acta 40:186-187

de Duve C, Lazarow PB, Poole B (1974) Biogenesis and turnover of rat-liver perox-
    isomes. In: Ceccarelli B, Clementi F, Meldolesi J (eds) Advances in cytopharmacolo-
    gy, vol II. Raven Press, New York, pp 219-223
Essner E (1967) Endoplasmic reticulum and the origin of microbodies in fetal mouse
    liver. Lab Invest 17:71-87
Essner E (1969) Localization of peroxidase activity in microbodies of fetal mouse
    liver. J Histochem Cytochem 17:454-466
Essner E (1970) Observations on hepatic and renal peroxisomes (microbodies) in the
    developing chick. J Histochem Cytochem 18:80-92
Fahimi HD (1969) Cytochemical localization of peroxidatic activity of catalase in
    rat hepatic microbodies (peroxisomes). J Cell Biol 43:275-288
Fahimi HD, Gray BA, Herzog VK (1976) Cytochemical localization of catalase and per-
    oxidase in sinusoidal cells of rat liver. Lab Invest 34:192-201
Fowler S, Remacle J, Trouet A, Beaufay H, Berthet J, Wibo M, Hauser P (1976) Analyti-
    cal study of microsomes and isolated subcellular membranes from rat liver. V.
    Immunological localization of cytochrome b$_5$ by electron microscopy: methodology
    and application to various subcellular fractions. J Cell Biol 71:535-550
Goldman BM, Blobel G (1978) Biogenesis of peroxisomes: Intracellular site of synthe-
    sis of catalase and uricase. Proc Natl Acad Sci USA 75:5066-5070
Gonzalez E, Beevers H (1976) Role of the endoplasmic reticulum in glyoxysome forma-
    tion in castor bean endosperm. Plant Physiol 57:406-409
Gulyas BJ, Yuan LC (1975) Microperoxisomes in the late pregnancy corpus luteum of
    Rhesus monkeys (Macaca mulatta). J Histochem Cytochem 23:359-368
Hicks L, Fahimi HD (1977) Peroxisomes (microbodies) in the myocardium of rodents
    and primates. Cell Tissue Res 175:467-481
Higashi T, Peters T Jr (1963) Studies on rat liver catalase. II. Incorporation of
    ^{14}C-leucine into catalase of liver cell fractions in vivo. J Biol Chem 238:3952-
    3954
Highfield PE, Ellis RJ (1978) Synthesis and transport of the small subunit of chlor-
    oplast ribulose bisphosphate carboxylase. Nature (London) 271:420-424
Hruban Z, Rechcigl M Jr (1969) Microbodies and related particles. Academic Press,
    London, New York.
Hruban Z, Swift H, Wissler RW (1963) Alterations in the fine structure of hepato-
    cytes produced by β-3-thienylanine. J Ultrastruct Res 8:236-250
Hruban Z, Swift H, Slesers A (1966) Ultrastructural alterations of hepatic micro-
    bodies. Lab Invest 15:1884-1901
Hruban Z, Vigil EL, Slesers A, Hopkins E (1972) Microbodies: Constituent organelles
    of animal cells. Lab Invest 27:184-191
Hruban Z, Gotoh M, Slesers A, Chou S-F (1974) Structure of hepatic microbodies in
    rats treated with acetylsalicylic acid, clofibrate, and dimethrin. Lab Invest 30:
    64-75
Jézéquel A-M, Arakawa K, Steiner JW (1965) The fine structure of the normal, neonatal
    mouse liver. Lab Invest 14:1894-1930
Kellems RE, Butow RA (1972) Cytoplasmic type 80S ribosomes associated with yeast
    mitochondria. I. Evidence for ribosome binding sites on yeast mitochondria. J Biol
    Chem 247:8043-8050
Kellems RE, Allison VF, Butow RA (1974) Cytoplasmic type 80S ribosomes associated
    with yeast mitochondria. II. Evidence for the association of cytoplasmic ribosomes
    with the outer mitochondrial membrane in situ. J Biol Chem 249:3297-3303
Kellems RE, Allison VF, Butow RA (1975) Cytoplasmic type 80S ribosomes associated
    with yeast mitochondria. IV. Attachment of ribosomes to the outer membrane of iso-
    lated mitochondria. J Cell Biol 65:1-14
Kindl H, Köller W, Frevert J (1980) Cytosolic precursor pools during glyoxysome bio-
    synthesis. Hoppe-Seyler's Z Physiol Chem 361:465-467
Köller W, Kindl H (1978) The appearance of several malate synthase-containing cell
    structures during the stage of glyoxysome biosynthesis. FEBS Lett 88:83-86
Lazarow PB (1978) Rat liver peroxisomes catalyze the β-oxidation of fatty acids. J
    Biol Chem 253:1522-1528
Lazarow PB (1980) Properties of the natural precursor of catalase: Implications for
    peroxisome biogenesis. Ann NY Acad Sci 343:293-303

Lazarow PB, De Duve C (1971) Intermediates in the biosynthesis of peroxisomal cata-
  lase in rat liver. Biochem Biophys Res Commun 45:1198-1204
Lazarow PB, De Duve C (1973a) The synthesis and turnover of rat liver peroxisomes.
  IV. Biochemical pathway of catalase synthesis. J Cell Biol 59:491-506
Lazarow PB, De Duve C (1973b) The synthesis and turnover of rat liver peroxisomes.
  V. Intracellular pathway of catalase synthesis. J Cell Biol 59:507-524
Lazarow PB, Fowler S, Hubbard AL (1979) Different composition of the peroxisomal
  and endoplasmic reticulum membranes of rat liver. J Cell Biol 83:263a
Legg PG, Wood RL (1970) New observations on microbodies: A cytochemical study on
  CPIB-treated rat liver. J Cell Biol 45:118-129
Leighton F, Poole B, Beaufay H, Baudhuin P, Coffey JW, Fowler S, De Duve C (1968)
  The large scale separation of peroxisomes, mitochondria and lysosomes from the
  livers of rats injected with Triton WR-1339. J Cell Biol 37:482-512
Leighton F, Coloma L, Koenig C (1975) Structure, composition, physical properties,
  and turnover of proliferated peroxisomes. A study of the trophic effects of Su-
  13437 on rat liver. J Cell Biol 67:281-309
Leskes A, Siekevitz P, Palade G (1971) Differentiation of endoplasmic reticulum in
  hepatocytes. I. Glucose-6-phosphatase distribution in situ. J Cell Biol 49:264-
  287.
Lord JM, Bowden L (1978) Evidence that glyoxysomal malate synthase is segregated
  by the endoplasmic reticulum. Plant Physiol 61:266-270
Maccecchini ML, Rudin Y, Blobel G, Schatz G (1979) Import of proteins into mitochondria: precursor forms of the extramitochondrially made F_1-ATPase subunits in
  yeast. Proc Natl Acad Sci USA 76:343-347
Novikoff AB, Goldfischer S (1969) Visualization of peroxisomes (microbodies) and
  mitochondria with diaminobenzidine. J Histochem Cytochem 17:675-680
Novikoff AB, Shin WY (1964) The endoplasmic reticulum in the Golgi zone and its
  relationship to microbodies, Golgi apparatus and autophagic vacuoles in rat liver
  cells. J Microsc 3:187-206
Novikoff AB, Novikoff PM, Davis C, Quintana N (1972) Studies on microperoxisomes.
  II. A cytochemical method for light and electron microscopy. J Histochem Cyto-
  chem 20:1006-1023
Novikoff PM, Novikoff AB, Quintana N, Davis C (1973) Studies on microperoxisomes.
  III. Observations on human and rat hepatocytes. J Histochem Cytochem 21:540-558
Palmiter RD, Gagnon J, Walsh KA (1978) Ovalbumin: A secreted protein without a tran-
  sient hydrophobic leader sequence. Proc Natl Acad Sci USA 75:94-98
Peters T Jr (1962) The biosynthesis of rat serum albumin. II. Intracellular phenom-
  ena in the secretion of newly formed albumin. J Biol Chem 237:1186-1189
Poole B (1971) The kinetics of disappearance of labeled leucine from the free leucine
  pool of rat liver and its effect on the apparent turnover of catalase and other
  hepatic proteins. J Biol Chem 246:6587-6591
Poole B, Leighton F, De Duve C (1969) The synthesis and turnover of rat liver peroxisomes. II. Turnover of peroxisome proteins. J Cell Biol 41:536-546
Poole B, Higashi T, De Duve C (1970) The synthesis and turnover of rat liver peroxisomes. III. The size distribution of peroxisomes and the incorporation of new
  catalase. J Cell Biol 45:408-415
Price VE, Sterling WR, Tarantola VA, Hartley RW Jr, Rechcigl M Jr (1962) The kinetics
  of catalase synthesis and destruction in vivo. J Biol Chem 237:3468-3475
Rachubinski RA, Verma DPS, Bergeron JJM (1980) Synthesis of rat liver microsomal
  cytochrome b_5 by free ribosomes. J Cell Biol 84:705-716
Reddy JK (1973) Possible properties of microbodies (peroxisomes) microbody proliferation and hypolipidemic drugs. J Histochem Cytochem 21:967-971
Reddy J, Svoboda D (1971) Microbodies in experimentally altered cells. VIII. Continuities between microbodies and their possible biologic significance. Lab Invest 24:74-81
Reddy J, Svoboda D (1973) Further evidence to suggest that microbodies do not exist
  as individual entities. Am J Pathol 70:421-432
Reddy JK, Azarnoff DL, Svoboda DJ, Prasad JD (1974) Nafenopin-induced hepatic microbody (peroxisome) proliferation and catalase synthesis in rats and mice: Absence
  of sex difference in response. J Cell Biol 61:344-358

Redman CM, Grab DJ, Irukulla R (1972) The intracellular pathway of newly formed rat liver catalase. Arch Biochem Biophys 152:496-501

Remacle J (1978) Binding of cytochrome b_5 to membranes of isolated subcellular organelles from rat liver. J Cell Biol 79:291-313

Remacle J, Fowler S, Beaufay H, Amar-Costesec A, Berthet J (1976) Analytical study of microsomes and isolated subcellular membranes from rat liver. VI. Electron microscope examination of microsomes for cytochrome b_5 by means of a ferritin-labeled antibody. J Cell Biol 71:551-564

Rhodin JAG (1963) An atlas of ultrastructure. WB Saunders Co, Philadelphia

Rigatuso JL, Legg PG, Wood RL (1970) Microbody formation in regenerating rat liver. J Histochem Cytochem 18:893-900

Robbi M, Lazarow PB (1978) Synthesis of catalase in two cell-free protein-synthesizing systems and in rat liver. Proc Natl Acad Sci USA 75:4344-4348

Scheele GA, Palade GE, Tartakoff AM (1978) Cell fractionation studies on the guinea pig pancreas. Redistribution of exocrine proteins during tissue homogenization. J Cell Biol 78:110-130

Schopfer D, Bajracharya D, Bergfeld R, Falk H (1976) Phytochrome-mediated transformation of glyoxysomes into peroxisomes in the cotyledons of mustard seedlings. Planta 133:73-80

Sternlieb I, Quintana N (1977) The peroxisomes of human hepatocytes. Lab Invest 36:140-149

Svoboda DJ, Azarnoff DL (1966) Response of hepatic microbodies to a hypolipidemic agent, ethyl chlorophenoxyisobutyrate (CPIB). J Cell Biol 30:442-450

Trelease RN, Becker WM, Gruber PJ, Newcomb EH (1971) Microbodies (glyoxysomes and peroxisomes) in cucumber cotyledons. Correlative biochemical and ultrastructural study in light- and dark-grown seedlings. Plant Physiol 48:461-475

Tsukada H, Mochizuki Y, Konishi T (1968) Morphogenesis and development of microbodies of hepatocytes of rats during pre- and postnatal growth. J Cell Biol 37:231-243

Vigil EL (1970) Cytochemical and developmental changes in microbodies (glyoxysomes) and related organelles of castor bean endosperm. J Cell Biol 46:435-454

Walk RA, Hock B (1978) Cell free synthesis of glyoxysomal malate dehydrogenase. Biochem Biophys Res Commun 81:636-643

Zimmerman M, Mumford RA, Steiner DF (eds) (1980) Precursor processing in the biosynthesis of proteins. Ann NY Acad Sci 343:1-443

Zimniak P, Hartter E, Ruis H (1975) Biosynthesis of catalase T during oxygen adaptation of *Saccharomyces cerevisiae*. FEBS Lett 59:300-304

Zimniak P, Hartter E, Woloszczuk W, Ruis H (1976) Catalase biosynthesis in yeast: Formation of catalase A and catalase T during oxygen adaptation of *Saccharomyces cerevisiae*. Eur J Biochem 71:393-398

Origin and Dynamics of Lysosomes

K. v. Figura, U. Klein, and A. Hasilik[1]

Introduction

The lysosomal apparatus, the endoplasmic reticulum, the Golgi apparatus
and the secretory and endocytotic vesicles form a system of intracellular organelles delimited by unit membranes. Specific functions can be
ascribed to the different elements of that system. Lysosomes are defined
functionally as membrane-bound vesicles, in which degradation of macro-
moleculars is accomplished [1]. To fulfil their function, lysosomes are
equipped with degradative enzymes and are accessible for intra- and
extracellular material destined for degradation. Both the transfer of
lysosomal enzymes and of substrates into lysosomes requires an inten-
sive communication of lysosomes with other parts of the system of intracellular membrane delimited structures [2, 3]. Within that system the
synthesis of lysosomal enzymes is completed, newly formed lysosomal enzymes are packaged and material destined for the degradation is transported. Fusion and fission of membrane-bound structures allows the
transfer of solutes, membrane parts and membrane-bound material among
that membraneous network [4]. Lysosomes are not formed within that membraneous network as complete organelles at once, but acquire their constituents — the lysosomal membrane, the complement of degradative enzymes and the macromolecular substrates — gradually and from various
sites.

The origin, transfer, and fate of lysosomal constituents will be the
subject of the following review.

Transfer of Extracellular Material into Lysosomes

Cells internalize extracellular macromolecules by endocytosis [5]. The
term endocytosis describes a process comprising the invagination of
plasma membrane segments that pinch off as vesicles carrying fluid and
solutes. Unless specific binding sites are present on the cell surface,
the uptake of a compound is directly related to its concentration in
the extracellular space. By this mechanism, called fluid endocytosis,
a variety of solutes is interiorized proportionately to its concentration in the medium. The efficiency of this uptake is limited by the
amount of fluid that can be taken up by a cell [6]. In fibroblasts
fluid endocytosis reaches values of about 0.1 µl/h and mg of cell protein [7, 8].

Far more effective and distinguished by selectivity is the uptake by
adsorptive endocytosis. In this process components of the medium are
recognized by high affinity binding sites on the cell surface prior to
the internalization [6]. Some of the binding sites have been identified

---

[1]Physiologisch-Chemisches Institut der Universität Münster, Waldeyerstraße 15,
 D 4400 Münster, FRG

as specific membrane proteins and classified as receptors, functioning in transport.

Besides the now classical example of clearance of asialoglycoproteins from circulation, which has been pioneered by Ashwell and coworkers (for review see [16 and 22]), adsorptive endocytosis has been shown to be involved in internalization of many other molecules as diverse as lysosomal enzymes, lipoproteins, sulfated proteoglycans, hormones, growth factors and transport proteins for vitamins.

Though our knowledge of the interaction of receptors with ligands, on internalization of their complexes and on the life cycle of the receptors is still far from complete, a number of data have been collected in recent years concerning all these aspects of adsorptive endocytosis. At present the best-studied system appears to be the LDL-uptake in fibroblasts. The following chapter therefore, will focus primarily on the LDL-uptake as worked out by Goldstein, Brown and Anderson (for a recent review see [9]).

LDL binds to cell surface receptors via apolipoprotein B [10, 11]. The LDL-receptors of bovine adrenal cortex and cultured human cells have recently been solubilized from membranes and partially characterized [12]. The binding of LDL to the cell surface of intact cells, to membrane preparations, or to the partially purified receptor shows a very high affinity (dissociation constant about 10^{-9} M), and depends on the presence of either calcium or manganese. Altering the charge of the lipoprotein affects the binding [13, 14].

The receptors for LDL are unevenly distributed on the cell surface. More than 50% are located in structures known as coated pits. These are specialized areas of the plasma membrane that comprise about 2% of the total cell surface in fibroblasts. Coated pits are morphologically identified as membrane depressions coated on the cytoplasmic side with a fuzzy material. A minor part of the LDL-receptors are randomly distributed. The receptors bind LDL irrespectively of their location. The internalization of the receptor-ligand complex, however, is restricted to the area of coated pits, that invaginate and form coated vesicles. The transfer of receptors from random sites to coated pits does not require the interaction with the ligand. It depends on a part of the receptor molecule called "internalization site", which presumably is located on the cytoplasmic side of the membrane.

Hypothetically the receptors are anchored in the coated pits by an interaction of the internalization site with clathrin, a 180,000 molecular weight protein, that is the major constituent of coated vesicles [11]. A mutation has been described in which the receptors are randomly distributed on the surface of cultured fibroblasts. The receptors bind LDL normally, but the complexes are not internalized. This mutation has been suggested to affect the internalization site.

At 4°C binding can be studied separated from the uptake. After washing the excess ligand off, the internalization is initiated by warming to 37°C. Within less than 10 min after the warming the bound LDL appears in lysosomes as shown by using ferritin-labeled LDL. The internalization is followed by a rapid degradation of the LDL components. Cholesterol, one of the ultimate products of degradation, triggers several regulatory responses in the endocytosing cell. The LDL-receptors seem to escape from degradation. Cultured fibroblasts contain some 20,000-50,000 receptors per cell, which correspond to the number of LDL-molecules interiorized per cell within less than 10 min after exposure to saturating levels of LDL. Since the rate of endocytosis is constant

for at least 6 h even in the absence of protein synthesis, a continuous
recycling of LDL-receptors has to be considered. It has been estimated
that a receptor spends during one cycle about 3 min in random positions
and about 6 min in coated pits on the cell surface [15].

Transport systems for other molecules have much in common with the LDL-
receptor system, though they differ in some details. The receptor-
ligand association constants are generally in the order of 10^{-8}-10^{-10} M
[16-21]. The recognition marker on the ligand is best characterized in
some glycoproteins, where it is located in the saccharide portion. Five
different receptors with specificity for galactose, L-fucose, N-acetylglucosamine, N-acetylglucosamine/mannose or mannose 6-phosphate are
known so far [22]. In general, these receptors are located only on certain cells, thereby allowing the direction of the uptake into specific
cell types. The galactose-specific receptor appears to be restricted
to hepatocytes, and that for N-acetylglucosamine/mannose to cells of
the reticulo-endothelial system [22-25]. The receptors specific for
galactose, N-acetylglucosamine, and mannose, as well as that for transcobalamin II, have been isolated and characterized [18, 22]. The solubilized and purified carbohydrate-specific receptors are water-soluble
proteins, whereas the LDL receptor precipitates in the absence of detergent [12].

Electronmicroscopic and light microscopic studies using fluorescence
labeled ligand indicate that concentration in the coated pits is a general feature of receptors involved in endocytosis. Besides receptors
for LDL this concerns those for α_2-macroglobulin [26-28], epidermal
growth factor [26, 29, 30] and insulin [26] in cultured cells. It is
a matter of controversy, however, whether localization of these receptors in coated pits depends on the interaction with the ligand [26,
28-30] or occurs spontaneously. Generally, the binding is rapidly followed by internalization. A detailed in vivo study in rat revealed,
however, that the two events are not necessarily coupled [31, 32].
After injecting into rats low doses of labeled asialotransferrin, only
binding to the liver galactose receptor was observed. The signal for
internalization could be generated by a second injection of higher dose
of asialotransferrin. Thus a critical ligand concentration may be necessary to induce the internalization. It may be envisaged that clustering
of receptors in coated pits depends on a critical ligand concentration.
Antibodies against a receptor-bound ligand can promote the clustering
of the receptors as was shown for epidermal growth factor and insulin
[33, 34].

Fate of Endocytotic Vesicles

In most cases endocytotic vesicles fuse rapidly with the lysosomal apparatus. This is evident from the short lag phase observed for the appearance of degradation products in the medium, such as ^{125}I-monojodo-
tyrosine derived from ^{125}I-labeled proteins [15] or $^{35}SO_4^{2-}$ derived
from ^{35}S-labeled proteoglycans [35, 36]. Morphological studies indicate
that endocytotic vesicles fuse with multivesicular bodies [37, 38] which
may be the source of mature lysosomes. In a few cases endocytotic vesicles have been observed to fuse with membrane-delimited structures other
than lysosomes, thus allowing internalized macromolecules to escape
from degradation. Examples are the maternal immunoglobulins delivered
into the fetal yolk sac [39], the transendothelial transport [40] and
the accumulation of the internalized nerve growth factor in the axon
tips [41].

Transport of Intracellular Material into Lysosomes

Levels of intracellular proteins may be regulated by modulating their degradation [42]. It has been widely suggested that degradation is located within the lysosomal apparatus of cells, mainly because enzymes capable of degrading proteins reside predominantly in lysosomes. Only little experimental evidence is available to support this concept. The line of evidence for a degradation of cellular, particularly cytosolic, proteins within lysosomes may be summarized as follows:

1. When inhibitors of the lysosomal cathepsins are introduced into rat liver lysosomes by means of liposomes, the rate of degradation of endogenous liver protein is reduced [43, 43a].

2. When liver after in vivo radiolabeling with leucine is perfused for 60 min and subfractionated, lysosomes are the only subcellular fraction in which a labeled TCA-soluble material is detectable [44].

3. Cytosolic proteins can be detected in lysosomes by immunological means. Cytosolic tryptophan pyrrolase has been found in liver tritosomes isolated from rats treated with hydrocortisone and chloroquine [45], labeled actomyosin peptides have been detected in lysosomes of rat muscle after an intramuscular injection of labeled N-ethylmaleinimide [46].

Since other mechanisms could account for the eventual localization of a cytosolic protein in lysosomes, demonstration of a direct intracellular transfer of a cytosolic protein into lysosomes is still needed.

Mechanisms of transfer of cytosolic proteins into lysosomes have not been elucidated up to date. Morphological studies indicate that cytosolic proteins and organelles such as mitochondria may end up in lysosomes by engulfment into autophagic vacuoles that fuse subsequently with lysosomes [47]. Entry by autophagocytosis would not provide selectivity in degradation. Yet there is evidence that protein catabolism displays specificity. The turnover rates of proteins have been observed to correlate with size [48], charge [49] and conformation [50]. The size and the charge of the protein subunit in turn are not independent of each other [50a]. Proteins with abnormal structure show an increased degradation rate [51, 52]. This selectivity may be due to a nonlysosomal proteolytic system [52]. A model has been proposed for the degradation of intracellular proteins which postulates an initial limited proteolysis extralysosomally, yielding products that enter lysosomes where degradation would be completed [53]. Both the extralysosomal proteolytic step and the mode of entry could contribute to the selectivity of protein catabolism.

Origin of the Lysosomal Membrane

Large amounts of plasma membrane are consumed during endocytosis [5]. Since plasma membrane-derived endocytotic vesicles eventually fuse with lysosomes to deliver their content, the plasma membrane could contribute to the formation of the lysosomal membrane. The lipid composition of the plasma membrane and the lysosomal membranes have in common the presence of cholesterol and sphingomyelin, which are low in other cytomembranes [54], whereas they differ considerably in other lipids [55]. The membrane proteins of plasma membrane are strikingly different [55]. The marker enzyme for plasma membrane such as Na^+, K^+, activated ATPase and alkaline phosphatase are absent in lysosomal membranes. Though 5'-nucleotidase, as well as leucine aminopeptidase, is found in both membranes, their properties in the two membrane preparations are different

and are probably due to different enzymes [56]. Solubilization of lysosomal membranes in sodium dodecyl sulfate and electrophoresis revealed
to presence of at least 16 protein bands, of which most stain also for
carbohydrate [57, 58]. Most of these proteins are exposed to the cytoplasmic side of lysosomes [57], whereas the neuraminic acid residues
appear to be mainly located on the luminal side [59]. The assymetric
localization could be the result of a preferential degradation of proteins exposed to the luminal side. The turnover rate of soluble and
membraneous lysosomal proteins is similar, although the turnover rate
of soluble lysosomal proteins themselves is heterogeneous [60]. It has
been suggested that lysosomal membrane proteins may function as receptors for cytosolic proteins [57]. Several lines of evidence point to
the possibility that after fusion of endocytotic vesicles with lysosomes
membrane proteins originally exposed on the outer surface of the cell
membrane are located on the cytoplasmic side of the lysosomal membrane
rather than on the luminal side where they would be expected. Anti-cell
surface immunoglobulins, which bind exclusively to the outer side of
the plasma membrane, were found to bind to the cytoplasmic surface of
lysosomes isolated from guinea pig polymorphonuclear leukocytes [61].
Ricin-binding sites, which are not found on the cytoplasmic side of the
plasma membrane, are present on the cytoplasmic surface of lysosomes
[62]. The hepatic receptor for asialoglycoproteins exposes its binding
site on the external side of the plasma membrane. After transfer into
lysosomes via endocytotic vesicles, functioning binding sites can only
be detected on the cytoplasmic side of the lysosomal membrane [22]. It
is unknown at present how the orientation of a glycoprotein is reversed
in a membrane.

The plasma membrane is not the only site wherefrom there is a membrane
flow to lysosomes. A close morphological relationship exists between
lysosomes, a specialized area of the smooth endoplasmic reticulum and
the trans side of the Golgi apparatus. This area, termed according to
its constituents GERL, has been postulated to be the site where lysosomes originate [3]. Since the endoplasmic reticulum itself is a membraneous system with highly specialized areas, one might anticipate
that the composition of lysosomal membrane differs from the overall
composition of endoplasmic reticulum and Golgi apparatus. Their lipid
composition was found to be different [54, 63] and the marker enzymes
of endoplasmic reticulum and Golgi apparatus are absent in lysosomal
membranes [56, 63]. Therefore, an extensive remodeling of cytomembranes
contributing to the lysosomal membrane such as plasma membrane, membranes of the endoplasmic reticulum, and Golgi apparatus is implicated
[4]. Further insight into the origin of lysosomal membrane awaits the
identification of a specific constituent of the lysosomal membrane and
the study of the metabolism of such a marker.

Transfer of Lysosomal Enzymes into Lysosomes

A particular feature of lysosomal enzymes synthesized by fibroblasts
is the presence of a recognition marker, for which specific receptors
are present on the cell surface of fibroblasts (for review see [64-67]).
The recognition marker contains mannose 6-phosphate residues [68] sensitive to alkaline phosphatase. Mannose 6-phosphate and mannose 6-phos-
phate containing mannans are competitive inhibitors for the recognition
and uptake of a variety of lysosomal enzymes [68-72]. The phosphorylated
oligosaccharides, mediating the recognition of human urine α-N-acetyl-
glucosaminidase, can be cleaved by endo-β-N-acetylglucosaminidase H and
are therefore presumably of the high mannose type [73-76]. Surprisingly
more than 70% of the phosphate groups located in endo-β-N-acetylgluco-
saminidase H-cleavable oligosaccharides isolated from β-hexosaminidase

and cathepsin D secreted by human skin fibroblasts are resistent to
alkaline phosphatase treatment [76a]. Mild acid hydrolysis (1 mM HCl,
100°C for 30 min) renders these oligosaccharides susceptible to alkaline phosphatase and increases electrophoretic mobility at pH 8.0.
These results suggest that phosphate groups are present in an intra-
or intermolecular diester linkage.

The binding of α-L-iduronidase to fibroblasts has been studied using
a sensitive fluorometric assay [19]. The binding follows saturation
kinetics with half maximal binding at 1.1×10^{-9} M enzyme. A single
fibroblast exposes 14,000 binding sites on the cell surface. The binding is inhibited by mannose 6-phosphate and is independent of divalent
cations such as Ca^{2+} or Mg^{2+}. In the absence of protein synthesis the
uptake remains unaltered for a period of several hours, indicating a
recycling of the cell surface binding sites. Thus, the complement of
the binding sites on the cell surface appears to be regenerated every
5 min. A receptor protein has so far not been solubilized from fibroblasts, but recently a mannose 6-phosphate specific binding protein
has been isolated from bovine liver [77].

Hickman and Neufeld first proposed in 1972 that the recognition marker
on lysosomal enzymes functions as a signal for transferring newly synthesized lysosomal enzymes into lysosomes [78]. They postulated that
newly synthesized lysosomal enzymes are first secreted and reach the
lysosome only by internalization via adsorptive endocytosis. This secretion-recapture mechanism could explain the pleiotropic effects of
the mutation in mucolipidosis II (I-cell disease) fibroblasts. These
cells show a multiple intracellular deficiency of lysosomal enzymes
accompanied by an accumulation of these enzymes extracellularly as a
result of the deficiency of the recognition marker on the lysosomal
enzymes. Recent studies have provided direct evidence that phosphorylation of lysosomal enzymes is severely reduced in mucolipidosis II
fibroblasts [75, 79, 80]. The availability of competitive inhibitors
for the recapture of secreted lysosomal enzymes allowed to test this
secretion-recapture hypothesis. Although the extracellular levels of
β-hexosaminidase were slightly increased by adding mannose 6-phosphate
to the culture medium in a concentration sufficient to prevent uptake,
the intracellular levels of this and other lysosomal enzymes were not
reduced even after prolonged incubation periods [81-83]. From these
results it was concluded that the secretion-recapture pathway accounts
only for the transfer of a minor part of a lysosomal enzyme. Nevertheless this minor fraction is the cause of the correction of defective
mucopolysaccharide catabolism observed in cell culture upon co-culti-
vation of fibroblasts from patients with different types of mucopoly-
saccharidoses [84].

The failure of inhibitors of uptake to alter the localization of lysosomal enzymes suggests that the bulk of newly formed lysosomal enzymes
is either transferred into lysosome via an intracellular route [66]
and/or via the cell surface in a receptor-bound form, that prevents
diffusion into the culture medium [81] (Fig. 1). In accordance with
the latter hypothesis lysosomal enzymes are present on the cell surface where they are detectable by immunofluorescence [85]. The amount
of β-hexosaminidase that can be solubilized with trypsin from cell surface of a single fibroblast corresponds to about 90,000 enzyme molecules. The principle of the original secretion-recapture hypothesis,
i.e., the function of the recognition marker as a signal for segregating lysosomal enzymes from glycoproteins destined for secretion, is
still preserved in current concepts of lysosomal enzyme transfer. It
has been demonstrated recently that cathepsin D enters the cisternae
of the endoplasmic reticulum in a co-translational manner [86]. The

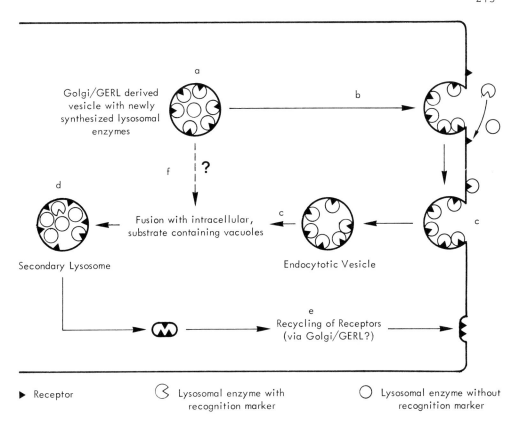

Fig. 1. Scheme of the transfer of lysosomal enzymes in cultured fibroblasts.
a After co-translational transfer into microsomes lysosomal enzymes are glycosylated
and phosphorylated. Packaging into vesicles equipped with the receptor occurs presumably in the GERL area. In these vesicles lysosomal enzymes are bound to the receptors, only enzymes that lack the recognition marker or that have a masked recognition marker are free. *b* These vesicles fuse with the cell membrane. The major part
of lysosomal enzymes remains bound to the receptor on the cell surface. Only such
lysosomal enzymes diffuse into the medium that lack the appropriate recognition marker or dissociate from the receptor. The latter may subsequently bind to free receptors
on the same or another cell, therefore allowing exchange of lysosomal enzymes between
different cells. *c* Internalization of cell surface bound lysosomal enzymes and transfer into lysosomes via endocytotic vesicles. *d* Dissociation of receptor ligand complex, which might require low pH or proteolytic processing of lysosomal enzymes. Lysosomal enzymes may loose their recognition marker, once they are in the lysosomes,
due to lysosomal phosphatases. This accounts for the fact that most of the lysosomal
enzymes extracted from tissues lack the recognition marker. *e* Recycling of the receptors for lysosomal enzymes to the cell surface. Part of the receptors may pass via
the site where newly synthesized lysosomal enzymes are packaged into vesicles (see
a). *f* It is unclear at present to which extent the vesicles containing the newly synthesized lysosomal enzymes may fuse directly with lysosomes, thus bypassing the cell
membrane

sequence of processing and packaging of this or any other hydrolase
into the lysosome remains a matter of speculation at present. It is
not known in which compartment of the cell and by what reactions lyso-

somal enzymes are phosphorylated, nor has it been shown whether the phosphorylated mannose residue is the only recognized structure during the segregation of lysosomal enzymes. The place of the segregation is not known either: does it occur within the network of the intracellular membranes (endoplasmic reticulum, GERL, Golgi apparatus) or at the plasma membrane? The two possibilities, however, are not mutually exclusive.

Lysosomal enzymes are located in fibroblasts in two species of lysosomal organelles, which differ by density [87]. The more dense organelles have the morphological appearance characteristic of lysosomes, whereas the more buoyant organelles are represented by multivesicular structures resembling the GERL network described in hepatocytes by Novikoff [3]. Lysosomal enzymes entering the cells by absorptive endocytosis appear first in the GERL-like organelles and later also in the more dense lysosomes. Substrates for lysosomal enzymes such as mucopolysaccharides are located and become degraded in both types of lysosomal organelles.

Generality of the Receptor-Mediated Transfer of Lysosomal Enzymes

Lysosomal enzymes that are recognized by fibroblasts via a mannose 6-phosphate-sensitive mechanism and hence contain a phosphorylated recognition marker are formed by platelets [68], bovine testes [76], smooth muscle cells [88], endothelial cells [88], and chondrocytes [89].

Mannose 6-phosphate specific receptors mediating uptake of lysosomal enzymes are present on human smooth muscle cells [88], Chinese hamster ovary cells [89a], rat chondrosarcoma cells [89b], and hepatocytes [8]. In hepatocytes the adsorptive endocytosis of lysosomal enzymes can be mediated also by galactose specific receptors [8], whereas in macrophages from various sources internalization of lysosomal enzymes via mannose N-acetylglucosamine specific receptors has been observed [23, 24, 89, 90]. Nothing is known about the involvement of these receptors in the transfer of newly synthesized lysosomal enzymes. Observations in various tissues and cells obtained from mucolipidosis II patients indicate that although the phosphorylation of lysosomal enzymes and the lysosomal functions are impaired in fibroblasts, lysosomes are functional in a number of other cell types (for review see [92]). In these cell types, as in endothelial cells that synthesize phosphorylated lysosomal enzymes but do not endocytose lysosomal enzymes, alternative mechanisms may be operating for the transfer of lysosomal enzymes.

Fate of Material Entering the Lysosomes

Macromolecules entering the lysosomal apparatus are subject to degradation. Among the final products of degradation there are monosaccharides, amino acids, small peptides, nucleotides, phosphate, and sulfate. Obviously, the storage capacity of the lysosomes for low molecular solutes is limited. Beside diffusion there may be mechanisms present in lysosomes, which govern the release of low molecular weight products (for discussion see [93]). Only a minor fraction of ^{125}I-monoiodotyrosine, the final radiolabeled product of endocytosed ^{125}I-labeled proteins, and of $^{35}SO_4{}^{2-}$ liberated from endocytosed ^{35}S-labeled sulfated polysaccharides is recovered in the cells. The major fraction appears in the culture medium.

It has been mentioned earlier that receptors for LDL and lysosomal enzymes apparently escape by recycling from degradation after delivering the ligands into lysosomes. They share this particular feature with receptors for asialoglycoproteins [94], transcobalamin II [95], and sulfated proteoglycans [96], whereas other receptors may become degraded after internalization. Thus epidermal growth hormone receptor becomes degraded for the most part after being internalized in the complex with the hormone [97, 98]. The recycling is considered to explain the discrepancies between the half-life of membrane proteins which for many is in excess of 50 h and the extensive internalization of the plasma membrane during endocytosis, e.g., mouse L-fibroblasts internalize up to 50% of their plasma membrane within 1 h [7]. Outer sugars of a membrane glycoprotein have been observed to have a much shorter half-life than the protein moiety. This may indicate a partial degradation and a reglycosylation during recycling [99].

Of particular interest is the fate of lysosomal enzymes. The catalytic activity of exogenous lysosomal enzymes taken up by endocytosis persists with half-lifes from less than two days up to more than two weeks (for review see [65]). The turnover rates as studied in rats treated with cycloheximide vary at least between 14 h (for cathepsin D) and 158 h (for β-glucuronidase) [99a]. Recently it has been observed that some lysosomal enzymes undergo a proteolytic processing. In fibroblasts lysosomal enzymes are synthesized in the form of larger precursors that are proteolytically processed to forms characteristic of purified tissue enzymes [100]. The final products are found exclusively intracellularly. In mucolipidosis II fibroblasts the processing is absent or incomplete. In the culture medium of fibroblasts only precursor molecules are present, persisting there over a period of several days. In vitro experiments suggest that lysosomal thiol proteases are responsible for the processing [101]. Synthesis of precursor forms and processing appears to be a general phenomenon for lysosomal enzymes and has now been documented for various lysosomal enzymes in fibroblasts, macrophages, kidney cells, smooth muscle cells, and endothelial cells [67, 86, 100, 102, 103] (Table 1). The function of the polypeptides cleaved off from the precursors is unknown. They may be involved in recognition and segregation of lysosomal enzymes. In some cases the precursors may be devoid of their potentially harmful hydrolytic activity.

Endocytosis of macromolecules and generation of degradation products seem to be involved in the metabolic control of the cell. Protein hormones interact with target cells through binding to cell surface receptors. The bound hormones become internalized and degraded in the lysosomes as was shown for epidermal growth hormone [97, 98], insulin [104], and chorionic gonadotropin [105]. Generally the number of cell surface receptors becomes reduced in the presence of effective concentrations of the corresponding hormone. Such a down-regulation of hormone receptors is explained in terms of endocytosis and degradation of hormone-receptor complexes [106]. So far it has not been proven that other effects of the hormones mentioned depend on their endocytosis and eventual degradation. There is some evidence for an endocytotic activation of the epidermal growth hormone, where internalization, but not binding, is correlated with stimulation of DNA synthesis [98]. A different mechanism seems to mediate the receptor down-regulation of the low density lipoprotein receptors. Internalized LDL is degraded and cholesterol is liberated by acid lipase. This cholesterol turns off the synthesis of the LDL receptor, thus protecting the cell against an overloading with cholesterol [107]. Feed-back regulation of 3-hydroxy-3-methyl-glutaryl coenzyme-A reductase, the rate-limiting enzyme in cholesterol biosynthesis and activation of acyl CoA-cholesterol acyltransferase [108, 109] turns off excessive endogenous synthesis of cholesterol and enhances the capacity of the cell to transform the cholesterol into its storage form, the ester.

Table 1. Molecular weight of precursor and processed forms of lysosomal enzymes

Enzyme	Cell type	Molecular weight ($\times 10^{-3}$) of		Reference
		Precursor	Processed forms	
β-Hexosaminidase	Human fibroblasts[a]			
α-chain		67	54	[100]
β-chain		63	29	[100]
α-Glucosidase	Human fibroblasts	95	79/76	[100]
α-L-Iduronidase	Human fibroblasts	80	72	[67]
α-N-Acetylglucosaminidase	Human fibroblasts	86	80	[103]
Arylsulfatase A	Human fibroblasts	64	62.5	[103]
Cathepsin D	Human fibroblasts[a]	53	31	[100]
Cathepsin D	Porcine kidney cells	46	44/30	[86]
β-Galactosidase	Mouse macrophages	82	63	[102]

[a]The precursor and processed forms in human smooth muscle cells and human endothelial cells were comparable to that found in human fibroblasts [88]

The observation that products of endocytosed material function as regulatory compounds in cellular metabolism strengthens the view that lysosomes participate in the control of cellular metabolism by extracellular factors. Lysosomes mediate the modulation of the number of some receptors on the cell surface and they may generate regulatory signals from endocytosed macromolecules.

Acknowledgments. Work from the authors cited in this review was supported by the Deutsche Forschungsgemeinschaft (SFB 104).

References

1. De Duve C (1969) In: Dingle JT, Fell HB (eds) Lysosomes in biology and pathology, vol I. North Holland Publishing Co, Amsterdam, pp 3-30
2. Novikoff AB (1973) In: Hers G, van Hoof F (eds) Lysosomes and storage diseases. Academic Press, London, New York, pp 1-41
3. Novikoff AB (1976) Proc Natl Acad Sci USA 73:2781-2787
4. Morré DJ, Kartenbeck J, Franke WW (1979) Biochim Biophys Acta 559:71-152
5. Silverstein SC, Steinman RM, Cohn ZA (1977) Annu Rev Biochem 46:669-722
6. Jacques PJ (1969) In: Dingle JT, Fell HB (eds) Lysosomes in biology and pathology, vol II. North Holland Publishing Co, Amsterdam, pp 395-420
7. Steinman RM, Silver JM, Cohn ZA (1978) In: Silverstein SC (ed) Transport of macromolecules. Berlin, Dahlem Konferenzen, pp 167-180
8. Ullrich K, Mersmann G, Weber E, von Figura K (1978) Biochem J 170:643-650
9. Brown MS, Goldstein JL (1979) Proc Natl Acad Sci USA 76:3330-3337
10. Mahley RW, Innerarity TL, Pitas RE, Weisgraber KH, Brown JH, Gross E (1977) J Biol Chem 252:7279-7287
11. Shireman R, Kilgore L, Fisher WR (1977) Proc Natl Acad Sci USA 74:5150-5154
12. Schneider WJ, Basu SK, McPhaul MJ, Goldstein JL, Brown MS (1979) Proc Natl Acad Sci USA 76:5577-5581
13. Weisgraber KH, Innerarity TL, Mahley RW (1978) J Biol Chem 253:9053-9093
14. Filipovic I, Schwarzmann G, Mraz W, Wiegandt H, Buddecke E (1979) Eur J Biochem 93:51-55
15. Goldstein JL, Anderson RGW, Brown MS (1979) Nature (London) 279:679-685
16. Ashwell G, Morell HG (1974) Adv Enzymol 41:99-128
17. Youngdahl-Turner P, Rosenberg LE, Allen RH (1978) J Clin Invest 61:133-141
18. Seligman PS, Allen RH (1978) J Biol Chem 253:1766-1772
19. Rome LH, Weissmann B, Neufeld EF (1979) Proc Natl Acad Sci USA 76:2331-2334
20. Figura von K, Kresse H (1974) J Clin Invest 53:85-90
21. Carpenter G, Lambach KJ, Morrison MM, Cohen S (1975) J Biol Chem 250:4297-4304
22. Neufeld EF, Ashwell G (1979) In: Lennarz W (ed) Biochemistry of proteoglycans and glycoproteins. Plenum Press, New York
23. Schlesinger PH, Doebber TW, Mandell BF, White R, De Schryver C, Rodman JS, Miller MJ, Stahl PD (1978) Biochem J 176:103-109
24. Ullrich K, Gieselmann V, Mersmann G, von Figura K (1979) Biochem J 182:329-335
25. Steer C, Clarenburg R (1979) J Biol Chem 254:4457-4461
26. Maxfield FR, Schlesinger J, Schechter Y, Pastan J, Willingham MC (1978) Cell 14:805-810
27. Maxfield FR, Willingham MC, Davies PJA, Pastan J (1979) Nature (London) 277:661-663
28. Willingham MV, Maxfield FR, Pastan JH (1979) J Cell Biol 82:614-625
29. Gordon P, Carpentier J, Cohen S, Orci L (1978) Proc Natl Acad Sci USA 75:5025-5029
30. McKanna JA, Haigler HT, Cohen S (1979) Proc Natl Acad Sci USA 76:5689-5693
31. Regoeczi E, Taylor P, Hatton MW, Wong K-L, Koj A (1978) Biochem J 174:171-178
32. Regoeczi E, Taylor P, Debaune MT, März L, Hatton MWC (1979) Biochem J 184:399-407
33. Schechter Y, Hernaez L, Schlesinger J, Cuatrecasas P (1979) Nature (London) 278:835-838

34. Schechter Y, Chang KJ, Jacobs S, Cuatrecasas P (1979) Proc Natl Acad Sci USA 76: 2720-2724
35. Neufeld EF, Cantz M (1973) In: Hers HG, van Hoof F (eds) Lysosomes and storage diseases. Academic Press, London, New York, pp 261-275
36. Kresse H, Tekolf W, von Figura K, Buddecke E (1975) Hoppe-Seyler's Z Physiol Chem 356:943-952
37. Friend DS (1969) J Cell Biol 41:269-279
38. Holtzman E (1976) Lysosomes: A survey. Springer, Wien
39. Roth TF, Cutting JA, Atlas SB (1976) J Supramol Struct 4:527-548
40. Palade GE, Simionescu M, Simionescu N (1978) In: Silverstein SC (ed) Transport of macromolecules in cellular systems. Berlin, Dahlem Konferenzen, pp 145-166
41. Bradshaw RA (1978) Annu Rev Biochem 47:191-216
42. Schimke RT (1970) In: Munro HN (ed) Mammalian protein turnover, vol IV. Academic Press, London, New York, pp 177-277
43. Dean RT (1975) Nature (London) 257:414-416
43a Ward WF, Chna BL, Li JB, Morgan HE, Mortimore GE (1979) Biochem Biophys Res Commun 87:92-98
44. Neely AN, Mortimore GE (1974) Biochem Biophys Res Commun 59:680-687
45. Rudek DE, Dien, PY, Schneider DL (1978) Biochem Biophys Res Commun 82:342-347
46. Gerard KW, Schneider DL (1979) J Biol Chem 254:11798-11805
47. De Duve C, Wattiaux R (1966) Annu Rev Physiol 28:435-492
48. Dice JF, Dehlinger PJ, Schimke RT (1973) J Biol Chem 248:4220-4228
49. Dice JF, Goldberg AL (1975) Fed Proc 34:651
50. Ballard FJ, Hopgood MF, Resheff L, Hanson RW (1974) Biochem J 140:531-538
50a Duncan WE, Offermann MK, Brown JS (1980) Arch Biochem Biophys 199:331-341
51. Goldberg AL, St John AC (1976) Annu Rev Biochem 45:747-803
52. Etlinger JD, Goldberg AL (1977) Proc Natl Acad Sci USA 74:54-58
53. Segal HL (1975) In: Dingle JT, Dean RT (eds) Lysosomes in biology and pathology, vol IV. North Holland Publishing Co, Amsterdam, pp 295-302
54. Henning R, Kaulen HD, Stoffel W (1970) Hoppe-Seyler's Z Physiol Chem 351:1191-1199
55. Henning R, Stoffel W (1973) Hoppe-Seyler's Z Physiol Chem 354:760-770
56. Kaulen HD, Henning R, Stoffel W (1970) Hoppe-Seyler's Z Physiol Chem 351:1555-1563
57. Schneider DL, Burnside J, Gorga FR, Nettleton CJ (1978) Biochem J 176:75-82
58. Milsom DW, Wynn CH (1973) Biochem Soc Trans 1:426-428
59. Henning R, Plattner H, Stoffel W (1973) Biochim Biophys Acta 330:61-75
60. Wang CC, Touster O (1975) J Biol Chem 250:4896-4902
61. Rikihisa Y, Mizumo D (1977) Exp Cell Res 110:103-110
62. Feigenson ME, Schnebli HP, Baggiolini M (1975) J Cell Biol 66:183
63. Thines-Sempoux (1973) In: Dingle JT (ed) Lysosomes in biology and pathology, vol III. North Holland Publishing Co, Amsterdam, pp 278-299
64. Neufeld EF, Sando GN, Gorvin J, Rome LH (1977) J Supramol Struct 6:95-101
65. Figura von K (1977) In: Harkness RA, Cockburn F (eds) The cultured cell and inherited metabolic disease. MTP Press, Lancaster, pp 105-119
66. Sly WS (1979) In: Svennerholm L, Mandel D, Dreyfus H, Urban PF (eds) Structure and function of gangliosides. Plenum Publishers, New York, pp 433-451
67. Neufeld EF (1980) In: Lowdon JA, Callahan JW (eds) Lysosomes and lysosomal storage diseases. Raven Press, New York, in press
68. Kaplan A, Achord DT, Sly WS (1977) Proc Natl Acad Sci USA 74:2026-2030
69. Kaplan A, Fischer D, Achord D, Sly WS (1977) J Clin Invest 60:1088-1093
70. Sando GN, Neufeld EF (1977) Cell 12:619-627
71. Ullrich K, Mersmann G, Weber E, von Figura K (1978) Biochem J 170:643-650
72. Kaplan A, Fischer D, Sly WS (1978) J Biol Chem 253:647-650
73. Figura von K, Klein U (1979) Eur J Biochem 94:347-354
74. Natowicz MR, Chi MMY, Lowry OH, Sly WS (1979) Proc Natl Acad Sci USA 76:4322-4326
75. Hasilik A, Neufeld EF (1980) J Biol Chem, in press
76. Distler J, Hieber V, Sahagian G, Schmickel R, Jourdian GW (1979) Proc Natl Acad Sci USA 76:4235-4239
76a Hasilik A, Klein U, von Figura K, unpublished

77. Sahagian G, Distler J, Jourdian GW (1980) Fed Proc, in press
78. Hickman S, Neufeld EF (1972) Biochem Biophys Res Commun 49:992-999
79. Hasilik A, Neufeld EF (1980) J Biol Chem, in press
80. Bach G, Bargal R, Cantz M (1979) Biochem Biophys Res Commun 91:976-981
81. Figura von K, Weber E (1978) Biochem J 176:943-950
82. Sly WS, Stahl P (1978) In: Silverstein S (ed) Transport of molecules in cellu-
    lar systems. Berlin, Dahlem Konferenzen, pp 229-244
83. Vladutiu G, Rattazi M (1979) J Clin Invest 63:595-601
84. Neufeld EF, Lim TW, Shapiro LJ (1975) Annu Rev Biochem 44:357-376
85. Figura von K, Voss B (1979) Exp Cell Res 121:267-276
86. Erickson AH, Blobel G (1979) J Biol Chem 254:11771-11774
87. Rome LH, Garvin AJ, Alietta MM, Neufeld EF (1979) Cell 17:143-153
88. Hasilik A, Voss B, von Figura K (1980) Hoppe-Seyler's Z Physiol Chem 361:262
89. Mittelviefhaus H, von Figura K (unpublished)
89a Robbins AL (1980) Proc Natl Acad Sci USA, in press
89b Rome L, Miller J (1980) Biochem Biophys Res Commun 92:986-993
90. Achord DT, Brot FE, Bell CE, Sly WS (1978) Cell 15:269-278
91. Stahl P, Rodman JS, Miller MJ, Schlesinger PH (1978) Proc Natl Acad Sci USA 75:
    1399-1403
92. McKusick V, Neufeld EF, Kelly   (1978) In: Stanbury JB, Wyngaarden JB, Frede-
    rickson DS (eds) The metabolic basis of inherited disease, 4th edn. McGraw-Hill,
    New York, pp 1282-1307
93. Reijngoud DJ, Tager JM (1977) Biochem Biophys Acta 472:419-449
94. Tanabe T, Pricer WE Jr, Ashwell G (1979) J Biol Chem 254:1038-1043
95. Youngdahl-Turner P, Mellman JS, Allen RH, Rosenberg LE (1979) Exp Cell Res 118:
    127-134
96. Prinz R, Schwermann J, Buddecke E, von Figura K (1978) Biochem J 176:671-676
97. Carpenter G, Cohen S (1976) J Cell Biol 71:159-171
98. Fox CF, Das M (1979) J Supramol Struct 10:199-214
99. Kreisel W, Volk B, Büchsel R, Reutter W (1980) Proc Natl Acad Sci USA, in press
99a Tersitore L, Curzio M, Cecchini G, Zuretti FM, Baccino FM (1979) Boll Soc Ital
    Biol Sper 55:60-64
100. Hasilik A, Neufeld EF (1980) J Biol Chem, in press
101. Frisch A, Neufeld EF (1980) Fed Proc, in press
102. Skudlarek MD, Swank RT (1979) J Biol Chem 254:9939-9942
103. Hasilik A, Waheed A, von Figura K (unpublished)
104. Terris S, Steiner DF (1975) J Biol Chem 250:8389-8398
105. Asoli M, Puett D (1978) J Biol Chem 253:4892-4899
106. Jarett L, Smith RM (1978) In: Silverstein SC (ed) Transport of macromolecules
     in cellular systems. Berlin, Dahlem Konferenzen, pp 117-132
107. Brown MS, Goldstein JL (1975) Cell 6:307-316
108. Brown MS, Dana SE, Goldstein JL (1973) Proc Natl Acad Sci USA 70:2162-2166
109. Goldstein JL, Dana SE, Brown MS (1974) Proc Natl Acad Sci USA 71:4288-4292

The Semliki Forest Virus Envelope: A Probe for Studies of Plasma Membrane Structure and Assembly

H. Garoff and K. Simons[1]

A number of animal viruses leave their host cells by budding through the host cell surface (Lenard and Compans 1974). The outer coat of these viruses consists of a segment of the host plasma membrane. The viruses are easy to purify. Their composition and structure are simpler than those of the cell surface membrane. Such enveloped viruses have therefore become widely used as probes for studies of plasma membrane structure and assembly. Some of these enveloped viruses shut off host protein synthesis and the infected cell becomes, in fact, a factory for virus production. This property greatly facilitates studies of membrane protein synthesis and assembly.

The best-studied membrane viruses include vesicular stomatitis virus (a rhabdovirus), influenza virus (a myxo virus) and alphaviruses. For recent reviews on membrane viruses, we refer to Kääriäinen and Renkonen (1977) Lenard (1978) and Patzer et al. (1979). We will in this review briefly summarize what is presently known of Semliki Forest virus, the virus we have been using for our studies. Semliki Forest (SF) virus is an alphavirus. Another well-studied member of the same virus is Sindbis virus. These viruses infect a wide variety of cells in culture both vertebrate and invertebrate. The virus consists of a nucleocapsid with the 42S RNA enclosed in a shell of protein and the membrane which the virus obtains when it buds out of the host cell (for an earlier review see Simons et al. 1978a).

Composition and Structure

Lipids

The lipid composition of SF virus reflects qualitatively and quantitatively that of the host cell plasma membrane (Renkonen et al. 1971). In 1963 Pfefferkorn and Hunter showed that the virus phospholipid is largely derived from preexisting cellular phospholipids. However, the question whether SF virus asserts any selectivity on the set of lipids it takes with it from the host cell or whether the virus lipids represent a random sample of the plasma membrane is still open.

An asymmetric distribution of the different phospholipids over the two leaflets of the membrane lipid bilayer has been best documented for the plasma membrane of the erythrocyte (see Op den Kamp 1979). Phospholipid localization studies on plasma membranes from nucleated eukaryotic cells are much more complicated and have been carried out mainly with derivatives of the plasma membrane, either using membrane viruses (right-side-out derivatives) or phagosomes (inside-out derivatives) (Sandra and Pagano 1978) which are formed through the uptake of latex beads by mouse LM cells. A condensed summary of the results obtained

---

[1]European Molecular Biology Laboratory, Postfach 10.2209, 6900 Heidelberg, FRG

Table 1. Phospholipid asymmetry in membrane structures derived from plasma membranes

Membrane system	Cells	Method	PC	SM	PS	PE
Semliki forest virus[A]	BHK-21	Exchange protein	52[a]			
		TNBS				22
		Phospholipases	51	33		20
Vesicular stomatitis virus[B+C]	BHK-21	TNBs				36-38
		Exchange protein	58			
Influenza virus[D]	MDBK	Phospholipase C	34-45	23	23-25	28-30
		Exchange proteins	43-48	15	14	
Phagosomes[E]	Mouse LM	Exchange proteins	48			
		TNBS				24-30

[a]The numbers are given as percentage of each individual phospholipid class being present in the outer membrane leaflet

[A]Van Meer et al. 1980; [B]FONG et al. 1976; [C]Shaw et al. 1979; [D]Rothman et al. 1976; [E]Sandra and Pagano 1978

using phospholipid exchange proteins, phosphilipases and labeling with trinitrobenzenesulphonate (TNBS) is shown in Table 1. Altogether the data suggest a rather uniform asymmetrical arrangements of the phospholipid classes in the membranes studied. Phosphatidylcholine (PC) is symmetrically distributed between the two leaflets. Of the sphingomyelin (SM) 15% - 33% is in the outer leaflets. Phosphatidylserine (PS) could not be studied in most cases but in influenza virus 14% - 25% was exposed on the outer surface, and of the phosphatidylethanolamine (PE) 20% - 38% was in the outer leaflet. The glycolipids found in the virus membranes are probably localized in the outer leaflet (Patzer et al. 1979). Cholesterol, which is a major component of the lipids in membrane viruses, is probably fairly equally distributed between the two leaflets, but this may vary (Patzer et al. 1979).

Proteins

The pioneering studies of Pfefferkorn and Clifford (1964) were the first to show that in contrast to the lipids the proteins in Sindbis virus did not pre-exist in the host cell and were presumably coded for by the virus genome. There are four species of polypeptides in SF virus (Garoff et al. 1974). Each polypeptide is present in about 240 copies per virus particle. Little host protein is incorporated into the virus particles. Strauss (1978) sets the limit at less than 0.2% - 0.5% of the total virus protein which amounts to no more than 2 - 5 host polypeptides assuming a (M_r = 30,000) per virus. Actin has been found in many other membrane viruses (Wang et al. 1976), but not in alphaviruses. The C-protein (M_r = 30,000) forms the protein shell in the nucleocapsid. The E1 (M_r = 49,000), E2 (M_r = 52,000) and E3 (M_r = 10,000) polypeptides are in the virus membrane and are all glycosylated. Each polypeptide contains one lactosamine-type of oligosaccharide unit consisting of approximately two sialosyl-N-acetyllactosamine branches (Mattila et al. 1976; Pesonen and Renkonen 1976). The branches are attached to a core tetrasacchride with the structure (Man α 1-3 [Man α

1-6] Man β 1-4 GlcNAc) (Pesonen et al. 1979). The core tetrasaccharide
is likely to be bound through a β 1-4 linkage to an N-acetylglucosamine
residue that is linked to asparagine in the polypeptide backbone. Terminal fucose is attached to the innermost N-acetylglucosamine unit.
E2 contains in addition to the lactosamine glycan a high-mannose type
of oligosaccharide, the precise structure of which is not known but
in general appears to be Asn-(GlcNAc)$_2$-(Man)$_{5-7}$ (Mattila and Renkonen
1978). However, the relative amounts of the lactosamine and high mannose glycans seem to vary in different E2 protein molecules.

Similar glycans have been found in both secretory and membrane proteins
made by the host cell. The virus proteins are apparently glycosylated
by the cellular glycosylation machinery. A number of studies have made
use of enveloped viruses as tools to study the details and mechanisms
of protein glycosylation (Robbins et al. 1977; Hunt et al. 1978; Tabas
et al. 1978).

Structure of the SF Virus Membrane Glycoproteins

The E1, E2, and E3 polypeptides form the spike-like projections on the
external surface of the SF virus membrane. Studies using detergent solubilization, protein cross-linking agents and specific antibodies have
shown that E1, E2, and E3 form a three-chain oligomer in which the
polypeptides are noncovalently bound to each other (Ziemiecki and
Garoff 1978). Each spike is thus probably made up of one copy of E1,
E2, and E3 each. Protease digestion has shown that about 90% of the
spike protein, including all of the protein-bound carbohydrate, is on
the external side of the lipid bilayer (Gahmberg et al. 1972). E3 is
completely on the outside, whereas E2 and E1 have hydrophobic tails
which attach them to the membrane (Utermann and Simons 1974). These
hydrophobic peptide tails have been mapped on the carboxy-terminal regions of the E1 and E2 peptide chains (Garoff and Söderlund 1978). The
proteins are thus oriented such that the N-terminal regions are external to the bilayer with carboxyterminal regions penetrating into
the bilayer. The same orientation has been found for other virus membrane glycoproteins (Skehel and Waterfield 1975; Gething et al. 1978)
and for glycophorin in the erythrocyte membrane (Marchesi et al. 1976)
and the H$_2$ and the HLA histocompatibility antigens (Henning et al.
1976; Springer and Strominger 1976). There is, however, another class
of surface membrane proteins found in the brush border of kidney and
intestinal epithelial cells. These proteins are oriented the other
way around with their amino-terminal regions in the membrane (Maroux
and Louvard 1976; Frank et al. 1978). No virus membrane proteins are
known with this orientation.

Further studies using protease digestion cross-linking agents, surface
labeling, and peptide mapping, have shown that the E2 peptide chain
spans the bilayer and a small carboxy-terminal segment of about 30
amino acids is located on the internal side of the membrane (Garoff
and Simons 1974; Garoff and Söderlund 1978). No indication was found
for E1 spanning the bilayer.

We have recently finished the complete nucleotide sequence of the
genes coding for the C, E1, E2, and E3 polypeptides from cloned DNA
(Garoff et al. 1980). The amino acid sequences deduced from the DNA
sequence give strong support to our studies of the SF virus spike protein structure. Only two hydrophobic sequences long enough to span the
bilayer as an α-helix have been found in the sequences of the virus
proteins using the criteria devised by Segrest and Feldmann (1974).
These are located in carboxy-terminal regions of E1 and E2 respectively

Glycophorin (Tomita and Marchesi, 1976)

Ala Gly

-Glu-Ile-Glu-Thr-Leu-Ile-Val-Phe-Gly-Val-Met-Ala-Gly-Val-Ile-Gly-Thr-Ile-Leu-Leu-Ile-Ser-Fyr-Gly-Ile-Arg-Arg-Leu-
-Ile-Lys-Lys-Ser-Pro-Ser-Asp-Val-Lys-Pro-Leu-Pro-Ser-Asp-Thr-Asp-Val-Pro-Leu-Ser-Ser-Val-Glu-Ile-Glu-Asn-Pro-Glu-
- Thr-Ser-Asp-Gln-COOH

Fowl plague virus haemagglutinin (Porter et al., 1979)

-Tyr-Lys-Asp-Val-Ile-Leu-Trp-Phe-Ser-Phe-Gly-Ala-Ser-Cys-Phe-Leu-Leu-Leu-Ala-Ile-Ala-Val-Gly-Leu-Val-Phe-Ile-Cys-Val-
-Lys-Asn-Gly-Asn-Met-Arg-Cys-Thr-Ile-Cys-Ile-COOH

Influenza A/Victoria/3/75 haemagglutinin (Min Jou et al., 1980)

-Tyr-Lys-Asp-Trp-Ile-Leu-Trp-Ile-Ser-Phe-Ala-Ile-Ser-Cys-Phe-Leu-Leu-Cys-Val-Val-Leu-Leu-Gly-Phe-Ile-Met-Trp-Ala-Cys-
-Gln-Lys-Gly-Asn-Ile-Arg-Cys-Asn-Ile-Cys-Ile-COOH

Semliki Forest virus El - protein (Garoff et al., 1980)

-Val-Gln-Lys-Ile-Ser-Gly-Gly-Leu-Gly-Ala-Phe-Ala-Ile-Gly-AlaIle-Leu-Val-Leu-Val-Val-Val-Thr-Cys-Ile-Gly-Leu-Arg-Arg-
-COOH

Semliki Forest virus E2 - protein (Garoff et al., 1980)

-Gly-Leu-Tyr-Pro-Ala-Ala-Thr-Val-Ser-Ala-Val-Val-Gly-Met-Ser-Leu-Leu-Ala-Leu-Ile-Ser-Ile-Phe-Ala-Ser-Cys-Tyr-Met-Leu-
-Val-Ala-Ala-Arg-Ser-Lys-Cys-Leu-Thr-Pro-Tyr-Ala-Leu-Thr-Pro-Gly-Ala-Ala-Val-Pro-Trp-Thr-Leu-Gly-Ile-Leu-Cys-Cys-
Ala-Pro-Arg-Ala-His-Ala-COOH

Fig. 1. Transmembrane sequences of glycophorin, membrane glycoproteins from influenza and Semliki Forest virus. The postulated hydrophobic transmembrane segment is *underlined*. The segments on the carboxy-terminal side probably form the internal domain of the protein

(Fig. 1). The hydrophobic sequence in E2 is followed by a segment of
31 amino acid residues before the carboxyterminus is reached. This segment of 31 residues must correspond to the internal peptide segment of
E2 previously postulated by our biochemical studies. There is a lysine
residue in this internal segment which is probably the one labeled by
$|^{35}S|$-formylmethionyl sulphate methylphosphate in the membrane host
preparations obtained from the virus by adding enough Triton X-100 to
lyse the membrane (Garoff and Simons 1974; Simons et al. 1980). This
lysine residue is probably also cross-linked to lysine residues in the
C-protein in the intact virus by diimidoesters (Garoff and Simons 1974).

Carbohydrate units in glycoproteins are known to appear either in Asn-
X-Ser or in Asn-X-Thr sequences (X represents a variable amino acid)
(Neuberger et al. 1972). A search for such sequences revealed only
one site in E1, located 141 residues from the amino terminus, two sites
in E2, located at residues 200 and 262, respectively, from the amino
terminus and two sites at residues 13 and 60 in E3. Since only one
glycan unit is found in E3, only one of the potential sites is probably
used. Not all such sequences can serve as a receptor for an oligosaccharide unit. Other factors must also be involved in determining where
carbohydrates attach (Neuberger et al. 1972).

Electron micrographs of negatively stained Sindbis virus particles
show that the glycoproteins are organized with trimer clustering in a
T = 4 icosahedral surface lattice (von Bonsdorff and Harrison 1975).
Von Bonsdorff and Harrison (1978) have furthermore shown that when
ghosts of the Sindbis virus membrane are released from the nucleocapside by low concentrations of Triton X-100, the spike glycoproteins
can form arrays, isomorphous in local packing to the viral surface
lattice. Thus the surface lattice is probably preserved after release
from the underlying nucleocapsid by reasonably strong lateral glyco-
protein-glycoprotein contacts. Experiments using phospholipases and
phospholipid exchange proteins have shown that the phospholipid head
groups in the external leaflet of the bilayer are shielded by the proteins, and are much more accessible after the spike proteins have been
removed by proteases (van Meer et al. 1980). Most of the carbohydrate
residues bound to the protein are not accessible to glycosidases in
the intact virus particle (McCarthy and Harrison 1977). All these findings suggest that the virus membrane is fairly rigidly organized. Each
spike protein is probably anchored through the membrane to one C-protein in the underlying nucleocapsid. External to the bilayer the spike
proteins are in close contact, perhaps mediated by their carbohydrate
moieties. The hydrophobic tails of the proteins are fairly distant
from each other, probably occupying less than 15% of the area in the
middle of the bilayer (Utermann and Simons 1974).

Assembly of the Virus Membrane

The virus 42S RNA serves as a messenger for translation of proteins
needed for new virus RNA synthesis (Strauss and Strauss 1976). Two
major RNA molecules are produced in the infected cell: new 42S RNA and
26S RNA molecules. The 26S RNA is homologous to the 3' end of the 42S
RNA, and it serves as a mRNA for the structural proteins of the virus
(Kääriäinen and Söderlund 1978). There is only one initiation sites
for protein synthesis on the 26S RNA (Clegg and Kennedy 1975; Glaville
et al. 1976). The virus proteins are translated sequentially from this
site in the order C, E3, and E1 (Clegg 1975; Garoff et al. 1978). Soon
after synthesis the C proteins bind to the 42S RNA and form the virus
nucleocapsid in the cytoplasm. The E polypeptides are inserted into
the membrane of the endoplasmic reticulum (Garoff et al. 1978) and are

transported to the surface of the cell (Richardson and Vance 1976).
Thus proteins made from the same 26S RNA are directed to two different
compartments in the cell. Moreover four polypeptides result from the
translation of this mRNA from only one initiation site.

Translation of the Virus Proteins

Together with B. Dobberstein we studied these questions in more detail
by making use of the in vitro translation system developed by Blobel
and Dobberstein (1975) for demonstrating translocation of secretory
proteins into microsomal membrane vesicles (Garoff et al. 1978). When
26S mRNA was used in the cell-free protein-synthesizing system derived
from HeLa cells two proteins were made: the C protein and a protein
with a M_r = 97,000 (97K protein). This latter protein is known to contain the sequences of E1, E2, and E3. When microsomes from dog pancreas
were added to the system, two new proteins were made in addition to
the C protein (65% of the protein made) and the 97K protein (2%). One
(8%) was the E1 protein and the other (25%) was the p62 protein. This
latter protein is the precursor for the E2 and E3 proteins. In cells
infected with SF virus the same four proteins are seen with identical
mobilities in SDS-gel electrophoreins. Protease digestion showed that
the C and the 97K protein were located on the "cytoplasmic" side of
the microsomal vesicles, whereas the p62 and E1 protein were inserted
in the correct asymmetric orientation into the microsomal membrane.
In p62 the carboxy-terminal region was cleaved off by protease treatment, whereas E1 was completely protected. A study of the time course
of translation and translocation showed that the C-protein was cleaved
from the growing polypeptide chain as soon as it was translated. So
was the p62 protein. Thus both proteolytic cleavages take place on the
nascent chain. The former cleavage proceeded in absence of microsomes,
whereas the latter was dependent on membrane insertion. A synchronized
translation of the 26S RNA showed that the microsomal vesicles had to
be added to the cell-free system after the C-protein had been translated, but before about 100 amino acids of the p62 protein had been
made. If translation proceeded, further insertion of the growing chain
into the microsomal vesicles was no longer possible and instead the
97K protein was synthesized. Thus there must be a signal sequence at
the amino terminus of p62, which directs the ribosome to the endoplasmic reticulum. Bonatti and Blobel (1979) have partially sequenced the
N-terminal region of the analog to p62 from Sindbis virus both made
in vitro and in the infected cell and the analog to the 97K protein
made in vitro. They found the sequences to be identical, showing that
the signal sequence is not cleaved during translation as is the case
for the amino terminal sequences of most secretory proteins. In this
case of the membrane glycoprotein of vesicular stomatitis virus which
was the first integral membrane protein to be made and assembled into
microsomal vesicles, kinetic experiments in vitro showed that the signal sequence is located within the first 80 amino residues of the polypeptide chain (Rothman and Lodish 1977) and later sequencing studies
revealed that a segment of 16 amino acid residues is cleaved from the
polypeptide chain during translation (Lingpappa et al. 1978; Irving
et al. 1979). Further experiments showed that the nascent membrane
glycoprotein of this virus competed with nascent preprolactin for the
membrane receptors on the microsomal vesicles (Lingpappa et al. 1978),
indicating that the signal peptides for secretory proteins and the
virus membrane proteins are functionally equivalent. The main difference lies in the carboxyterminal regions of the proteins. The secretory proteins are translocated in their entirety through the membrane
and are released into the lumen of the vesicle (Blobel et al. 1979),
whereas the glycoprotein of the vesicular stomatitis virus gets locked

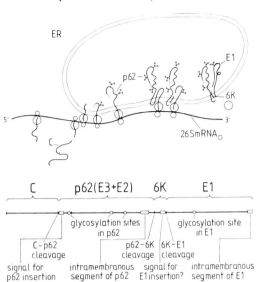

Synthesis of SFV proteins

Fig. 2. A scheme for the assembly of the Semliki Forest virus membrane glycoprotein into the membrane of the endoplasmic reticulum (Garoff et al. 1980). It is not exactly known when and how the 6K peptide is excised during translation

into the membrane during translocation, leaving the carboxyterminal region of the protein spanning the membrane (Katz and Lodish 1979). Translocation of the p62 peptide chain stops after the transmembrane sequence of E2 has been inserted into the membrane (see Fig. 1).

After the p62 protein has been assembled in the membrane how is E1 inserted? Our sequence studies revealed an interesting feature previosly not known. We found a region of 180 base pairs interposed between the genes for E2 and E1. Welch and Sefton (1980) have recently detected a peptide with a M_r = 6000 (6K peptide) in cells infected with SF virus. This peptide is found in infected cells bound to the rough endoplasmic reticulum. Preliminary amino acid sequence data by W.J. Welsh and B.M. Sefton (pers. comm.) show that the gene coding for the 6K peptide corresponds to this 180 nucleotide segment. Most likely the 6K peptide functions as a signal peptide for E1. Hashimoto et al. (1980) have found a temperature-sensitive mutant of SF virus in which the cleavage between the C and the p62 proteins is blocked. In cells infected with this mutant E1 is inserted in the correct orientation into the endoplasmic reticulum at the non-permissive temperature. The uncleaved protein containing the C and p62 protein sequences has a M_r of 87,000 and is left in the cytosol. These findings suggest that a functional signal peptide exists for E1 and that it is coded for in the 6K region. We have therefore tentatively positioned the signal peptide for E1 in the 6K peptide before the amino terminus of E1. Transfer of the E1 polypeptide through the membrane of the endoplasmic reticulum stops at the transmembrane sequence located at its carboxy-terminus. The hydrophobic peptide segment ends two arginine residues before the chain-terminating codon UAA is reached. The core saccharide units are added to the polypeptide chains during translation (Garoff et al. 1978). Glycosylation of the membrane proteins is not required, however, for correct insertion and cleavage of the p62 and the E1 proteins (Garoff and Schwartz 1978). If cells are infected with SF virus in the presence of the glycosylation inhibitor tunicamycin, completely unglycosylated forms of the proteins are formed. Analysis of protease-treated micro-

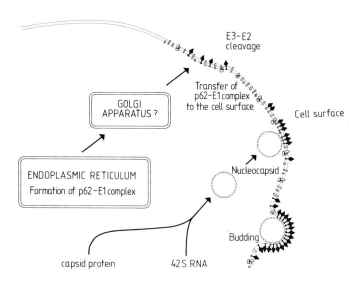

Fig. 3. Schematic picture of the intracellular glycoproteins to the cell surface and of the budding of the nucleocapsid through the plasma membrane

somes from such cells shows that the unglycosylated p62 and E1 are correctly inserted in the membrane.

Figure 2 summarizes the events that lead to segregation of the C-protein into the cytoplasm and the E proteins into the endoplasmic reticulum. It is evident that the virus mRNA manages to direct the synthesis of these proteins with different distributions within the cell by interposing a signal peptide between the regions coding for C-protein and the E proteins. The ribosome translating the 26S mRNA first becomes membrane-bound when this signal peptide has been translated and this initiates translocation of the p62 protein through the membrane. The 97K protein is an aberrant protein which is formed if the ribosome fails to bind to the endoplasmic reticulum before the critical length of 100 amino acids of the p62 protein has been made. The putative signal peptide in the 6K peptide apparently does not function efficiently enough to completely inhibit the formation of the 97K protein since this protein can be detected in infected cells. One interesting feature becomes apparent when one scrutinizes the known transmembrane sequences of plasma membrane and virus glycoproteins (Fig. 1). The hydrophobic stretch lacks Lys, Arg, Glu, Asp, His, Gln, Asn, and Pro residues, is at least 23 amino acid residues long, and it ends with a cluster of at least two arginine or lysine residues. Perhaps it is this combination of a hydrophobic segment ending in a basic cluster that is important for locking the transmembrane segment in place during the co-translational translocation through the membrane.

Intracellular Transport of Membrane Protein to the Cell Surface

Soon after synthesis the p62 and the E1 polypeptides form a complex in the endoplasmic reticulum (Ziemiecki et al. 1980). This two-chain complex is transported from the endoplasmic reticulum to the surface of the cell (Fig. 3). The first complexes reach the plasma membrane within about 30 min after synthesis. The protein-bound carbohydrate units are modified during intracellular transport (Robbins et al. 1977). Glucose and mannose residues are removed and then other monosaccharides are added in a stepwise fashion to produce the lactosamine type of glycan units found in viral membrane proteins. The p62-E1 complex presumably has to be transported to the Golgi apparatus where some of the enzymes responsible for these glycosylations are known to be localized. However, so far the Golgi apparatus has not been convincingly demonstrated as an intermediate station for surface glycoproteins en route to the plasma membrane.

A number of studies indicate that the carbohydrate moieties do not play a decisive role in directing secretory proteins or surface glycoproteins to their destinations (see Roth et al. 1979). Other posttranslationals modifications include fatty acid acylation (Schmidt et al. 1979) (see, however, Simons et al. 1978b) and sulphation (Pinter and Compans 1975) but their functional significance is not known.

How the surface proteins find their way from the endoplasmic reticulum to the plasma membrane is not known. Rothman and Fine (1980) have implicated coated vesicles as the transport vehicles for the vesicular stomatitis virus membrane glycoprotein from the endoplasmic reticulum to the Golgi apparatus and from there to the cell surface, but their results are far from conclusive. This is an area of research which, undoubtedly, will attract a lot of attention in the coming years.

When the p62-E1 complex reaches the plasma membrane the cleaveage of p62 to E2 and E3 takes place (Ziemiecki et al. 1980). When antibodies reacting specifically with either E1 or E2 are added to the extracellular medium, the cleavage of p62 is inhibited. This cleavage generates the final three-chain structure of the SF virus glycoproteins. Ziemiecki et al. (1980) have shown the E2 protein in the plasma membrane spans the membrane with a 30 amino acid residue segment from the carboxyterminal end protruding into the cytosol. Thus the three-domain structure of virus glycoproteins (external, transmembrane and internal) is generated during the co-translational assembly into the membrane of the endoplasmic reticulum, and this structure is maintained during intracellular transport.

Proteolytic Cleavages During SF Virus Morphogenesis

There are at least four proteolytic cleavages involved in the formation of the four virus polypeptides E, E1, E2, and E3 (Fig. 4). The first of these between the C and the E3 proteins has been found to take place on the nascent chain in the absence of membranes in all eukaryotic protein-synthesizing in vitro systems tested. Aliperti and Schlesinger (1978) have suggested that the cleavage is a function of the C-protein itself, but this needs confirmation. A ribosomal protease could also be involved (Langer et al. 1979). The two following cleavages release the 6K peptide. These cleavages also occur on the nascent chain, and require microsomal membranes to take place. The cleavage between the carboxy-terminus of 6K peptide and the amino-terminus of E1 could be due to the signal peptidase. The fourth cleavage, which cleaves the precursor p62 into E2 and E3, probably occurs at the plasma membrane. The cleavage site contains an arginine pair located before the amino terminus of E1. Many prohormones and proproteins have been shown to be processed to their final forms via specific cleavages at sites bearing pairs of basic amino acid residues (lysine and arginine). This is the case for proinsulin (Kemmler et al. 1971), progastrin (Gregory and Tracy 1972) proglucagon (Tager and Steiner 1973), proalbumin (Russell and Geller 1975), proparathyroid hormone (Habener et al. 1977) and the common precursor for melanotropins, corticotropin, lipotropin, and endorphins (Nakanishi et al. 1979). These cleavages, which probably involve a combined trypsin-like and carboxypeptidase B like

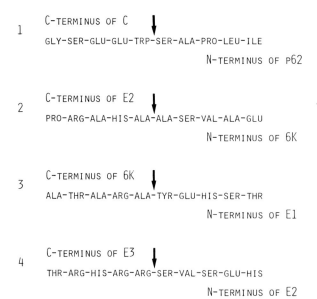

1   C-TERMINUS OF C
    GLY-SER-GLU-GLU-TRP-SER-ALA-PRO-LEU-ILE
                    N-TERMINUS OF P62

2   C-TERMINUS OF E2
    PRO-ARG-ALA-HIS-ALA-ALA-SER-VAL-ALA-GLU
                    N-TERMINUS OF 6K

3   C-TERMINUS OF 6K
    ALA-THR-ALA-ARG-ALA-TYR-GLU-HIS-SER-THR
                    N-TERMINUS OF E1

4   C-TERMINUS OF E3
    THR-ARG-HIS-ARG-ARG-SER-VAL-SER-GLU-HIS
                    N-TERMINUS OF E2

Fig. 4. Proteolytic cleavages involved in Semliki Forest virus morphogenesis

activity, occur late in secretion shortly before extracellular release. Also the hemagglutinin of fowl plague virus is split late during intracellular transport to the plasma membrane by a host protease into two polypeptide chains by two cleavages involving the excision of a (-Lys-Arg-Glu-Lys-Arg-) connecting peptide (Porter et al. 1979). It is tempting to speculate that the secreted proportions and the virus membrane glycoproteins are processed by the same family of host proteases, part of the post-Golgi pathway to the cell surface.

Budding Mechanism

The transmembrane structure of the spike proteins is likely to play a major role in the final stage in SF virus morphogenesis which is the budding of the nucleocapsid through the plasma membrane. We have earlier postulated that the budding process is mediated by transmembrane binding of the spike glycoproteins to the nucleocapsid (Garoff and Simons 1974). The nucleocapsid is bound to the cytoplasmic (internal) domains of the spike proteins and budding proceeds by more spike proteins diffusing into the budding site and being trapped by the budding. The virus glycoproteins are known to move by lateral diffusion in the host cell membrane (Birdwell and Strauss 1974). When all the binding sites have been occupied (spike protein-capsid protein stochiometry in the mature virus is 1:1) the virus particle is released. The budding may be faciliated by lateral interactions between the spike glycoproteins (Von Bonsdorff and Harrison 1978). Host proteins which have no affinity for the nucleocapsid or for the virus glycoproteins may simply be excluded from the budding site for steric reasons. This would explain the selectivity of the budding process, the lipids being derived from the host membrane and the proteins specified by the virus genome.

This self-assembly model of Semliki Forest virus assembly is shown in a schematic form in Figure 3. Experiments done by Waite et al. (1970, 1972) suggest that metabolic energy is not needed for the budding process. Treatment with cyanide, sodium fluoride, sodium azide, or iodoacetic acid does not affect the release of virus from cells restored to normal conditions after hypotonic arrest of the budding process.

One important question that needs to be resolved is why budding takes place mostly at the cell surface. If the virus spike glycoproteins are being transported from the endoplasmic reticulum and through the Golgi apparatus to the cell surface, why do the nucleocapsids not bud into the intracellular compartments? The last proteolytic cleavage in alphavirus maturation seems to occur at the cell surface: this cleavage may be a prerequisite for budding (Jones et al. 1974; Keranen and Kääriäinen 1975). It could result in a conformational change of the spike protein facilitating the formation of lateral contacts between the spike proteins. The low concentrations of spike glycoproteins in the intracellular compartments is probably another reason for the absence of intracellular budding (Saraste et al. 1980). Since the spike proteins are in transit through the intracellular compartments, their concentration may not reach the critical level necessary for budding. It is also possible that a sizable proportion of the cytoplasmic tails of the spike glycoproteins in transit are covered by proteins involved in intracellular transport and therefore not available for interactions with the nucleocapsid. Saraste et al. (1980) have described a temperature-sensitive mutant of SFV with an intracellular transport defect. In cells infected with this mutant the spike glycoproteins do not reach the cell surface and accumulate in intracellular membranes at the non-permissive temperature. The temperature defect is reversible. By shifting down to the permissive temperature, the spike proteins presumably resume their native conformation and within 10 min binding of the nucleocapside to intracellular membranes is observed followed by budding into the intracellular compartments. Under these conditions the concentration of spike proteins accumulated in the intracellular membranes might be high enough to allow budding to occur. By 60 min after shift down intracellular budding is diminished and budding is observed at the cell surface.

Richardson and Vance (1978a,b) have envisaged a direct role of the cytoskeleton in the budding of SF virus. More data are needed before this claim can be substantiated. They found that the drugs colchicine and dibucaine lower virus formation. Their interpretation of the kinetics led to the conclusion that the spike proteins accumulate at the surface of the cell. However, a crude purification method that did not allow precise differentiation between intracellular smooth membranes and surface membranes was used. Their finding that the p62 cleavage was hindered by the drugs could also be interpreted to mean that p62 reaches the surface more slowly under these conditions in which the cytoskeletal elements have been disrupted.

References

Aliperti G, Schlesinger MJ (1978) Evidence for an autoprotease acitivity of Sindbis virus capsid protein. Virology 90:366-369

Birdwell CR, Strauss JH (1974) Replication of Sindbis virus. IV Electron microscope study of the insertion of viral glycoproteins into the surface of infected chick cells. J Virol 14:366-374

Blobel G, Dobberstein B (1975) Transfer of proteins across membranes I. J Cell Biol 67:852-862

Blobel G, Walter P, Chang CN, Goldman B, Ericson HH, Lingappa VR (1979) Translocation of proteins across membranes. Soc Exp Biol 33:9-36

Bonatti S, Blobel G (1979) Absence of a cleavable signal sequence in Sindbis virus glycoprotein PE$_2$. J Biol Chem 254:12261-12264

Bonsdorff von CH, Harrison SC (1975) Sindbis virus glycoproteins form a regular icosahedral surface lattice. J Virol 16:141-145

Bonsdorff von CH, Harrison SC (1978) Hexagonal glycoprotein arrays from Sindbis virus membranes. J Virol 28:578-583

Clegg JCS (1975) Sequential translation of capsid and membrane protein genes of alphaviruses. Nature (London) 254:454-455

Clegg JCS, Kennedy SIT (1975) Initiation of synthesis of the structural proteins of Semliki Forest virus. J Mol Biol 97:401-411

Fong BS, Hunt RC, Brown JC (1976) Asymmetric distribution of phosphatidylethanol-amine in the membrane of vesicular stomatitis virus. J Virol 20:658

Frank G, Brunner J, Hanser K, Wacker H, Semenzi G, Zuber H (1978) The hydrophobic anchor of small-intestinal sucrose isomaltase. FEBS Lett 96:183-188

Gahmberg CG, Utermann G, Simons K (1972) The membrane proteins of SFV have a hydro-phobic part attached to the viral membrane. FEBS Lett 28:179-182

Garoff H, Schwartz R (1978) Glycosylation is not necessary for membrane insertion and cleavage of Semliki Forest virus membrane proteins. Nature (London) 247:487-490

Garoff H, Simons K (1974) Location of the spike glycoproteins in the Semliki Forest virus membrane. Proc Natl Acad Sci USA 71:3988-3992

Garoff H, Söderlund H (1978) The amphiphilic membrane glycoproteins of Semliki Forest virus are attached to the lipid bilayer by their COOH-terminal ends. J Mol Biol 124:535-549

Garoff H, Simons K, Renkonen O (1974) Isolation and characterization of the mem-brane proteins of Semliki Forest virus. Virology 61:493-504

Garoff H, Simons K, Dobberstein B (1978) Assembly of the Semliki Forest virus mem-brane glycoproteins in the membrane of the endoplasmic reticulum in vitro. J Mol Biol 124:587-600

Garoff H, Frishauf A-M, Simons K, Lehrach H, Delius H (1980) Nucleotide sequence of the cDNA encoding for the Semliki Forest virus glycoproteins. Submitted

Gething MJ, White JM, Waterfield MD (1978) Purification of the fusion protein of Sendai virus: analysis of the NH_2-terminal sequence generated during precursor activation. Proc Natl Acad Sci USA 76:2737-2740

Glanville N, Ranki M, Morser J, Kääriäinen L, Smith AE (1976) Initiation of trans-lation directed by 42S and 26S RNA from Semliki Forest virus in vitro. Proc Natl Acad Sci USA 73:3059-3063

Gregory RA, Tracy HJ (1972) Isolation of two big gastrims from Zollinger-Ellison tumour tissue. Lancet 11:797-799

Habener JF, Chang HT, Potts JJT (1977) Enzyme processing of propromethyroid hormone by cell-free extracts of parathyroid glands. Biochemistry 16:3910-3917

Hashimoto K, Erdei S, Kääriäinen L, Keranen S (1980) Evidence for a separate signal sequence for carboxyterminal enevelope glycoprotein E1 of Semliki Forest virus. Submitted

Henning R, Milner RJ, Reske K, Cunningham BA, Edelman GM (1976) Subunit structure, cell surface orientation and partial amino acid sequences of murine histocompati-bility antigens. Proc Natl Acad Sci USA 73:118-122

Hunt LA, Etchison JB, Summers DF (1978) Oligosaccharide chains are trimmed during synthesis of the envelope glycoproteins of vesicular stomatitis virus. Proc Natl. Acad Sci USA 75:754-758

Irving RA, Toneguzzo F, Rhee SH, Hofmann T, Ghosh HP (1979) Synthesis and assembly of membrane glycoproteins: Presence of leader peptide in nonglycosylated precursor of membrane glycoprotein of vesicular stomatitis virus. Proc Natl Acad Sci USA 76:570-574

Jones JK, Waite MRF, Bose HR (1974) Cleavage of a viral envelope precursor during the morphogenesis of Sindbis virus. J Virol 13:809-817

Kamp Op den JAF (1979) Lipid asymmetry in membranes. Annu Rev Biochem 48:47-71

Kääriäinen L, Renkonen O (1977) Envelopes of lipid-containing viruses as models for membrane assembly. In: Poste G, Nicolson GL (eds) The synthesis, assembly and turnover of cell surface components. North Holland, Amsterdam, pp 741-801

Kääriäinen L, Söderlund H (1978) Structure and replication of alphaviruses. Curr Top Microbiol Immunol 82:15-69

Katz FN, Lodish HF (1979) Transmembrane biogenesis of the vesicular stomatitis virus glycoprotein. J Cell Biol 80:416-426

Kemmler W, Peterson JD, Steiner OF (1971) Studies on the conversions of proinsulin to insulin. J Biol Chem 246:6786-6791

Keranen S, Kääriäinen L (1975) Proteins synthesized by Semliki Forest virus and its 16 temperature-sensitive mutants. J Virol 16:388-396

Langner J, Wiederlanders B, Ansorge S, Bohley P, Kirschke H (1979) The ribosomal serine proteinase; cathepsin R. Acta Biol Med Germ 38:1527-1538

Lenard H (1978) Virus envelopes and plasma membranes. Annu Rev Biophys Bioeng 7: 139

Lenard J, Compans RW (1974) The membrane structure of lipid-containing viruses. Biochim Biophys Acta 344:51-94

Lingappa VR, Katz F, Lodish HF, Blobel G (1978) A signal sequence for the insertion of a transmembrane glycoprotein. J Biol Chem 253:8667-8670

Marchesi VT, Furthmayer H, Tomita M (1976) The red cell membrane. Annu Rev Biochem 45:667-698

Maroux S, Louvard D (1976) On the hydrophobic part of aminopeptidase and maltases, which bind the enzyme to the intestinal brush border membrane. Biochim Biophys Acta 419:189-195

Mattila K, Renkonen O (1978) Separation of A- and B-type glycopeptides of Semliki Forest virus by convalin A affinity chromatography and preliminary characterization of the B-type glycopeptides. Virology 91:508-510

Mattila K, Luukkonen A, Renkonen O (1976) Protein-bound oligosaccharides of Semliki Forest virus. Biochim Biophys Acta 419:435-444

Meer van G, Simons K, Op den Kamp JAF, van Deenen LLM (1980) Biochemistry submitted

McCarthy M, Harrison SC (1977) Glycosidase susceptibility: a probe for the distribution of glycoprotein oligosaccharides in Sindbis virus. J Virol 23:61-73

Min Jou WM, Verhoeyen M, Devos R, Saman E, Fang R, Huylebroeck D, Fiers W, Ihrelfall G, Barber C, Carey N, Emtage S (1980) Complete structure of the hemagglutinin gene from the human influenza A/Victoria/3/75 (H3N2) strain as determined from cloned DNA. Cell 19:683-696

Nakanishi S, Inoue A, Kita T, Nakamura M, Chang ACY, Cohen SN, Numa S (1979) Nucleotide sequence of cloned DNA for bovine corticotropin-β-lipoprotein precursor. Nature (London) 278:423-427

Neuberger A, Gottschalk A, Marshall RD, Spiro RG (1972) Carbohydrate peptide linkages in glyocproteins and methods for their elucidation. In: Gottschalk A (ed) The glycoproteins: Their composition, structure and function. Elsevier Publishing Co, Amsterdam, pp 450-490

Patzer EJ, Wagner RR, Duboui EJ (1979) Viral membranes: model systems for studying biological membranes. Crit Rev Biochem 6:165-217

Pesonen M, Renkonen O (1976) Sequence and anomeric configuration of monosaccharides in type A glycopeptides of Semliki Forest virus. Biochim. Biophys Acta 455:510-525

Pesonen M, Haahtela K, Renkonen O (1979) Core tetrasaccharide liberated by endo-β-D-N-acetylglucosaminidase D from lactosamine-type oligosaccharides of Semliki Forest virus. Biochem Biophys Acta 588:102-112

Pfefferkorn ER, Hunter HS (1963b) The source of the ribonucleic acid and phospholipid of Sindbis virus. Virology 20:433-445

Pfefferkorn ER, Clifford RL (1964) The origin of the protein of Sindbad virus. Virology 23:217-223

Pinter A, Compans RW (1975) Sulfated components of enveloped viruses. J Virol 16: 859-866

Porter HG, Barber C, Carey NH, Hallelwell RA, Threlfall G, Emtage JC (1979) Complete nucleotide of an influenza virus hemagglutinin gene from cloned DNA. Nature (London) 282:471-477

Renkonen O, Kääriäinen L, Simons K, Gahmberg CO (1971) The lipid class composition of Semliki Forest virus and of plasma membrane of host cell. Virology 46:318

Richardson CD, Vance DE (1976) Biochemical evidence that Semliki Forest virus obtains its envelope from the plasma membrane of the host cell. J Biol Chem 251: 5544-5550

Richardson CD, Vance DE (1978a) Chemical cross-linking of proteins of Semliki Forest virus: Virus particles and plasma membrane from BHK-2 cells treated with colchicine or dibucaine. J Virol 28:193-198

Richardson CD, Vance DE (1978b) The effect of colchicine and dibucaine on the morphogenesis of Semliki Forest virus. J Biol Chem 253:4584-4589

Robbins PW, Hubbard SC, Turco SJ, Wirth DF (1977) Proposal for a common oligosaccharide intermediate in the synthesis of membrane glycoproteins. Cell 12:893

Roth MG, Fitzpatrick JP, Compans RW (1979) Polarity of influenza and vesicular stomatitis virus maturation in MDCK cells: Lack of a requirement for glycosylation of viral glycoproteins. Proc Natl Acad Sci USA 76:6430-6434

Rothman JE, Fine (1980) Coated vesicles transport newly synthesized membrane glycoproteins from endoplasmic reticulum to plasma membranes in two successive stages. Proc Natl Acad Sci USA 77:780-784

Rothman JE, Lodish HF (1977) Synchronized transmembrane insertion and glycosylation of a nascent membrane protein. Nature (London 296:775-779

Rothman JE, Tsai DK, Dawidowicz EA, Lenard J (1976) Transbilayer phospholipid asymmetry and its maintenance in the membrane of influenza virus. Biochemistry 15:2361

Russell JH, Geller DM (1975) The structure of rat proalbumin. J Biol Chem 250:3409-3413

Sandra A, Pagano RE (1978) Phospholipid asymmetry in LM cell plasma membrane derivatives: polar headgroup and acyl chain distributions. Biochemistry 17:332

Saraste J, von Bonsdorff C-H, Hashimoto K, Kääriäinen L, Keranen S (1980) Semliki Forest virus mutants with temperature-sensitive transport defect of envelope proteins. Virology 100:229-245

Schmidt MFG, Bracha M, Schlesinger MJ (1979) Evidence for covalent attachment of fatty acids to Sindbis virus glycoproteins. Proc Natl Acad Sci USA 76:1687-1691

Segrest JP, Feldman RJ (1979) Membrane Proteins: Amino acid sequence and membrane penetration. J Mol Biol 87:853-858

Shaw JM, Morre NF, Patzer EF, Correa-Freire M, Thompson TE, Wagner RR (1979) Transmembrane movement and asymmetry of phosphatidylcholine in the membrane of vesicular stomatitis virus. Biochemistry 18:538-543

Skehel JJ, Waterfield MD (1975) Studies on the primary structure of the influenza virus hemagglutinin. Proc Natl Acad Sci USA 72:93-97

Simons K, Garoff H, Helenius A, Ziemiecki A (1978a) The structure and assembly of the membrane of Semliki Forest virus. In: Pullman B (ed) Frontiers in physiochemical biology. Academic Press, London New York, pp 387-407

Simons K, Garoff H, Helenius A (1980) Alphavirus proteins. In: Schlesinger W (ed) Togaviruses. Academic Press, London New York, in press

Springer TA, Strominger JL (1976) Detergent-soluble HLA antigens contain a hydrophilic region at the COOH-terminus and a penultimate hydrophobic region. Proc. Natl. Acad. Sci USA 73:2481-2485

Strauss EG (1978) Mutants of Sindbis virus III. Host polypeptides present in purified HR and ts 103 virus peptides. J Virol 28:466-474

Strauss JH, Strauss EG (1976) Togaviruses: In: Nayak DP (ed) The molecular biology of animal viruses. Marcel Dekker, New York, in press

Tabas I, Schlesinger S, Kornfeld S (1978) Processing of high mannose oligosaccharides to form complex type oligosaccharides on newly synthesized polypeptides of vesicular stomatitis virus protein and the IgG heavy chain. J Biol Chem 253:716-722

Tager HS, Steiner DF (1973) Isolation of a glucagon-containing peptide: primary structures of a possible fragment of proglucagon. Proc Natl Acad Sci USA 70:2321-2325

Tomita H, Marchesi VT (1975) Amino-acid sequence and oligosaccharide attachment sites of human erythrocyte glycophorin. Proc Natl Acad Sci USA 72:2964-2968

Utermann G, Simons K (1974) Studies on the amphiphilic nature of the membrane proteins of Semliki Forest virus. J Mol Biol 85:569-587

Waite MRF, Pfefferkorn ER (1970) Inhibition of Sindbis virus prodcution by media of low ionic strength: Intracellular events and requirements for reversal. J Virol 10:537-544

Waite NRF, Brown DT, Pfefferkorn ER (1972) Inhibition of Sindbis virus release by media of low ionic strength: a electron microscope study. J Virol 10:537-544

Wang E, Wolf BA, Lamb RA, Choppin PW, Goldberg AR (1976) The presence of actin in enveloped viruses. In: Goldman R, Pollard T, Rosenbaum J (eds) Cell motility. Cold Spring Harbor Press, Cold Spring Harbor, pp 589-600

Welch WJ, Sefton BM (1980) Characterization of a small, nonstructural viral polypeptide present late during infection of BHK cells by Semliki Forest virus. J Virol 33:230-237

Ziemiecki A, Garoff H (1978) Subunit composition of the membrane glycoprotein complex of Semliki Forest virus. J Mol Biol 122:259-269

Ziemiecki A, Garoff H, Simons K (1980) Formation of the Semliki Forest virus membrane glycoprotein complexes in the infected cell. J Gen Virol, in press

Assembly of Membrane Proteins in *Escherichia coli.* A Genetic Approach

M. Schwartz[1]

Introduction

Each type of membrane present in a eukaryotic cell contains a specific set of proteins. With the exception of some of the proteins present in mitochondria or chloroplasts, and formed within these organelles, all membrane proteins are synthesized in the cytoplasm. There seems to be a consensus that these proteins can follow either of two routes from their site of formation to their final location. The first route was originally uncovered for secreted proteins. It starts by a synthesis on polysomes bound to the rough endoplasmic reticulum (RER). It concerns proteins which end up in the RER itself, in the Golgi apparatus, in lysosomes, in the plasma membrane, and perhaps in other organelles as well (Palade 75; Rothman and Lenard 1977; Katz et al. 1977; Rothman and Lodish 1977). The other route is that followed by several of the proteins which end up in mitochondria or chloroplasts. These are synthesized on free polysomes, like the cytoplasmic proteins, and they reach directly their final location (Chua and Schmidt 1978, 1979; Highfield and Ellis 1978; Schatz 1979; Raymond and Shore 1979; Poyton and McKemmie 1979).

The first route has been the most thoroughly documented, and a model, termed the "signal hypothesis", has been proposed to account for the first step which it involves (Blobel and Sabatini 1971; Milstein et al. 1972; Blobel and Dobberstein 1975). According to this model, proteins destined to follow the RER route are synthesized as precursors containing an extra amino acid sequence at their NH2-terminus. This extra sequence, termed the signal sequence, is composed predominantly of hydrophobic amino acids. It initiates the binding of the translation complex to the endoplasmic reticular membrane, and thereby allows the transfer of the protein across the membrane, concomitant with translation. The signal sequence is eliminated during or after the translocation step by an enzyme located on the luminal side of the membrane.

Although much experimental evidence has accumulated in support of the signal hypothesis, many questions still remain to be solved. Even though most of the proteins following the RER route are synthesized as precursors, is it true that the extra sequence present in these precursors plays a critical role in the initiation of transmembrane transfer? Assuming that this is the case, is this role that predicted by the signal hypothesis? In fact another model, recently proposed by Wickner (1979) suggests that the role of the signal sequence is different. Does the discrimination between proteins destined to follow the RER route and the others only rests on the possession of a signal sequence? Could it not be, on the contrary, that there exists severe constraints as to which kind of polypeptide can be "threaded through" a

[1]Unité de Génétique Moléculaire, Département de Biologie Moléculaire, Institut Pasteur, 25 rue du Dr. Roux, 75724 Paris Cedex 15, France

membrane? Crucial as they are for understanding the process of membrane biogenesis, these questions are rather difficult to answer in the case of eukaryotic cells. Bacteria, on the other hand, may help to solve these questions.

It may seem strange to test the signal hypothesis in bacteria since these organisms have no RER, no Golgi apparatus, and none of the other organelles present in a eukaryotic cell. However they are still faced with problems of protein export. Indeed a bacterium like *Escherichia coli* comprises four compartments: the cytoplasm, the inner or cytoplasmic membrane, the periplasmic space, and the outer membrane. Each of these compartments contains a specific set of proteins, which are all synthesized in the cytoplasm. Several lines of evidence suggest that the mechanism whereby proteins are exported to the periplasmic space or the outer membrane bears a close resemblance with the initial step of protein secretion in the eukaryotic cell (Silhavy et al. 1979; Davis and Tai 1980). Periplasmic and outer membrane proteins were shown to be synthesized on polysomes bound to the cytoplasmic membrane, which would then play the same role as the RER membrane in eukaryotic cells (Randall and Hardy 1977; Smith et al. 1977; Varenne et al. 1978). They are synthesized in the form of precursors containing a typical signal sequence at their NH_2-terminal end (Inouye and Beckwith 1977; Inouye et al. 1977; Randall et al. 1978; Sutcliffe 1978). In addition the transfer of some proteins through the cytoplasmic membrane was shown to be cotranslational (Smith et al. 1977). In view of these results the signal hypothesis has been extended to the case of prokaryotic cells (Chang et al. 1979). There, however, it can be more easily tested by using genetics.

By developing one particular example, that of an outer membrane protein in *E. coli*, I wish to demonstrate how genetics can be used to study protein export. The results will be discussed in their bearing to the signal hypothesis.

In addition, in a much shorter section of this presentation I will provide preliminary results regarding the penetration of a colicin into bacteria. The rationale for doing so is that this process bears some formal resemblance to the penetration of proteins into mitochondria or chloroplasts, i.e., with the "second route" referred to at the beginning of this Introduction. In this case again, genetics may be of some help.

Export of the *lamB* Protein to the Outer Membrane

The *lamB* Protein

Almost since I started in research I have been interested in the maltose system of *E. coli*. One of the strange observations pertaining to this system was that many Mal⁻ mutants were also resistant to phage λ, i.e., unable to adsorb this phage (Lederberg 1955). As an attempt to understand this observation, Linda Randall and myself decided several years ago to identify the receptor for phage λ (Randall-Hazelbauer and Schwartz 1973). We demonstrated that a 50K protein, which could be extracted from the outer membrane of λ-sensitive cells, neutralized phage λ in vitro. This protein was absent in many λ-resistant mutants. Later studies have shown that this protein is not only a phage receptor. It also has another role, more beneficial to the cell, which is to facilitate the transport of maltose and maltodextrins across the outer membrane (Szmelcman and Hofnung 1975; Wandersman et al. 1979). The structural gene for this protein is called *lamB*. It is located in one of the three operons which constitute the "maltose regulon" (Raibaud et al. 1979).

The *lamB* protein is synthesized on membrane-bound polysomes (Randall et al. 1978). When it is synthesized in vitro a polypeptide slightly longer than the mature protein is obtained (Randall et al. 1978; Marchal et al. 1980). It contains an extra sequence of 24 amino acids at its NH2-terminal end (Hedgpeth et al. 1980). The question which we asked, as stated in the Introduction, concerned the role of this extra sequence. Is it necessary for the export of *lamB* protein? Would this sequence lead any polypeptide out of the cytoplasm? More generally, we were interested in identifying the genetic regions involved in the export of the *lamB* protein, in locating what could be called the "export determinants". The approach which we used, in a collaborative work with the groups of J. Beckwith and T. Silhavy at Harvard Medical School, and of M. Hofnung at the Institut Pasteur, was based on a genetic technique set up by M. Casadaban, and which I shall now describe.

Casadaban's Technique of Gene Fusion

The technique of Casabadan (1976) allows selection of "fusion strains" of *E. coli* in which the beginning of essentially any gene, called X, can be fused to *lacZ*, the gene coding for β-galactosidase, a cytoplasmic enzyme (Fig. 1). The amount of gene X present in the hybrid gene is different in the different fusion strains. On the other hand the amount of *lacZ* is essentially the same in all strains and corresponds to the whole gene (1021 codons) with the exception of the first 20 to 30 codons. The proteins coded by the hybrid genes are still endowed with the enzymatic activity characteristic of the *lacZ* product, so that fusion strains are able to grow on lactose.

The rationale behind using this technique to study protein localization (Silhavy et al. 1977) was that, if gene X coded for an envelope protein, the product of some of the gene X-*lacZ* hybrid genes might be exported to the cell envelope. By correlating the cellular location of the hybrid protein with the amount of gene X present in the hybrid gene it might then be possible to locate more precisely the "export determinants", presumably located in gene X.

Isolation of *lamB-lacZ* Fusion Strains

By using Casadaban's technique we were able to isolate 25 different *lamB-lacZ* fusion strains (Silhavy et al. 1977; Hall, Schwartz and Silhavy in prep.). In these strains the synthesis of hybrid protein is induced by maltose, as is normally that of *lamB* protein. We shall focus on three fusions, called 61-4, 52-4, and 42-1, which so far yielded the most information (Fig. 2). According to genetic tests, hybrid genes 61-4 and 52-4 contained only the very beginning of gene *lamB*, while 42-1 contained between 1/3 and 1/2 of this gene (150 to 250 codons). Accordingly the polypeptides coded by the first two hybrid genes are approximately of the same size as authentic *lacZ* polypeptide (116,000) while that coded by 42-1 is significantly larger. In the first two cases the hybrid protein was found located in the cytoplasm. In the third case about half of the protein was membrane-bound, and half of that (1/4 of the total) was in the outer membrane, i.e., at the normal location of the *lamB* protein. From these results we could conclude:
1. that the grafting of about the first half of *lamB* to *lacZ* does change to a significant extent the location of β-galactosidase in the cell.
2. that at least some of the "export determinants" responsible for this

238

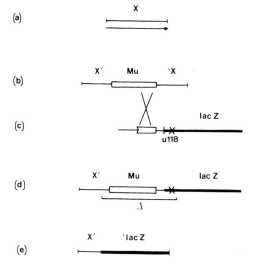

(a)

(b)

(c)

(d)

(e)

Fig. 1a-e. Casadaban's technique of gene fusion. (a) gene X is shown, with its direction of transcription indicated by an *arrow*; (b) phage Mu lysogenizes *E. coli* by inserting its DNA essentially at random in the bacterial chromosome. Cells which have a Mu prophage inserted in gene X can be recognized because they lost gene X function. A thermoinducible mutant of phage Mu has been used; (c) Casadaban constructed λ-transducing phages carrying the *lac* operon and a portion of Mu DNA. The particular phage used in this work carried an ochre mutation (U118) at the position corresponding to amino acid 17 in β-galactosidase. Only the relevant part of this phage DNA is shown here. This phage can integrate its DNA into the chromosome of Mu lysogens by using Mu DNA homology; (d) Such an integration yields a strain in which *lacZ* is located in the vicinity of the beginning of gene X. This strain is thermosensitive, because it contains a thermoinducible Mu prophage. It is also Lac⁻, because of the U118 mutation, and also because the *lac* operon present on Casadaban's phages does not have its promoter. Deletion events such as shown (Δ) yield strains which are thermoresistant (they lost the Mu prophage) and which are Lac⁺ as long as the deletion removed the U118 mutation (codon 17 in *lacZ*) but did not penetrate too far into *lacZ* (the limit seems to be around codon 30); (e) Many of the thermoresistant and Lac⁺ derivatives carry a hybrid gene which has the structure shown. The amount of gene X present in this hybrid gene (designated *X'* on the figure) depends upon the site of Mu insertion (see b) and upon the size of the deletion Δ creating the fusion (see d)

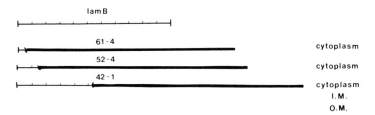

Fig. 2. Structure of three *lamB-lacZ* fusion strains. The divisions shown in gene *lamB* correspond to genetic intervals determined by deletion mapping. The structure of three *lamB-lacZ* hybrid genes is shown under the map, the *lacZ* part being represented by a *heavy bar*. *lamB* and *lacZ* DNA are not represented at the same scale (the *lacZ* polypeptide is 116 K while the *lamB* polypeptide is about 50 K). The cellular location of the proteins coded by the three hybrid genes is indicated: *IM* inner membrane; *OM* outer membrane

change are located within *lamB* in the region which is present in hybrid gene 42-1 but absent in the others.

The continuation of this work was greatly facilitated by an unexpected property of the strain carrying hybrid gene 42-1. When maltose was added to cultures of this strain, the cells stopped dividing, and eventually lysed. This "maltose-sensitive" phenotype is due to the synthesis of large amounts of the *lamB-lacZ* hybrid protein. Indeed mutations preventing the synthesis of hybrid protein, for instance nonsense mutations in the hybrid gene, automatically restore a maltose-resistant phenotype. In view of the cellular distribution of the 42-1 hybrid protein, it seemed likely that the lethal effect was related to the incompleteness of the export of this protein. If this were so, mutations preventing the very initiation of this export might "cure" the fusion strain of its maltose-sensitive phenotype.

Mutations Affecting the Signal Sequence of the *lamB* Protein Precursor

Most of the maltose-resistant derivatives obtained from the 42-1 strain lost the ability to synthesize the hybrid protein, and were therefore unable to grow on lactose. However, rare mutants were obtained which were still Lac[+]. Twenty-six such mutants were analyzed in detail (Emr et al. 1978; Emr and Silhavy 1980). The mutations had little effect on the amount of hybrid protein synthesized, but this protein was now entirely cytoplasmic. The mutations were found to map at the very beginning of the *lamB* portion of the hybrid gene.

Genetic studies demonstrated that 12 of the mutations were small deletions, while 14 behaved like point mutations. These mutations were transduced into a strain carrying an otherwise wild-type *lamB* gene. The resulting transductants had a phenotype characteristic of *lamB* mutants, (phage λ resistance in particular). They had no detectable *lamB* protein in their outer membrane. On the other hand they contained a new polypeptide in their cytoplasm. In the case of the point mutants this polypeptide was slightly larger than mature *lamB* protein, and was shown to correspond to the uncleaved precursor of this protein. The polypeptide found in the cytoplasm of the deletion mutants was somewhat smaller, in accordance with the known extent of the deletions. The exact location of 15 of these mutations was determined by using DNA sequencing techniques (Emr et al. 1980). All of them lead to alterations in the DNA region coding for the signal sequence (Fig. 3). Of the 15 mutants, 8 lead to a change in hydrophobic residues which were converted to charged amino acids. The other mutations were deletions. Three of these were internal to the portion of DNA coding for the signal sequence while the others also extended outside of this region.

It is apparent from this work that the export of the *lamB* protein is totally prevented by mutations affecting its signal sequence. When such mutations are present, this protein remains in the cytoplasm, in the form of an uncleaved precursor. Very similar results have been obtained by Bassford et al. in a work performed in parallel on the periplasmic maltose-binding protein (*malE* product) (Bassford et al. 1979; Bassford and Beckwith 1979; Bedouelle et al. 1980). By using the same techniques these authors also obtained mutations affecting the signal sequence of the *malE* protein, and these mutations have the same characteristics as those described above. These studies argue strongly for a critical role of the signal sequence in the initiation of protein export. They also strongly suggest that the hydrophobic character of these sequences is essential for their function. Further studies on signal sequence mutations should help define the relative functional importance of the different amino acids present in this sequence. In

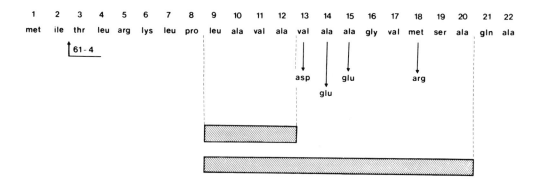

Fig. 3. Mutations affecting the export of *lamB* protein. The amino terminal sequence of the *lamB* protein precursor, as shown here, was deduced from the DNA sequence (Hedgpeth et al. 1980). The residues indicated in italics are present in the mature form of the protein while the others correspond to the signal sequence. The amino acid alterations corresponding to point mutations which prevent the export of *lamB* protein were established by DNA sequencing (Emr et al. 1980). They altered residues 13, 14, 15, and 18, as shown. The extent of some deletions resulting in the same phenotype was also determined. The most significant are shown here as hatched bar. Finally the position of the fusion joint in hybrid genes 61-4 and 52-4 was determined by sequencing the aminoterminal portion of the hybrid proteins (Moreno et al. 1980)

this respect it is worth mentioning that a mutation altering the hydrophobic portion of the signal sequence of *E. coli* lipoprotein had little effect on the export of this protein to the outer membrane (Lin et al. 1978).

Is the Signal Sequence Sufficient to Lead a Polypeptide Out of the Cytoplasm?

According to the signal hypothesis, any polypeptide possessing a signal sequence at its amino terminal end should leave the cytoplasmic compartment of the cell. The results obtained with the 42-1 fusion strain (Silhavy et al. 1977) already suggested that this is not strictly the case since about half of the hybrid protein synthesized by this strain remains in the cytoplasm. However, the conclusions which can be obtained with this strain are somewhat obscured by the fact that it does not grow normally. Much clearer results have been obtained with the two other fusion strains, 61-4 and 52-4, which grow normally in the presence of maltose. As mentioned earlier, the hybrid proteins synthesized by these strains are entirely (90% at least) located in the cytoplasm. These proteins were purified, and their NH2-terminal sequence was analyzed (Moreno et al. 1980). The 61-4 protein contains only two amino acids from the *lamB* protein precursor at its NH2-terminal end; therefore, its cytoplasmic location is not surprising. On the other hand the 52-4 protein contains 39 amino acids from the *lamB* protein precursor, i.e., the complete signal sequence (24 amino acids)

plus 15 residues from the mature form of *lamB* protein (Fig. 3). Therefore, at least in this case, grafting a signal sequence to the amino
terminal end of a cytoplasmic polypeptide has not been sufficient to
provoke the removal of this polypeptide from the cytoplasmic compartment of the cell.

Several interpretations can be offered for this finding. Dr. Wickner
will probably argue that it provides evidence for his "membrane trigger
hypothesis" (Wickner 1979) which implies that the membrane recognition
step involves a large fraction of the polypeptide destined to be exported. However, other interpretations must also be considered. The
true "signal" sequence, the only element required for membrane recognition in the "signal hypothesis", may only be slightly longer than
the "extra" sequence present in the precursor of exported proteins.
Alternatively some rules may exist according to which polypeptides can
be "threaded" through a membrane. The *lacZ* polypeptide might not follow
these rules. The transfer of the hybrid protein through the membrane
would then stop at an early stage, and the protein, only weakly bound
to the membrane, would detach spontaneously as synthesis proceeds.
Evidence obtained with the 42-1 *lamB-lacZ* fusion strain, and with some
malE-lacZ fusion strains tends to support the latter explanation.

The Notion of "Export Determinants" Reconsidered

We mentioned earlier that the 42-1 *lamB-lacZ* fusion strain is maltose-
sensitive, and incompletely exports the hybrid protein. A possible explanation of this phenotype would be that the "threading" of the hybrid
protein through the cytoplasmic membrane occurs normally as long as
lamB sequences are concerned, and then becomes very inefficient when
lacZ sequences are encountered. This could result in an accumulation
of hybrid proteins in the cytoplasmic membrane, and a possible "jam-
ming" of the export routes. This could prevent the further export of
hybrid protein, but also of other proteins essential for cell division,
hence the lethal effect of hybrid protein synthesis.

The results obtained with *malE-lacZ* fusion strains indicate even more
strongly that the *lacZ* polypeptide cannot be transferred efficiently
through the cytoplasmic membrane (Bassford et al. 1979). In this case
the hybrid protein is never exported to the periplasmic space, which
represents the normal location of the *malE* protein. When most of gene
malE is present in the hybrid gene the hybrid protein is found associated
with the cytoplasmic membrane. As in the case of the 42-1 *lamB-lacZ* fusion this strain is sensitive to the addition of maltose.

The failure of the hybrid protein to be efficiently transferred through
the cytoplasmic membrane suggests that there may be rather stringent
rules regarding the structure of polypeptides which can be threaded
through a membrane. Some determinants present all along an exported
polypeptide may play an active role in the export process. Alternatively sequences present in a cytoplasmic polypeptide such as β-galactosidase may play an inhibitory role. It may seem strange that some of a
lamB-lacZ hybrid protein can be exported to the outer membrane, while
none of the *malE-lacZ* proteins is transferred to the periplasm. A model
was proposed to account for this difference (Silhavy et al. 1979). It
implies the notion that the *lamB* polypeptide contains an additional
signal which prevents its complete transfer through the cytoplasmic
membrane. Interactions between the inner and outer membranes would
then allow the transfer of the polypeptide to its final location.

In conclusion the work performed with *lamB-lacZ* and *malE-lacZ* fusion
strains demonstrates that the signal sequence plays a critical role

in the initiation of protein export in bacteria, but that additional features must be present in a polypeptide in order for it to be transferred through the cytoplasmic membrane, and directed to its final location.

The Penetration of Colicin E3 into Sensitive Cells

Colicins are antibacterial proteins produced by strains of *Escherichia coli* which carry plasmids called colicinogenic factor (Hardy 1975; Holland 1975). Different colicins kill sensitive bacteria by different mechanisms. Colicin E3, which is one of the best characterized (Holland 1976) induces a specific cut in the 16S RNA of the 30S ribosomal subunit, and thereby destroys the protein-synthesizing machinery of the cell. A single molecule of colicin can kill a bacterium. For many years it was believed that colicin E3, and the other colicins was well, remained at the surface of the target cells and exerted their lethal action by some indirect mechanism. This way of thinking had to be radically revised when Boon (1971) and Bowman et al. (1971) demonstrated that purified colicin E3 induced the specific cut on 16S RNA when it was added to purified ribosomes. The conclusion then became inescapable that colicin E3, or at least part of it, had to penetrate into the target cell to exert its lethal action. The question then became "how does it do it?". Answering this question could have some relevance to the post-translational transfer of cytoplasmically made proteins into mitochondria or chloroplasts, as occurs in the eukaryotic cell.

Michèle Mock and myself have started a genetic approach of this problem (Mock and Schwartz 1978, 1980). Once again the idea is to locate the genetic determinants — which could be called the "import" determinants — responsible for the penetration of colicin E3 into sensitive cells. We mutagenized a colicinogenic strain, and looked for mutants which failed to produce an active colicin. After screening 41,400 colonies we obtained 69 mutants with such a property. Only three of them synthesized a polypeptide of the same molecular weight as colicin E3. Two of these mutant colicins were inactive in vitro as well as in vivo and were apparently altered in their C-terminal portion, known to carry the catalytic activity. On the other hand the third mutant colicin had the same activity as wild-type colicin in vitro. In addition it adsorbed like wild-type colicin to the specific outer membrane receptor. Since it was totally inactive in vivo it must be specifically impaired in its ability to cross the bacterial envelope. By using DNA sequencing techniques we should be able to locate the amino acid change resulting from the mutation. It is hoped that the study of this and similar mutations might uncover the structural characteristics which enable the colicin to penetrate into bacteria.

Conclusions

From the work summarized here it seems that the use of genetics should help in understanding how proteins cross membranes in bacteria. However is it reasonable to assume, as was somehow implicit all along in this presentation, that solving this problem in bacteria will shed light on the biogenesis of organelles in eukaryotic cells? Was it justified, in other words, to place this lecture in a section called "Model Systems"? On the one hand it may seem very speculative to assume that the sorting out of proteins in the Golgi apparatus in a eukaryotic cell has anything to do with that of proteins destined to the different layers of the envelope in bacteria. On the other hand, however, the fact that ovalbumin is secreted into the periplasmic space when its structural gene is introduced into *E. coli* cells (Fraser and Bruce 1978) argues that the "Zip code" directing the proteins to their proper cellular location, may be, at least in part, as universal as the genetic code.

References

Bassford P, Beckwith J (1979) Escherichia coli mutants accumulating the precursor of a secreted protein in the cytoplasm. Nature (London) 277:538-541

Bassford PJ Jr, Silhavy TJ, Beckwith J (1979) Use of gene fusion to study secretion of maltose-binding protein into *Escherichia coli* periplasm. J Bacteriol 139:19-31

Bedouelle H, Bassford PJ Jr, Fowler AV, Zabin I, Beckwith J, Hofnung M (1980) The nature of mutational alterations in the signal sequence of the maltose binding protein of *Escherichia coli*. Nature (London) 285:78-81

Blobel G, Dobberstein B (1975) Transfer of proteins across membranes. I. Presence of proteolytically processed and unprocessed nascent immunoglobulin light chains on membrane bound ribosomes of murine myeloma. J Cell Biol 67:835-851

Blobel G, Sabatini DD (1971) Ribosome-membrane interaction in eukaryotic cells. In: Manson LA (ed) Biomembranes, vol II. Plenum Publ Corp, New York, p 193-195

Boon T (1971) Inactivation of ribosomes in vitro by colicin E3 and its mechanism of action. Proc Natl Acad Sci USA 68:2421-2425

Bowman CM, Sidikaro J, Nomura M (1971) Specific inactivation of ribosomes by colicin E3 in vitro and mechanism of immunity in colicinogenic cells. Nature New Biol (London) 234:133-137

Casadaban MJ (1976) Transposition and fusion of the *lac* genes to selected promoters in *E. coli* using bacteriophages λ and Mu. J Mol Biol 104:541-555

Chang CN, Model P, Blobel G (1979) Membrane biogenesis: cotranslational integration of the bacteriophage f1 coat protein into an *Escherichia coli* membrane fraction. Proc Natl Acad Sci USA 76:1251-1255

Chua NH, Schmidt G (1978) Post translational transport into chloroplasts of a precursor to the small subunit of ribulose-1,5-bisphosphate carboxylase. Proc Natl Acad Sci USA 75:6110-6114

Chua NH, Schmidt G (1979) Transport of proteins into mitochondria and chloroplasts. J Cell Biol 81:461-483

Davis BD, Tai PC (1980) The mechanism of protein secretion across membranes. Nature (London) 283:433-438

Emr SD, Silhavy TJ (1980) Mutations affecting localization of an *Escherichia coli* outer membrane protein, the bacteriophage λ receptor. J Mol Biol 141:63-90

Emr SD, Schwartz M, Silhavy TJ (1978) Mutations altering the cellular localization of the phage λ receptor, an *Escherichia coli* outer membrane protein. Proc Natl. Acad Sci USA 75:5802-5806

Emr SD, Hedgpeth J, Clément JM, Silhavy TJ, Hofnung M (1980) Sequence analysis of mutations that present export of λ receptor, an *Escherichia coli* outer membrane protein. Nature (London) 285:82-85

Fraser TH, Bruce BJ (1978) Chicken ovalbumin is synthesized and secreted by *Escherichia coli*. Proc Natl Acad Sci USA 75:5936-5940

Hardy KG (1975) Colicinogeny and related phenomena. Bacteriol. Rev. 39:464-515

Hedgpeth J, Clément JM, Marchal C, Perrin D, Hofnung M (1980) Proc Natl Acad Sci USA 77:2621-2625

Highfield PE, Ellis RT (1978) Synthesis and transport of the small subunit of chloroplast ribulose bisphosphate carboxylase. Nature (London) 271:420-424

Holland IB (1975) Physiology of colicin action. Adv Microb Physiol 12:55-139

Holland IB (1976) Colicin E3 and related bacteriocins: penetration of the bacterial surface and mechanism of ribosomal inactivation. In: Cuatracasas P (ed) Receptors and recognition, vol I. Chapman and Hall, London, pp 99-127

Inouye H, Beckwith J (1977) Synthesis and processing of an *Escherichia coli* alkaline phosphatase precursor in vitro. Proc Natl Acad Sci USA 74:1440-1444

Inouye S, Wang S, Sekizawa J, Halegoua S, Inouye M (1977) Amino acid sequence for the peptide extension on the prolipoprotein of the *Escherichia coli* outer membrane. Proc Natl Acad Sci USA 84:1004-1008

Katz FN, Rothman JE, Lingappa VP, Blobel G, Lodish JF (1977) Membrane assembly in vitro: synthesis, glycosylation, and asymmetric insertion of a transmembrane protein. Proc Natl Acad Sci USA 74:3278-3282

Lederberg EM (1955) Pleiotropy for maltose fermentation and phage resistance in *E. coli* K12. Genetics 40:580-581

Lin JJC, Kanazawa H, Ozols J, Wu HC (1978) An *Escherichia coli* mutant with an amino acid alteration within the signal sequence of outer membrane prolipoprotein. Proc Natl Acad Sci USA 75:4891-4895

Marchal C, Perrin D, Hedgpeth J, Hofnung M (1980) Synthesis and maturation of the λ receptor in *E. coli* K12: in vivo and in vitro expression of gene *lamB* under *lac* promoter control. Proc Natl Acad Sci USA 77:1491-1495

Milstein C, Brownlee GG, Harrison TM, Mathews MB (1972) A possible precursor of immunoglobin light chains. Nature (London) New Biol 239:117-120

Mock M, Schwartz M (1978) Mechanism of colicin E3 production in strains harboring wild-type or mutant plasmids. J Bacteriol 136:700-707

Mock M, Schwartz M (1980) Mutations which affect the structure and activity of colicin E3. J Bacteriol 142:384-390

Moreno F, Fowler AV, Hall M, Silhavy TJ, Zabin I, Schwartz M (1980) A signal sequence is not sufficient to lead β-galactosidase out of the cytoplasm. Nature (London) 286:356-359

Palade GE (1975) Intracellular aspects of the process of protein secretion. Science 189:347-358

Poyton RO, McKemmie E (1979) A polyprotein precursor to all four cytoplasmically translated subunits of cytochrome C oxydase from *Saccharomyces cerevisiae*. J Biol Chem 254:6763-6771

Raibaud O, Roa M, Braun-Breton C, Schwartz M (1979) Structure of the *malB* region in *Escherichia coli* K12 I Genetic map of the *malK-lamB* operon. Mol Gen Genet 174:241-248

Randall LL, Hardy SJS (1977) Synthesis of exported proteins by membrane bound polysomes from *Escherichia coli*. Eur J Biochem 75:43-53

Randall LL, Hardy SJS, Josefsson LG (1978) Precursors of three exported proteins in *Escherichia coli*. Proc Natl Acad Sci USA 75:1209-1212

Randall-Hazelbauer L, Schwartz M (1973) Isolation of the bacteriophage lambda receptor from *Escherichia coli*. J Bacteriol 116:1436-1446

Raymond Y, Shore GC (1979) The precursor for carbamyl phosphate synthetase is transported to mitochondria via a cytosolic route. J Biol Chem 254:9335-9338

Rothman JE, Lenard J (1977) Membrane assymmetry. Science 195:743-755

Rothman JE, Lodish HF (1971) Synchronized transmembrane insertion and glycosylation of a nascent membrane protein. Nature (London) 269:775-780

Schatz G (1979) How mitochondria import proteins from the cytoplasm. FEBS Lett 103:203-211

Silhavy TJ, Casadaban MJ, Shuman HA, Beckwith J (1976) Conversion of β-galactosidase to a membrane-bound state by gene fusion. Proc Natl Acad Sci USA 73:3423-3427

Silhavy TJ, Shuman HA, Beckwith J, Schwartz M (1977) Use of gene fusions to study outer membrane protein localization in *Escherichia coli*. Proc Natl Acad Sci USA 74:5411-5415

Silhavy TJ, Bassford PJ Jr, Beckwith JR (1979) A genetic approach to the study of protein localization in *Escherichia coli*. In: Inouye (ed) Bacterial outer membranes: biogenesis and function. John Wiley and Sons Inc, New York, p 203-254

Smith WP, Tai PC, Thompson RC, Davis BD (1977) Extracellular labelling of nascent polypeptides traversing the membrane of *Escherichia coli*. Proc Natl Acad Sci USA 74:2830-2834

Sutcliffe JG (1978) Nucleotide sequence of the ampicillin resistance gene of Escherichia coli plasmid pBR322. Proc Natl Acad Sci USA 75:3737-3741

Szmelcman S, Hofnung M (1975) Maltose transport in *Escherichia coli* K12. Involvement of the bacteriophage lambda receptor. J Bacteriol 124:112-118

Varenne S. Piovant M, Pages JM, Lazdunski C (1978) Evidence for synthesis of alkaline phosphatase on membrane-bound polysomes in *Escherichia coli*. Eur J Biochem 86:603-606

Wandersman C, Schwartz M, Ferenci T (1979) *Escherichia coli* mutants impaired in maltodextrin transport. J Bacteriol 140:1-13

Wickner W (1979) The assembly of proteins into biological membranes: the membrane trigger hypothesis. Annu Rev Biochem 48:23-45

Studies of the Path of Assembly of Bacteriophage M13 Coat Protein Into the *Escherichia coli* Cytoplasmic Membrane

P. Hearne, M. Nokelainen, A. Ponticelli, Y. Hirota, K. Ito, and W. Wickner[1]

Viruses have aided in the study of many biological processes such as DNA synthesis and transcriptional regulation. The capsid protein of coliphage M13 offers several distinct advantages for the study of membrane biogenesis:

1. It spans the host cell plasma membrane at each stage of virus infection with its N-terminus on the outer (periplasmic) surface (Wickner 1975, 1976) and its C-terminus exposed to the cytoplasm (Webster and Cashman 1978).

2. Its amino acid sequence and that of its precursor form have been independently determined by the methods of DNA (von Wezenbeek et al. pers. commun.), RNA (Sugimoto et al. 1977), and protein sequencing (Asbeck et al. 1969; Nakashima and Konigsberg 1974). It entirely lacks four amino acids.

3. It is made in great abundance by the infected cell, and can account for up to 1/3 of the membrane protein synthesis (Smilowitz et al. 1972).

4. Its unusually small size allows it to be directly assayed by SDS polyacrylamide gel electrophoresis of unfractionated, infected cells after pulse labeling with radioactive amino acids (Ito et al. 1980).

5. It can readily be isolated in *gram* quantities (Knippers and Hoffmann-Berling 1966; Woolford and Webster 1975).

6. Antibodies can be prepared which are quite specific for the N-terminal 8 residues of coat protein (Wickner 1975, 1976) and which will cross-react with procoat protein as well (Ito et al. 1980).

7. An amber mutant is available in the coat protein gene as well as in each of the other 7 virus genes (Henry and Pratt 1969).

In Vivo Studies
===============

We began our studies of assembly by assaying the orientation of the coat protein synthesized in infected cells (Wickner 1975, 1976). Antibody to purified coat protein, which was shown to react only with the eight extreme NH_2-terminal residues, bound only to the outer surface of the plasma membranes of infected cells. Chemical modification studies (Webster and Cashman 1978) have shown that the C-terminus is exposed on the inner surface of the membrane. Thus the coat protein spans the membrane whith its basic N-terminus on the outer surface of the membrane, its hydrophobic central region in contact with the apolar fatty acyl phase of the bilayer, and its acidic C-terminus exposed to the cytoplasm. We found that the coat protein which came to the cell with the infecting virus had this same orientation across the membrane;

[1]Department of Biological Chemistry and The Molecular Biology Institute, University of California, Los Angeles, California 90024, USA

this suggested that membrane protein orientation is less a matter of
the pathway of assembly than of the thermodynamics of protein-membrane
interaction.

Coat protein was predicted by mRNA sequencing (Sugimoto et al. 1977)
and by in vitro protein synthesis (Chang et al. 1978) to be made as a
precursor (procoat) with 23 additional N-terminal residues. To study
its synthesis and metabolism in vivo, we have performed a series of
pulse labeling experiments (Ito et al. 1979, 1980; Date et al. 1980a,b)
with M13-infected cells. [^{3}H]-proline was chosen for these labeling
studies since it is found in procoat and coat but not in *E. coli* lipoprotein or prolipoprotein, two major protein species which migrate at
the same R_f on SDS polyacrylamide gels. Pulse labeling, combined with
"chase" with nonradioactive amino acids and analysis by SDS gels and
fluorography, showed that procoat is the initial gene product and that
it rapidly chases to coat protein. Pulse-labeled procoat is soluble,
as determined by immunoassay or SDS gel electrophoresis, and after
cell lysis by any of a variety of techniques. This agrees well with
the fact that procoat is made exclusively by polysomes which are not
membrane-bound (Ito et al. 1979). Procoat is found bound to the membrane at intermediate chase times; protease mapping showed that this
procoat is entirely on the inner bilayer surface. The next major species to appear during the chase is coat protein spanning the bilayer
in its final conformation.

The ability to kinetically separate procoat synthesis, binding to the
membrane, insertion across the bilayer, and cleavage to coat protein
has allowed us to examine the energy requirements of each stage of the
pathway (Date et al. 1980a,b). These experiments involved pulse labeling
M13-infected cells and "chasing" with nonradioactive amino acids plus
poisons. When no poison was present during the chase, procoat rapidly
chased to coat. However, a variety of agents which abolished the transmembrane electrochemical potential (cyanide, azide, dinitrophenol, and
carbonyl cyanide m-chlorophenylhydrazone [CCCP]) blocked the post-
translational conversion of procoat to coat. CCCP was chosen for further study, since it alone did not inhibit the isolated leader peptidase. The procoat which persisted in the presence of CCCP bound to the
plasma membrane on the inner surface bud did not cross the bilayer and
was not cleaved by leader peptidase. Isolation of a spontaneous CCCP-
resistant *E. coli* mutant allowed the effect of this uncoupler on conversion of procoat to coat to be assigned directly to energy metabolism.
In contrast to CCCP, arsenate had no effect on the membrane potential
bud dissipates the nucleoside triphosphate pool, thereby almost immediately blocking protein synthesis. Thus polypeptide chain elongation
is not involved in procoat crossing the bilayer, but the transmembrane
potential is required.

In Vitro Reconstitutions

Procoat can also be synthesized in a cell-free incubation. This procoat is also initially soluble and, as with its in vivo counterpart,
is found as a 5s oligomeric species (Wickner et al. 1978). It will then
bind to added membrane or even to protein-free liposomes. Protease mapping studies showed that this procoat, added to the outer face of the
liposomes, gained across to internal (trapped) protease (Wickner et
al. 1978).

Chang, Blobel, and Model have also studied procoat metabolism with the
aid of in vitro protein synthesis (Chang et al. 1978, 1979). They observed that procoat was processed to coat in synthesis reactions with

E. coli inner membranes present from the start. When synthesis of procoat was allowed to proceed for one hour, followed by the addition of
membranes, very little procoat was post-translationally processed to
coat protein. This experiment was interpreted to indicate that procoat
insertion and processing could only occur if membranes were present
during translation, i.e., if they were present to receive the growing
procoat polypeptide as it emerged from the ribosome. We have reproduced
these experiments (manuscript in preparation) and found them to be correct; however, additional experiments along similar lines have completely altered their interpretation. Labeled procoat synthesis was allowed
to proceed for only 5 - 10 min, then the label was "chased" with nonradioactive amino acids and polypeptide chain elongation was terminated
with puromycin. Thirty seconds later, membranes were added. After additional incubation, analysis by SDS polyacrylamide gel electrophoresis
and fluorography showed that very efficient (75% complete) conversion
of procoat to coat had occurred. How was this distinctly *post*-transla-
tional assembly to be rationalized with the results of Chang et al.?
We found that when procoat synthesis in the absence of membranes was
performed for 5, 10, 20, 40, or 60 min, followed by a post-translational incubation with membranes, a dramatically decreasing proportion
of the procoat was converted to coat after the longer incubations.
Thus, post-translational assembly *does* occur in this system unless the
procoat is incubated in this crude synthesis extract for nonphysio-
logical periods (1 h) without membranes. In sum, there is no conflict
in the data in the literature, and soluble procoat is efficiently converted to coat *after* translation is complete both in vivo and in vitro.

Purification of the Components for Assembly

We have purified an *E. coli* leader peptidase 6000-fold from uninfected
cells (Zwizinski and Wickner 1980). Enzyme activity was assayed by the
post-translational conversion of procoat (synthesized in the cell-free
reaction) to coat protein. Purified enzyme cleaved procoat at the
proper position in the sequence, and even did so when the labeled procoat substrate was purified away from the other components of the trans-
cription/translation reaction. Leader peptide was clearly detected
when purified procoat was digested by purified leader peptidase, and
a soluble factor was found which may hydrolyze the liberated leader
peptide.

We observed that the conversion of procoat to coat is delayed in cells
infected with any of several M13 assembly gene mutants (Ito et al.
1979). This observation has led to the isolation of chemically pure
procoat protein (Silver et al. in prep.), raising the exciting possibility that each step of the conversion of procoat to coat might be
reproduced in a cell-free reaction.

The Genetics of Membrane Assembly

In addition to the approaches discussed above, it seems important to
seek mutants in membrane biogenesis. Specifically, mutants might be
found in the gene of any protein which was required to catalyze a step
of procoat metabolism. Such mutants would show a delayed or completely
blocked conversion of procoat protein to coat protein. If, as seems
likely, the mutant protein is also needed for the assembly of vital
host proteins into the membrane, then only conditionally lethal mutants
would be viable.

We have screened 1300 independently isolated temperature-sensitive
mutants of *E. coli* for the conversion of M13 procoat to coal. In these

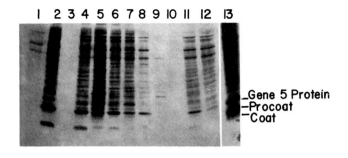

Fig. 1. Independently isolated temperature-sensitive *E. coli* mutants screened for conversion of M13 procoat to coat

_Gene 5 Protein
—Procoat
—Coat

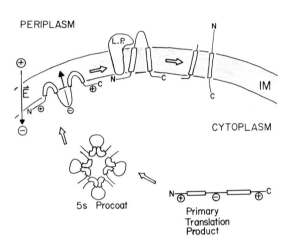

Fig. 2. Model formation of M13 coat protein

PERIPLASM

L.P.

IM

CYTOPLASM

5s Procoat

Primary
Translation
Product

experiments, cells were made male, grown and infected with M13 at a permissive temperature (32°C) and shifted to a nonpermissive temperature (42°C). They were then pulse labeled with $[^3H]$-proline, briefly "chased" with nonradioactive amino acids, and analyzed by SDS polyacrylamide gel electrophoresis and fluorography (Fig. 1). While some strains proved to be incapable of supporting M13 infection (lanes 5 and 12) or of supporting protein synthesis at 42°C (lanes 1, 3, 9, and 10), many of the strains showed synthesis of both gene 5 protein and coat protein (lanes 2, 4, 6, 7, 8, and 11). In no case, however, was there gene 5 protein and either no coat protein or persistent procoat. In lane 13 is a control experiment in which M13 amber 7 phage was used for infection; cells infected by this virus show a delayed conversion of procoat to coat and, indeed, procoat is seen in this lane (but no others).

Of course, this essentially negative result does not prove that no such mutants exist; rather, it indicates that they may not be abundant Clearly, at least one enzyme *is* known to be involved in assembly, namely leader peptidase, and it would be useful to have mutants in its gene. Others have sought mutants in membrane biogenesis functions and each of their approaches holds considerable promise.

Conclusions

Our current working model of how M13 coat protein is made is illustrated in Figure 2. While almost none of the salient features (such as a soluble precursor and the requirement for a trans-bilayer electrochemical potential) was anticipated at the start of this study, more surprises may emerge from further studies. Questions yet to be answered include:

1. What is the structure of the soluble precursor form of procoat?

2. Are there "topographic catalysts" which help to insert the procoat across the bilayer?

3. Which side of the bilayer has the active site of leader peptidase?

4. What is the fate of the leader peptide?

5. Precisely how does the electrochemical potential function in procoat assembly?

References

Asbeck v F, Beyreuther K, Köhler H, von Wettstein G, Braunitzer G (1969) Hoppe-Seyler's Z Physiol Chem 350:1047-1066
Chang CN, Blobel G, Model P (1978) Proc Natl Acad Sci USA 75:361-365
Chang CN, Model P, Blobel G (1979) Proc Natl Acad Sci USA 76:1251-1255
Date T, Zwizinski C, Ludmerer S, Wickner W (1980a) Proc Natl Acad Sci USA 77:827-831
Date T, Goodman JM, Wickner W (1980b) Proc Natl Acad Sci USA in press
Henry T, Pratt D (1969) Proc Natl Acad Sci USA 62:800-804
Ito K, Mandel G, Wickner W (1979) Proc Natl Acad Sci USA 76:1199-1203
Ito K, Date T, Wickner W (1980) J Biol Chem 255:2123-2130
Knippers R, Hoffmann-Berling H (1966) J Mol Biol 21:281-292
Nakashima Y, Konigsberg W (1974) J Mol Biol 88:598-600
Silver P, Nikelainen M, Hearne P, Wickner W in preparation
Smilowitz H, Carson J, Robbins P (1972) J Supramol Struct 1:8-18
Sugimoto K, Sugisaki H, Takanami M (1977) J Mol Biol 110:487-507
Webster RE, Cashman JS (1978) In: Denhardt DT, Dressler D, Ray DS (eds) The single-stranded DNA phages. Cold Spring Harbor Laboratories, New York, pp 557-569
Wezenbeek P, Hulsebos T, Schoenmakers JGG, personal communication
Wickner W (1975) Proc Natl Acad Sci USA 72:4749-4753
Wickner W (1976) Proc Natl Acad Sci USA 73:1159-1163
Wickner W, Mandel G, Zwizinski C, Bates M, Killick T (1978) Proc Natl Acad Sci USA 75:1754-1758
Woolford JL, Webster RE (1975) J Biol Chem 250:4333-4339
Zwizinski C, Wickner W (1980) J Biol Chem submitted

Subject Index

International
Cell Biology 1980-1981

Editor: H. G. Schweiger

1981. Approx. 650 figures.
Approx. 1200 pages
ISBN 3-540-10475-5

Contents: Opening Lecture. – Genomes and Gene Expression. – Cytoskeleton. – Pathology and Pathogenicity. – Differentiation and Development. – Membranes and Cell Surfaces. – Functional Organization.

International Cell Biology 1980–1981 contains contributions presented at the Second International Congress on Cell Biology held in West Berlin, August 31 – September 5, 1980. The authors of the contributions were selected as speakers for the Congress for their leading role in their respective fields. The topics cover a uniquely broad range of research areas, providing an excellent reflection of the present status of cell biology. This book will remain a useful source of information to biologists, medical researchers and biochemists for years to come.

Springer-Verlag
Berlin
Heidelberg
New York

Chloroplasts

Editor: J. Reinert
With contributions by numerous experts

1980. 40 figures, 11 tables. XI, 240 pages
(Results and Problems in Cell Differentiation,
Volume 10)
ISBN 3-540-10082-2

Contents: Types of Plastids: Their Development and Interconversions. – The Continuity of Plastids and the Differentiation of Plastid Populations. – Plastid DNA – The Plastome. – RNA and Protein Synthesis in Plastid Differentiation. – Biosynthesis of Thylakoids and the Membrane-Bound Enzyme Systems of Photosynthesis. – Fraction I Protein. – Factors in Chloroplast Differentiation. – The Survival, Division and Differentiation of Higher Plant Plastids Outside the Leaf Cell. – Subject Index.

Springer-Verlag
Berlin
Heidelberg
New York

Research on chloroplasts has undergone a remarkable expansion during the last years, especially as far as their role in photosynthesis is concerned. Closely connected with this expansion is our increasing knowledge of the role and the participation of chloroplasts as autonomous cell constituents in the development, in particular the developmental biochemistry of plant cells. This volume compiles recent experimental facts and theoretical points in the field of chloroplast research and their bearing upon plant development.